Applied Mathematical Sciences | Volume 17

L. Collatz
W. Wetterling

Optimization Problems

Translated by:
P. Wadsack

Springer-Verlag New York · Heidelberg · Berlin
1975

L. Collatz
Institut für Angewandte Mathematik
Universität Hamburg
2000 Hamburg 13
Rothenbaumchaussee 41
West Germany

W. Wetterling
T. H. Twente
Enschede
Netherlands

Translator:
P. Wadsack
University of Wisconsin-Madison
Mathematics Department
Madison, Wisconsin

AMS Classifications
Primary: 49-01
Secondary: 65Fxx, 90C05, 90D40

Library of Congress Cataloging in Publication Data

Collatz, Lothar, 1910–
Optimization problems.

(Applied mathematical sciences, v. 17)
Translation of Optimierungsaufgaben.
Bibliography: p.
Includes index.
1. Mathematical optimization. 2. Game theory.
1. Wetterling, Wolfgang W. E., 1932– joint author.
II. Title. III. Series.
QA1.A647 vol. 17 [QA402.5] 510′.8s [519.7]
75-15795

Printed in the United States of America

ISBN 0-387-90143-4 Springer-Verlag New York Heidelberg Berlin
ISBN 3-540-90143-4 Springer-Verlag Berlin Heidelberg New York

PREFACE

The German edition of this book, first published in 1966, has been quite popular; we did not, however, consider publishing an English edition because a number of excellent textbooks in this field already exist. In recent years, however, the wish was frequently expressed that, especially, the description of the relationships between optimization and other subfields of mathematics, which is not to be found in this form in other texts, might be made available to a wider readership; so it was with this in mind that, belatedly, a translation was undertaken after all.

Since the appearance of the German edition, the field of optimization has continued to develop at an unabated rate. A completely current presentation would have required a total reworking of the book; unfortunately, this was not possible. For example, we had to ignore the extensive progress which has been made in the development of numerical methods which do not require convexity assumptions to find local maxima and minima of non-linear optimization problems. These methods are also applicable to boundary value, and other, problems. Many new results, both of a numerical and a theoretical nature, which are especially relevant to applications, are to be found in the areas of optimal contol and integer optimization.

Although these and many other new developments had to be ignored, we hope that the book continues to satisfy the goals set forth in the preface to the German edition.

Finally, we want to take this opportunity to express our gratitude, to Peter R. Wadsack for a careful translation, and to Springer Verlag for kind cooperation.

FROM THE PREFACE TO THE GERMAN EDITION

With this book we would like to provide an introduction to a field which has developed into a great new branch of knowledge in the last thirty years. Indeed, it continues to be the object of intensive mathematical research. This rapid development has been possible because there exists a particularly close contact between theory and application.

Optimization problems have appeared in very different applied fields, including such fields as political economics and management science, for example, where little use was formerly made of mathematical methods. It also has become apparent that questions from very different areas of numerical mathematics may be regarded as examples of optimization. Thus, many types of initial value and boundary value problems of ordinary and partial differential equations, as well as approximation problems, game theoretic questions, and others, reduce to optimization problems. As this field has grown in importance, the number of texts has increased. Thus some justification for yet another text might be required. Now most existing texts deal with some subfield, whether linear or non-linear optimization, game theory, or whatever. So it became our intention to provide a certain overview of the entire field with this book, while emphasizing the connections and interrelations among different fields and subfields, including those previously mentioned. Since it is also our impression that these new fields -- for example, the beautiful general theorems on systems of equations and inequalities -- are not yet generally known, even in mathematical circles, we want to use this book to provide a general, easily comprehensible, and for the practitioner, readily accessible, introduction to this varied field, complete with proofs and unobscured by excessive computational detail. Thus, several deeper concepts, such as the theory of optimal processes (due to Pontrjagin), for one example, or the theory of dynamic optimization (due to Bellman), for another, are not discussed.

The book resulted from a number of courses in the subject given by the authors at the Universität Hamburg. In addition, one of the authors included the theorems of the alternative for systems of equations and inequalities, up to the duality theorem of linear optimization (§5 of this book) in an introductory course on "Analytic Geometry and Algebra"; for these theorems may be presented in a few hours as an immediate sequel to matrix theory and the concept of linear independence of vectors. It seems desirable that the young student become familiar with these things. In some countries they already are covered in high school seminars, for which they are well suited. They contribute to the dissemination of mathematics into other sciences and thus their significance will certainly grow in the future.

TABLE OF CONTENTS

I. LINEAR OPTIMIZATION

§1. Introduction

Using simple applications as examples, we will develop the formulation of the general linear optimization problem in matrix notation.

1.1. The Fundamental Type of Optimization Problem

Example 1. First we discuss a problem in production planning whose mathematical formulation already contains the general form of a linear optimization problem.

A plant may produce q different products. Production consumes resources, specifically, m different types of resources, such as labor, materials, machines, etc., each of limited availability. The production of one unit of the kth product yields a net profit of p_k, $k = 1,\ldots,q$. Thus, if x_1 units of the first product, x_2 units of the second, and generally, x_k units of the kth product are produced, the total profit will be $\sum_{k=1}^{q} p_k x_k$. Our problem is to devise a production plan which maximizes the total profit. In doing so, we must bear in mind that the jth resource is available only up to some maximal finite quantity, b_j, and that the production of one unit of the kth product consumes a quantity a_{jk} of the jth resource. The x_k must be chosen, therefore, to satisfy the inequalities $\sum_{k=1}^{q} a_{jk} x_k \le b_j$, $j = 1,\ldots,m$, and naturally must satisfy also the requirements $x_k \ge 0$.

We can formulate this problem as a linear optimization problem in the following manner.

Let there be given the (always real!) numbers p_k, b_j, a_{jk}, $j = 1,\ldots,m$, $k = 1,\ldots,q$. Find numbers x_k such that

$$Q(x_1,\ldots,x_q) = \sum_{k=1}^{q} p_k x_k = \text{Max!}, \tag{1.1}$$

i.e., is as large as possible, subject to the constraints

$$\sum_{k=1}^{q} a_{jk} x_k \le b_j \quad (j = 1,\ldots,m) \tag{1.2}$$

and the positivity constraints

$$x_k \ge 0 \quad (k = 1,\ldots,q). \tag{1.3}$$

The notation $Q(x_1,\ldots,x_q) = \text{Max!}$, resp. Min!, will be used henceforth. It instructs

1. check whether the function Q possesses a maximum, resp. minimum, subject to the given constraints; and if it does,

2. determine the extreme value and the values of

the variables $x_1,\ldots,x_q$ for which Q attains this extreme.

In particular, the notation $Q(x_1,\ldots,x_q) = \text{Max!}$ makes no claim about the existence of a maximum. It should be interpreted as merely a statement of the problem.

In the context of linear optimization, we consider problems of the type just described: find the maximum of a function Q (the <u>objective function</u>), which is linear in the variables x_k, where the x_k satisfy a system of linear inequalities and are non-negative. The following variations from this fundamental type also occur.

1. The objective function $Q(x_1,\ldots,x_q)$ of form (1.1) is to be minimized. A switch to $-Q(x_1,\ldots,x_q)$ reduces this case to the one described above.

2. The inequalities read $\geq$ instead of $\leq$. Multiplication by -1 converts such inequalities to form (1.2).

3. In the constraints (1.2) we have $=$ instead of $\geq$. The introduction of <u>slack variables</u>, as we shall see below, reduces form (1.2) to this type, where the contraints form a system of equations.

4. The positivity constraints (1.3) may be omitted (or perhaps are already contained in the constraints (1.2)).

5. Various combinations are also possible. The constraints may consist partially of equations and partially of inequalities. Positivity constraints may be prescribed for some, but not all, of the x_k.

<u>Example 2</u>. With this concrete, if highly idealized, example, we will further clarify, and illustrate graphically,

the formulation of the linear optimization problem.

An agricultural cooperative raises cows and sheep. The cooperative has 50 stalls for cows, and 200 for sheep. It also has 72 acres of pasture. One acre is needed to sustain one cow, while a sheep requires 0.2 acres. To care for the animals, the cooperative can provide up to 10,000 hours of labor per year. A cow requires 150 hours annually, and a sheep, 25. The annual profit that is realized is \$250 per cow and \$45 per sheep.

The cooperative would like to determine the number, x_1, of cows, and x_2, of sheep, which it should keep to maximize the total profit.

Formulated mathematically, this becomes the linear optimization problem

$$\left.\begin{aligned} Q(x_1,x_2) = 250x_1 + 45x_2 &= \text{Max!} \\ x_1 &\le 50 \\ x_2 &\le 200 \\ x_1 + 0.2x_2 &\le 72 \\ 150x_1 + 25x_2 &\le 10000 \\ x_1 \ge 0,\ x_2 &\ge 0. \end{aligned}\right\} \quad (1.4)$$

Figure 1.1 graphically illustrates this problem.

Those points, whose coordinates (x_1,x_2) satisfy all of the inequalities (1.4), are precisely the points of the shaded, six-sided polygon M, boundary points included. Now, $Q(x_1,x_2) = c$ determines a family of parallel lines dependent on the parameter c. Problem (1.4) thus can be formulated as follows. From among all lines of the family

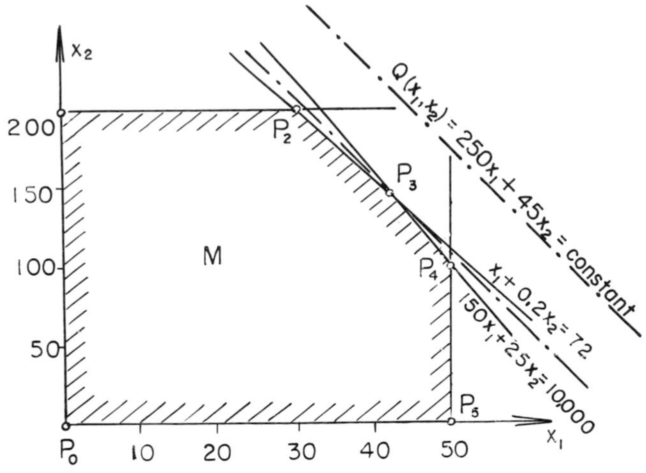

Figure 1.1

which contain points of M, find that line for which c is maximal. Let that line be denoted by $Q(x_1,x_2) = c^*$. Then each point which this line has in common with M, and no other, yields a solution to problem (1.4). It is intuitively clear (and will later be proven in general) that at least one corner point of M must be among these points of intersection of the line $Q(x_1,x_2) = c$ with the polygon M. The case where the intersection contains two corner points occurs only when the constants in $Q(x_1,x_2)$ are adjusted so as to cause a whole side of M to lie in the line $Q(x_1,x_2) = c^*$. In either case, it suffices to compute the values of $Q(x_1,x_2)$ at all corner points of M. The largest value obtained in this way is simultaneously the maximum of $Q(x_1,x_2)$. The coordinates of the corresponding corner point solve the optimization problem. We obtain:

corner	x_1	x_2	$Q(x_1,x_2)$
P_0	0	0	0
P_1	0	200	9000
P_2	32	200	17000
P_3	40	160	17200
P_4	50	100	17000
P_5	50	0	12500

So we see that the maximal profit of \$17,200 is attained by keeping 40 cows and 160 sheep.

1.2. The Fundamental Type in Matrix Notation

The fundamental type of linear optimization problem, described by (1.1), (1.2), and (1.3), will now be reformulated in a more concise notation by the introduction of vectors and matrices. The p_k's, b_j's, and x_k's are collected in (column) vectors,

$$\underset{\sim}{p} = \begin{pmatrix} p_1 \\ p_2 \\ \cdots \\ p_q \end{pmatrix}, \quad \underset{\sim}{b} = \begin{pmatrix} b_1 \\ b_2 \\ \cdots \\ b_m \end{pmatrix}, \quad \underset{\sim}{x} = \begin{pmatrix} x_1 \\ x_2 \\ \cdots \\ x_q \end{pmatrix} \tag{1.5}$$

the a_{jk}'s into the matrix

$$\underset{\sim}{A} = \begin{pmatrix} a_{11} & a_{12} & \cdots & a_{1q} \\ a_{21} & a_{22} & \cdots & a_{2q} \\ \cdots & \cdots & \cdots & \cdots \\ a_{m1} & a_{m2} & \cdots & a_{mq} \end{pmatrix}. \tag{1.6}$$

The transpose matrix of $\underset{\sim}{A}$ will be denoted by $\underset{\sim}{A}'$:

$$\underset{\sim}{A}' = \begin{pmatrix} a_{11} & a_{21} & \cdots & a_{m1} \\ \cdots & \cdots & \cdots & \cdots \\ a_{1q} & a_{2q} & \cdots & a_{mq} \end{pmatrix},$$

and correspondingly, a column vector, say $\underset{\sim}{p}$, is transposed to the row vector

$$\underset{\sim}{p}' = (p_1, p_2, \ldots, p_q).$$

The linear optimization problem now reads as follows.

Let $\underset{\sim}{p}$ and $\underset{\sim}{b}$ be given real vectors, as in (1.5), and let $\underset{\sim}{A}$ be a given real matrix, as in (1.6). Find the real vector $\underset{\sim}{x}$ for which

$$Q(\underset{\sim}{x}) = \underset{\sim}{p}'\underset{\sim}{x} = \text{Max!} \tag{1.1a}$$

subject to the <u>constraints</u>

$$\underset{\sim}{A}\underset{\sim}{x} \leq \underset{\sim}{b} \tag{1.2a}$$

and the <u>positivity</u> <u>constraints</u>

$$\underset{\sim}{x} \geq \underset{\sim}{0}. \tag{1.3a}$$

Here $\underset{\sim}{0}$ is the zero vector. The relation $\geq$ or $\leq$ between vectors means that the corresponding relation holds for each component. By introducing a dummy vector, $\underset{\sim}{y} = \underset{\sim}{b} - \underset{\sim}{A}\underset{\sim}{x}$, the inequalities (1.2a) may be transformed into equations. Instead of (1.2a) we have the equations

$$\underset{\sim}{A}\underset{\sim}{x} + \underset{\sim}{y} = \underset{\sim}{b} \tag{1.2b}$$

and to (1.3a) we add the further positivity constraints

$$\underset{\sim}{y} = \begin{bmatrix} y_1 \\ \cdots \\ y_m \end{bmatrix} \geq \underset{\sim}{0} \tag{1.3b}$$

A vector with non-negative components which is used in this manner to transform inequalities into equations is called a slack variable vector, and its components are called slack variables. (But notice that this process does not reduce the total number of inequalities, as new constraints have been added.)

Now set

$$\left.\begin{aligned} &\hat{\underset{\sim}{x}} = \begin{pmatrix} \underset{\sim}{x} \\ \underset{\sim}{y} \end{pmatrix} = \begin{bmatrix} x_1 \\ \cdots \\ x_q \\ y_1 \\ \cdots \\ y_m \end{bmatrix}, \quad \tilde{\underset{\sim}{p}} = \begin{pmatrix} \underset{\sim}{p} \\ \underset{\sim}{0} \end{pmatrix} = \begin{bmatrix} p_1 \\ \cdots \\ p_q \\ \left.\begin{matrix} 0 \\ \cdots \\ 0 \end{matrix}\right\} m \text{ components} \end{bmatrix}, \qquad \tilde{\underset{\sim}{b}} = \underset{\sim}{b}, \\ &\tilde{\underset{\sim}{A}} = (\underset{\sim}{A}, \underset{\sim}{E}_m) = \begin{pmatrix} a_{11} & \cdots & a_{1q} & 1 & 0 & \cdots & 0 \\ a_{21} & \cdots & a_{2q} & 0 & 1 & \cdots & 0 \\ \cdots & \cdots & \cdots & \cdots & \cdots & \cdots & \cdots \\ a_{m1} & \cdots & a_{mq} & 0 & 0 & \cdots & 1 \end{pmatrix}, \end{aligned}\right\} \tag{1.7}$$

$$n = q + m \tag{1.8}$$

where $\underset{\sim}{E}_m$ is the m-dimensional identity matrix. Then (1.1a), (1.2a), and (1.3a) become equivalent to

$$Q(\tilde{\underset{\sim}{x}}) = \tilde{\underset{\sim}{p}}'\tilde{\underset{\sim}{x}} = \text{Max!}, \tag{1.1c}$$

$$\tilde{\underset{\sim}{A}}\tilde{\underset{\sim}{x}} = \tilde{\underset{\sim}{b}}, \tag{1.2c}$$

$$\tilde{x} \geq \underset{\sim}{0}. \tag{1.3c}$$

In this way, we obtain a linear optimization problem
in n = m + q variables, where the constraints are given
by linear equations. Conversely, if we are given a linear
optimization problem of the type of (1.1c), (1.2c), and
(1.3c), with equations as constraints (and where the x, p,
and A do not necessarily have the special form of (1.7)),
then the reverse transformation into a linear optimization
problem with inequalities as constraints can be carried out
trivially. For (1.2c) is equivalent to

$$\left. \begin{array}{c} \tilde{A}\tilde{x} \leq \tilde{b} \\ -\tilde{A}\tilde{x} \leq -\tilde{b} \end{array} \right\}. \tag{1.2d}$$

In sample applications the auxiliary constraints
typically are given in the form of inequalities. For
theoretical considerations however, it is more effective as
a rule to consider the case where the constraints are in the
form of equations. The two cases are equivalent, as we just
demonstrated. Therefore in the following we shall consider
primarily linear optimization problems of the type

$$Q(\underset{\sim}{x}) = \underset{\sim}{p}'\underset{\sim}{x} = \text{Max! (or Min!)},$$
$$A\underset{\sim}{x} = \underset{\sim}{b}, \qquad \underset{\sim}{x} \geq \underset{\sim}{0} \tag{1.9}$$

where A is a matrix with m rows and n columns, $\underset{\sim}{p}$ and
$\underset{\sim}{x}$ are vectors with n components, and $\underset{\sim}{b}$ is a vector
with m components. Also, we require that

$$n > m \tag{1.10}$$

(as is the case, for example, when n satisfies (1.8)). The number of rows in the matrix $\underset{\sim}{A}$ therefore must be smaller than the number of columns. If instead we had $n < m$, the theory of linear equations would tell us that (at least) one of the following three cases applies.

1. $\underset{\sim}{x}$ is uniquely determined by $\underset{\sim}{A}\underset{\sim}{x} = \underset{\sim}{b}$; this $\underset{\sim}{x}$ is then the solution of the optimization problem if $\underset{\sim}{x} \geq \underset{\sim}{0}$; if $\underset{\sim}{x} \geq \underset{\sim}{0}$ does not hold, (1.1) has no solution.

2. The equations $\underset{\sim}{A}\underset{\sim}{x} = \underset{\sim}{b}$ are inconsistent. The optimization problem has no solution.

3. Some of the equations are linearly dependent on others, and therefore dispensable.

Other problems, though in appearance quite distinct from example 1, nevertheless lead to a problem of type (1.9), as the following example shows.

Example 3. (Transportation Problem, due to W. Knödel, 1960)

We have seven sugar factories F_j producing a_j, $j = 1,\dots,7$, tons of sugar per month. We also have 300 localities G_k, $k = 1,\dots,300$, each consuming r_k tons of sugar monthly. Therefore, $\sum_{j=1}^{7} a_j = \sum_{k=1}^{300} r_k$. The transportation costs per ton of sugar, from F_j to G_k, are c_{jk}. Our task is to find the distribution scheme -- the number of tons, x_{jk}, to be transported from F_j to G_k -- which minimizes the total cost of transportation. This leads us to the problem

$$\left.\begin{array}{l} \sum\limits_{j=1}^{7} x_{jk} = r_k, \quad \sum\limits_{k=1}^{300} x_{jk} = a_j, \quad x_{jk} \ge 0 \\ \\ Q(x_{jk}) = \sum\limits_{j,k} c_{jk} x_{jk} = \text{Min!} \end{array}\right\} \tag{1.11}$$

Set

$$\underset{\sim}{x} = (x_{1,1},\ldots,x_{1,300},x_{2,1},\ldots,x_{7,300})'$$

$$\underset{\sim}{p} = (c_{1,1},\ldots,c_{1,300},c_{2,1},\ldots,c_{7,300})'$$

$$\underset{\sim}{b} = (a_1,\ldots,a_7,\ r_1,\ldots,r_{300})'$$

writing the column vectors as transposed row vectors with the same components in order to save space; we will use this more compact notation from now on. Also set

$$\underset{\sim}{A} = \underbrace{\left(\begin{array}{c|c|c|c} \overbrace{\begin{matrix} 1 & 1 & \cdots & 1 \\ 0 & 0 & \cdots & 0 \\ & & 0 & \\ & & & \end{matrix}}^{300 \text{ columns}} & \overbrace{\begin{matrix} 0 & 0 & \cdots & 0 \\ 1 & 1 & \cdots & 1 \\ & & 0 & \\ & & & \end{matrix}}^{300 \text{ columns}} & \cdots & \overbrace{\begin{matrix} & & & \\ & & 0 & \\ & & & \\ 1 & 1 & \cdots & 1 \end{matrix}}^{300 \text{ columns}} \\ \hline \begin{matrix} 1 & & & \\ & 1 & & 0 \\ & & \ddots & \\ & 0 & & 1 \end{matrix} & \begin{matrix} 1 & & & \\ & 1 & & 0 \\ & & \ddots & \\ & 0 & & 1 \end{matrix} & \cdots & \begin{matrix} 1 & & & \\ & 1 & & 0 \\ & & \ddots & \\ & 0 & & 1 \end{matrix} \end{array}\right)}_{2100 \text{ columns}} \begin{array}{l} \left.\begin{matrix} \\ \\ \\ \\ \end{matrix}\right\} 7 \text{ rows} \\ \left.\begin{matrix} \\ \\ \\ \\ \end{matrix}\right\} 300 \text{ rows} \end{array} \tag{1.12}$$

and then problem (1.11) takes exactly the form

$$Q(\underset{\sim}{x}) = \text{Min!}, \quad \underset{\sim}{A}\underset{\sim}{x} = \underset{\sim}{b}, \quad \underset{\sim}{x} \ge \underset{\sim}{0}.$$

This method was applied to a real problem in Austria

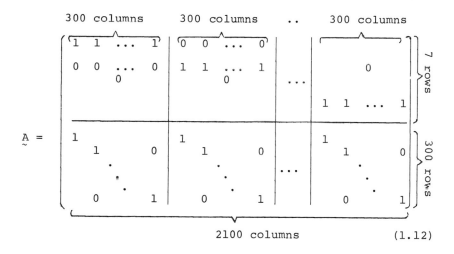

and the solution was computed from the above form. As a result, a saving of about ten percent of the cost of the pre-existing distribution scheme was attained. This amortised the costs of programming, computer time, etc. in ten days.

§2. Linear Optimization and Polyhedra

In this section we present the foundations of the theory of polyhedra, at least insofar as it applies to the elementary theory of optimization and to the simplex method. The less theoretically oriented reader may skip theorem 7 and its proof. The examples of §4 will clarify the present approach, and make it seem quite obvious. An entirely different approach, in §5.5 and §5.6 below, avoids the somewhat tedious considerations of polyhedra, but loses all motivation of the simplex method.

2.1. Feasible Points and Minimal Points

In Example 2, we saw that the set of all points, whose coordinates satisfied the constraints, formed the six-sided polygon pictured in Figure 1.1. The maximum of the objective function was attained at a vertex of the polygon. We now want to demonstrate that analogous conclusions hold in general for problems of the type of (1.9). We write this problem as a minimum problem:

$$Q(\underset{\sim}{x}) = \underset{\sim}{p}'\underset{\sim}{x} = \text{Min!}, \tag{2.1}$$

$$\underset{\sim}{A}\underset{\sim}{x} = \underset{\sim}{b}, \quad \underset{\sim}{x} \geq \underset{\sim}{0}. \tag{2.2}$$

Here, $\underset{\sim}{A}$ is a real matrix, with m rows and n columns, $\underset{\sim}{p}$ and $\underset{\sim}{b}$ are real vectors, with n and m components, respectively; the latter may also be written as $\underset{\sim}{p} \in R^n$, $\underset{\sim}{b} \in R^m$. R^n denotes n-dimensional, real vector space, and R^m similarly. Also, $m < n$. We are to find a vector $\underset{\sim}{x} \in R^n$, satisfying (2.2), which realizes the minimum in (2.1). Instead of "vector" we will frequently use the term "point". $\underset{\sim}{x} \in R^n$ is called a <u>feasible vector</u> or a <u>feasible point</u>, if $\underset{\sim}{x}$ satisfies the equations and inequalities of (2.2). The set of all feasible points will be denoted by M.

<u>Definition 1</u>: Let $\underset{\sim}{x}^1,\ldots,\underset{\sim}{x}^k$ be vectors in R^n and let $\alpha_1,\ldots,\alpha_k$ be real numbers. Then $\underset{\sim}{x} = \sum_{j=1}^{k} \alpha_j \underset{\sim}{x}^j$ is called a <u>convex combination</u> of $\underset{\sim}{x}^1,\ldots,\underset{\sim}{x}^k$ if $\alpha_j \geq 0$ $(j = 1,\ldots,k)$ and $\sum_{j=1}^{k} \alpha_j = 1$. If in addition, each $\alpha_j > 0$, $\underset{\sim}{x}$ is a <u>strict convex combination</u> of $\underset{\sim}{x}^1,\ldots,\underset{\sim}{x}^k$.

<u>Definition 2</u>: A point-set K is <u>convex</u> if every convex combination of any two points, $\underset{\sim}{x}^1$ and $\underset{\sim}{x}^2$, of K, is also in K; equivalently, the line segment connecting $\underset{\sim}{x}^1$ and $\underset{\sim}{x}^2$ is in K for all x^1, x^2 in K.

The set M of feasible points for problem (2.1), (2.2) is convex, since $\underset{\sim}{A}\underset{\sim}{x}^i = \underset{\sim}{b}, \underset{\sim}{x}^i \geq \underset{\sim}{0}$, $\alpha_i \geq 0$ $(i = 1,2)$, $\alpha_1 + \alpha_2 = 1$ implies that $\underset{\sim}{A}(\alpha_1\underset{\sim}{x}^1 + \alpha_2\underset{\sim}{x}^2) = \alpha_1\underset{\sim}{b} + \alpha_2\underset{\sim}{b} = \underset{\sim}{b}$, $\alpha_1\underset{\sim}{x}^1 + \alpha_2\underset{\sim}{x}^2 \geq \underset{\sim}{0}$. Furthermore, M clearly is a closed point-set in R^n.

<u>Definition 3</u>: A point $\underset{\sim}{x} \in M$ is a <u>vertex</u> of M if $\underset{\sim}{x}$ cannot be represented as a strict convex combination of

two distinct points of M.

Now let $\underset{\sim}{a}^1,\ldots,\underset{\sim}{a}^n$ be the column vectors of matrix $\underset{\sim}{A}$. Instead of $\underset{\sim}{A}\underset{\sim}{x} = \underset{\sim}{b}$, we may write

$$\sum_{k=1}^{n} \underset{\sim}{a}^k x_k = \underset{\sim}{b}. \tag{2.3}$$

If $\underset{\sim}{x} \in M$, the components x_k satisfy this equation.

<u>Theorem 1</u>: $\underset{\sim}{x} \in M$ is a vertex of M if and only if the column vectors of $\underset{\sim}{A}$ which correspond to positive components x_k are linearly independent.

<u>Proof</u>: I. Let $\underset{\sim}{x}$ be a vertex of M. Without loss of generality, we may assume that precisely the first r components of $\underset{\sim}{x}$ are positive: $x_k > 0$ $(k = 1,\ldots,r)$, $x_k = 0$ $(k = r+1,\ldots,n)$. Here $0 \leq r \leq n$. If $r = 0$, the corresponding set of column vectors is empty, and therefore linearly independent by definition. If $r > 0$, (2.3) becomes $\sum_{k=1}^{r} \underset{\sim}{a}^k x_k = \underset{\sim}{b}$. Suppose that $\underset{\sim}{a}^1,\ldots,\underset{\sim}{a}^r$ are linearly dependent. Then there are numbers $d_1,\ldots,d_r$, not all zero, such that $\sum_{k=1}^{r} \underset{\sim}{a}^k d_k = 0$. Since $x_k > 0$, there are sufficiently small $\delta > 0$ such that $x_k \pm \delta d_k > 0$, $k = 1,\ldots,r$.

Furthermore, $\sum_{k=1}^{r} \underset{\sim}{a}^k (x_k \pm \delta d_k) = \underset{\sim}{b}$. Thus the vectors $\underset{\sim}{x}^1, \underset{\sim}{x}^2$ with components

$$\begin{aligned} x_k^1 &= x_k + d_k \\ x_k^2 &= x_k - d_k \end{aligned} \quad (k = 1,\ldots,r),$$
$$x_k^1 = x_k^2 = 0 \quad (k = r+1,\ldots,n)$$

both belong to M and are distinct. $\underset{\sim}{x} = \frac{1}{2}(\underset{\sim}{x}^1 + \underset{\sim}{x}^2)$ is a strict convex combination of $\underset{\sim}{x}^1$ and $\underset{\sim}{x}^2$ and consequently not a vertex of M. This contradiction followed from the assumption of dependence; so $a^1,\ldots,a^r$ are linearly independent.

II. As before, precisely the first r components x_k are positive. Let $a^1,\ldots,a^r$ be linearly independent. Suppose that x may be represented as a strict convex combination of two distinct points of M: $x = \alpha x^1 + (1-\alpha)x^2$ $(0 < \alpha < 1)$. Since $x_k = 0$ $(k = r+1,\ldots,n)$, $x \geq 0$, $x^j \geq 0$ $(j = 1,2)$, it follows that $x_k^j = 0$ $(j = 1,2;\ k = r+1,\ldots,n)$. x^1 and x^2 lie in M, so that $Ax^1 = Ax^2 = b$. This implies that $\sum_{k=1}^{r} a^k(x_k^1 - x_k^2) = 0$. Since $a^1,\ldots,a^r$ are linearly independent, it follows that $x_k^1 = x_k^2$ $(k = 1,\ldots,r)$ and therefore, $x^1 = x^2$. Our assumption has lead to a contradiction, so it must be that x is a vertex of M.

<u>Corollary</u>: If x is a vertex of M, then x has at most m positive components. The remaining components are zero.

Ordinarily, a vertex x has exactly m positive components. Such a vertex is called <u>regular</u>. A vertex with fewer than m positive components is called <u>degenerate</u>.

We now prove several theorems about properties of M, the set of feasible points, which are intuitively obvious for two- or three-dimensional space. For arbitrary finite dimension, however, the proof of the theorems becomes rather tedious, as we shall see.

<u>Theorem 2</u>: M has only finitely many vertices.

<u>Proof</u>: Each vertex is uniquely determined by a cor-

responding set of column vectors of $\underset{\sim}{A}$. But there are only finitely many subsets of linearly independent column vectors of size less than or equal to m.

Theorem 3: If M is not empty, then the set of vertices of M also is not empty.

Proof: For $\underset{\sim}{x} \in M$ we define a function $\rho(\underset{\sim}{x})$ by the number of non-zero components of $\underset{\sim}{x}$. We have $0 \leq \rho(\underset{\sim}{x}) \leq n$. If M is not empty, the function $\rho(\underset{\sim}{x})$ attains its minimum, ρ_0, on M. Suppose $\rho(\underset{\sim}{\bar{x}}) = \rho_0$. We will show that $\underset{\sim}{\bar{x}}$ is a vertex of M. If $\rho_0 = 0$, then $\underset{\sim}{\bar{x}} = \underset{\sim}{0}$ also, and $\underset{\sim}{\bar{x}}$ is a vertex, since the set of column vectors corresponding to positive components is empty, and therefore linearly independent by definition. If $\rho_0 > 0$, we may assume that $\underset{\sim}{\bar{x}} = (\bar{x}_1,\ldots,\bar{x}_{\rho_0},0,\ldots,0)'$. We suppose $\underset{\sim}{\bar{x}}$ is not a vertex and derive a contradiction. For then the column vectors $\underset{\sim}{a}^1,\ldots,\underset{\sim}{a}^{\rho_0}$ are linearly dependent, and there exist numbers, $d_1,\ldots,d_{\rho_0}$, not all zero, such that $\sum_{k=1}^{\rho_0} \underset{\sim}{a}^k d_k = \underset{\sim}{0}$. For those indices k, for which $d_k \neq 0$, consider $\bar{x}_k/|d_k|$ and find the smallest of these numbers. We may assume that $\lambda = \bar{x}_1/|d_1| \leq \bar{x}_k/|d_k|$ $(k = 1,\ldots,\rho_0)$ and $d_1 > 0$. The point

$$\underset{\sim}{\bar{\bar{x}}} = (\bar{x}_1-\lambda d_1,\ldots,\bar{x}_{\rho_0}-\lambda d_{\rho_0},0,\ldots,0)'$$

belongs to M because $\underset{\sim}{A}\underset{\sim}{\bar{\bar{x}}} = \underset{\sim}{A}\underset{\sim}{\bar{x}} - \lambda \sum_{k=1}^{\rho_0} \underset{\sim}{a}^k d_k = \underset{\sim}{b}$ and $\underset{\sim}{\bar{\bar{x}}} \geq \underset{\sim}{0}$, and has fewer than ρ_0 components because $\bar{x}_1 - \lambda d_1 = 0$, contrary to the definition of ρ_0. Therefore $\underset{\sim}{\bar{x}}$ is a vertex.

There are three possible cases:

(1) M is the empty set. The constraints (2.2) are inconsistent.

(2) M is a non-empty, bounded subset of R^n.

(3) M is an unbounded subset of R^n.

In case (2), M is called a (convex) polyhedron. In this case, the continuous function $Q(\underset{\sim}{x})$ assumes its minimum on the closed and bounded set M. In theorem 6 we will see that this occurs at a vertex of M. In case (1) the linear optimization problem has no solution. In case (3) there are two possibilities:

(a) $Q(\underset{\sim}{x})$ is bounded below on M and assumes its minimum. (That the minimum is actually assumed will not become clear until §5.6.)

(b) $Q(\underset{\sim}{x})$ is not bounded below on M and has no minimum. The optimization problem has no solution.

<u>Definition 4</u>: A point $\underset{\sim}{x}^o \in M$ is a <u>minimal point</u> if $Q(\underset{\sim}{x}^o) \leq Q(\underset{\sim}{x})$ for all $\underset{\sim}{x} \in M$.

It is easily seen that every convex combination of minimal points is again a minimal point. This proves

<u>Theorem 4</u>: The set of minimal points of a linear optimization problem is convex.

2.2. Further Results on Vertices and Minimal Points

<u>Theorem 5</u>: If M is a convex polyhedron, then every point of M can be written as a convex combination of finitely many vertices of M.

Proof: If $\underset{\sim}{x} \in M$, (2.3) holds, so $\sum_{k=1}^{n} \underset{\sim}{a}^k x_k = \underset{\sim}{b}$, and $x_k \geq 0$ $(k = 1,\ldots,n)$. Let r be the number of positive x_k. The proof proceeds by induction on r. If $r = 0$, $\underset{\sim}{x}$ is a vertex by theorem 1. So let $r > 0$ and assume the theorem holds for $0, 1,\ldots,r-1$ positive x_k. Let Z be the subset of indices k for which $x_k > 0$. If the $\underset{\sim}{a}^k$ with $k \in Z$ are linearly independent, theorem 1 again shows that $\underset{\sim}{x}$ is a vertex. Alternatively, if the $\underset{\sim}{a}^k$ with $k \in Z$ are linearly dependent, we can find numbers d_k, $k \in Z$, not all zero, such that $\sum_{k \in Z} \underset{\sim}{a}^k d_k = \underset{\sim}{0}$. Let $\underset{\sim}{x}(\lambda)$ be the vector with components $x_k + \lambda d_k$ for $k \in Z$, and 0 for $k \notin Z$. Because M is convex, closed, and bounded, there exist numbers $\lambda_1 < 0$, $\lambda_2 > 0$ such that the point $\underset{\sim}{x}(\lambda)$ is in M if and only if $\lambda_1 \leq \lambda \leq \lambda_2$. For $k \notin Z$, the component $x_k(\lambda_i) = 0$ $(i = 1,2)$. In addition, for $k \in Z$, there is at least one component $x_k(\lambda_1) = 0$ (otherwise, there would exist a $\lambda < \lambda_1$ with $\underset{\sim}{x}(\lambda) \in M$). Similarly, there is a $k \in Z$ with $x_k(\lambda_2) = 0$. The induction hypothesis implies that the points $\underset{\sim}{x}(\lambda_1)$ and $\underset{\sim}{x}(\lambda_2)$ are each a convex combination of vertices of M; consequently, so is $\underset{\sim}{x}$.

Theorem 6: If M is a convex polyhedron, then $Q(\underset{\sim}{x})$ attains its minimum at at least one vertex of M.

Proof: We have already seen that, for a polyhedron M, there exists a minimal point, $\underset{\sim}{x}^0$. Thus we need only show that one of the vertices, $\underset{\sim}{x}^1,\ldots,\underset{\sim}{x}^p$, is a minimal point. By theorem 5, there exist numbers $\alpha_j \geq 0$ such that $\sum_{j=1}^{p} \alpha_j = 1$ and $\underset{\sim}{x}^0 = \sum_{j=1}^{p} \alpha_j \underset{\sim}{x}^j$. Since $Q(\underset{\sim}{x})$ is linear in $\underset{\sim}{x}$, we have $Q(\underset{\sim}{x}^0) = \sum_{j=1}^{p} \alpha_j Q(\underset{\sim}{x}^j)$. Since $\underset{\sim}{x}^0$ is minimal, $Q(\underset{\sim}{x}^0) \leq Q(\underset{\sim}{x}^j)$. Now there is at least one index k such that $\alpha_k > 0$. If for this k, $Q(\underset{\sim}{x}^k) > Q(\underset{\sim}{x}^0)$, then $Q(\underset{\sim}{x}^0) < \sum_{j=1}^{p} \alpha_j Q(\underset{\sim}{x}^j)$. Therefore, $Q(\underset{\sim}{x}^k) = Q(\underset{\sim}{x}^0)$, and vertex $\underset{\sim}{x}^k$ is a minimal point.

If the set M of feasible points is unbounded, $Q(\underset{\sim}{x})$ is not necessarily bounded below, and thus has no minimum on M. But we do have the following theorem.

Theorem 7: If M is unbounded and $Q(\underset{\sim}{x})$ attains its minimum on M, then at least one vertex of M is a minimal point.

Proof: Let $\underset{\sim}{x}^0 \in M$ be a minimal point, but not a vertex. Let $\underset{\sim}{x}^1, \ldots, \underset{\sim}{x}^p$ be the vertices of M. Then $\underset{j=0,1,\ldots,p}{\text{Max}} \left(\sum_{k=1}^{n} x_k^j \right) = C \geq 0$. If $C = 0$, then $\underset{\sim}{x} = \underset{\sim}{0}$ is tho only vertex of M and thus the minimal point. Let $C > 0$. To the constraints (2.2) we add the equation $x_1 + \ldots + x_n + x_{n+1} = 2C$ and the inequality $x_{n+1} \geq 0$. Consider the linear optimization problem $\tilde{Q}(\tilde{\underset{\sim}{x}}) = \tilde{\underset{\sim}{p}}'\tilde{\underset{\sim}{x}} =$ Min!, $\tilde{\underset{\sim}{A}}\tilde{\underset{\sim}{x}} = \tilde{\underset{\sim}{b}}$, $\tilde{\underset{\sim}{x}} \geq \underset{\sim}{0}$, where

$$\tilde{x} = \begin{pmatrix} x_1 \\ \cdots \\ x_n \\ x_{n+1} \end{pmatrix}, \quad \tilde{p} = \begin{pmatrix} p_1 \\ \cdots \\ p_n \\ 0 \end{pmatrix}, \quad \tilde{b} = \begin{pmatrix} b_1 \\ \cdots \\ b_n \\ 2C \end{pmatrix},$$

$$\tilde{A} = \left(\begin{array}{c|c} & 0 \\ A & \vdots \\ & 0 \\ \hline 1 \quad 1 \ \ldots \ 1 & 1 \end{array}\right).$$

For this optimization problem, the set $\tilde{M}$ of feasible points is bounded, since $\tilde{x} \in \tilde{M}$ implies $0 \leq x_j \leq 2C$, $j = 1,\ldots,n+1$. There is a one-to-one correspondence between the points $\tilde{x} \in \tilde{M}$ and those points $x \in M$ for which $\sum_{j=1}^{n} x_j \leq 2C$, which is defined by setting $x_{n+1} = 2C - \sum_{j=1}^{n} x_j$. Let $\tilde{x}^0,\ldots,\tilde{x}^p$ be the points this correspondence assigns to $x^0,\ldots,x^p$, respectively. If $\tilde{x} \in \tilde{M}$ and $x \in M$ are corresponding points, then $\tilde{Q}(\tilde{x}) = Q(x)$. Since $Q(x)$ attains its minimum with respect to M at x^0, $\tilde{Q}(\tilde{x})$ attains its minimum with respect to $\tilde{M}$ at $\tilde{x}^0$, and the minimal value is the same.

Partition the vertices of $\tilde{M}$ into two classes:

(1) those where $x_{n+1} > 0$, and

(2) those where $x_{n+1} = 0$.

Let $\tilde{x}$ be a vertex of class (1). Theorem 1 and the form of $\tilde{A}$ show that the column vectors of A which correspond to positive components $x_1,\ldots,x_n$ are linearly independent. Hence the point of M corresponding to $\tilde{x}$ is a vertex of M. Conversely, every vertex of M corresponds to a vertex of class (1). Thus the vertices of class (1) are precisely $\tilde{x}^1, \ldots, \tilde{x}^p$.

Let the vertices of class (2) be $\tilde{x}^{p+1}, \ldots, \tilde{x}^r$. As $\tilde{M}$ is bounded, theorem 5 applies, and we have $\tilde{x}^0 = \sum_{j=1}^{\tilde{r}} \alpha_j \tilde{x}^j$, where $\alpha_j \geq 0$, $\sum_{j=1}^{r} \alpha_j = 1$. Since $x_{n+1}^0 \geq C$ and $x_{n+1}^j = 0$, $j = p+1, \ldots, r$, $\alpha_j > 0$ for at least one $j \leq p$. As in the proof of theorem 6, we conclude that one of the vertices $\tilde{x}^1, \ldots, \tilde{x}^p$ is a minimal point of $\tilde{Q}(\tilde{x})$ with respect to $\tilde{M}$. The corresponding vertex of M is then a minimal point of $Q(x)$ with respect to M.

2.3. The Basis of a Vertex

Our conclusions so far are valid for matrices A in (2.2) whose row number m is less than the column number n, but which are otherwise arbitrary. For the following considerations, we make the additional assumption that

$$\text{rank of } A = m = \text{row number of } A. \tag{2.4}$$

This condition proves productive and yet is not an essential restriction. For if the rank of A is less than m, either the system of equations $Ax = b$ in (2.2) has no solution, or some of the equations are dependent on others, and therefore dispensable. If we eliminate the dependent equations, we obtain a new system, $\tilde{A}\tilde{x} = \tilde{b}$, for which rank of $\tilde{A}$ equals the row number, and to which, therefore, the following considerations apply.

Example: In example 3, (1.12) shows that matrix A has a rank of at most 306, since the sum of rows 1 through 7 is equal to the sum of rows 8 through 307. And so too, the extended matrix $(\underset{\sim}{A}|\underset{\sim}{b})$ has a rank of at most 306, since, by construction, the sum of the first seven components of $\underset{\sim}{b}$ is equal to the sum of the 8th through 307th. Consequently, at least one row of the system is redundant, and we may as well drop the first row. We now may convince ourselves rather easily that the matrix $\tilde{\underset{\sim}{A}}$ obtained in this way has rank 306; e.g., the determinant of the matrix formed from columns 1 through 300 plus 301, 601,...,1801, is one. Let $\tilde{\underset{\sim}{b}}$ be the vector obtained from $\underset{\sim}{b}$ by eliminating the first component. The system $\tilde{\underset{\sim}{A}}\underset{\sim}{x} = \tilde{\underset{\sim}{b}}$ satisfies (2.4).

Now let $\underset{\sim}{x}$ be a vertex of M, the set of feasible vectors for the optimization problem (2.1), (2.2). By the corollary to theorem 1, $\underset{\sim}{x}$ has at most m positive components x_k. Let Z' be the set of indices of these components, so that $x_k > 0$ for $k \in Z'$ and $x_k = 0$ for $k \notin Z'$. By theorem 1, the column vectors $\underset{\sim}{a}^k$ of $\underset{\sim}{A}$ with $k \in Z'$ are linearly independent.

<u>Theorem 8</u>: For the above vertex $\underset{\sim}{x}$ we can find m linearly independent column vectors $\underset{\sim}{a}^k$, $k \in Z$, in the matrix $\underset{\sim}{A}$, which include the given vectors $\underset{\sim}{a}^k$, $k \in Z'$; i.e. $Z' \subset Z$.

<u>Proof</u>: If $\underset{\sim}{x}$ is a regular (= not degenerate) vertex, the conclusion follows immediately from theorem 1, and $Z' = Z$. If $\underset{\sim}{x}$ is a degenerate vertex, we have $r < m$ linearly independent column vectors $\underset{\sim}{a}^k$, $k \in Z'$, and by a well-known theorem on matrices, there are $m-r$ additional

column vectors $\underset{\sim}{a}^k$ which allow us to complete the system to m linearly independent vectors.

Definition. A system of m linearly independent column vectors of the matrix $\underset{\sim}{A}$, which has been assigned to a vertex $\underset{\sim}{x}$, in accordance with theorem 8, is called a basis for the vertex $\underset{\sim}{x}$.

Although a regular vertex has a uniquely determined basis, a degenerate vertex generally has several bases.

Applying theorem 8, the polyhedral case of the linear optimization problem (2.1), (2.2) now may be solved (theoretically) in the following manner. Form all possible systems of m column vectors $\{\underset{\sim}{a}^k;\ k \in Z\}$ from the n vectors $\underset{\sim}{a}^1,\ldots,\underset{\sim}{a}^n$. There are $\binom{n}{m}$ such systems. Now eliminate all systems for which the $\underset{\sim}{a}^k$ are linearly dependent. For the remaining systems, compute the numbers t_k in the equations $\sum_{k\in Z} \underset{\sim}{a}^k t_k = \underset{\sim}{b}$. Next eliminate any system for which at least one t_k is negative. For the remaining systems set $x_k = t_k$ if $k \in Z$, and set $x_k = 0$, if $k \notin Z$. The vector $\underset{\sim}{x}$ with components x_k is a vertex of M, by theorem 1; conversely, every vertex of M is obtained in this way, by theorem 8. If M is a polyhedron, $Q(\underset{\sim}{x})$ attains its minimum at a vertex, by theorem 6. For every vector $\underset{\sim}{x}$ obtained by this method, compute $Q(\underset{\sim}{x})$. The smallest of these numbers is the minimal value, and the corresponding vertex solves the optimization problem.

In practice, this method is rarely applicable, because $\binom{n}{m}$ grows very rapidly. Even for small numbers, say $n = 20$ and $m = 10$, we already get a large result, since

$\binom{20}{10} = 184{,}756$. We really need a more selective method, one which will pick out the vertices $\underset{\sim}{x}$ for which $Q(\underset{\sim}{x})$ is minimal with greater efficiency. One such process, the simplex method, will be described in the following sections.

§3. Vertex Exchange and the Simplex Method

§3.1 describes the process of vertex exchange. The computational basis for the process rests in formulas (3.4) and (3.5). §3.3 presents a careful discussion of the case where the exchange leads to a degenerate vertex. Again, this is necessarily somewhat tedious, and the reader who is interested primarily in practical applications may skip this section.

3.1. Vertex Exchange

We consider a linear optimization problem of the type of (2.1), (2.2); i.e.,

$$Q(\underset{\sim}{x}) = \underset{\sim}{p}'\underset{\sim}{x} = \text{Min!}, \quad \underset{\sim}{A}\underset{\sim}{x} = \underset{\sim}{b}, \quad \underset{\sim}{x} \geq 0.$$

The rank of matrix $\underset{\sim}{A}$ is equal to the row number, m, and less than the column number, n.

Let $\underset{\sim}{x}^0$ be a vertex of the set M of feasible vectors, and form a basis for this vertex from the linearly independent column vectors $\underset{\sim}{a}^k$, $k \in Z$, of matrix $\underset{\sim}{A}$. Here, as in §2, Z is a subset of the index set, $k = 1,\ldots,n$, containing exactly m of the indices. Since $\underset{\sim}{x}^0 \in M$ and $x_k^0 = 0$ for $k \notin Z$, we have

$$\sum_{k\varepsilon Z} x_k^0 a^k = b. \tag{3.1}$$

Because the vectors a^k, $k \varepsilon Z$, are linearly independent, they form a basis for R^m; in particular, every column vector of matrix A can be written as a linear combination of these vectors:

$$a^i = \sum_{k\varepsilon Z} c_{ki} a^k \quad (i = 1,\ldots,n). \tag{3.2}$$

If $j \varepsilon Z$, $c_{kj} = \delta_{kj}$, where $\delta_{kj} = 0$ if $k \neq j$ and $\delta_{jj} = 1$.

For the present, we assume that x^0 is a regular vertex, so that we have $x_k^0 > 0$, $k \varepsilon Z$. We show that if one of the numbers c_{ki}, $k \varepsilon Z$ and $i \notin Z$, is positive (say $c_{\hat{k}j} > 0$), then we can find a new vertex, x^1, with a basis consisting of the vector a^j together with all of the vectors a^k, $k \varepsilon Z$, save one.

For $\delta \geq 0$, let $x(\delta)$ be the vector with components

$$\left.\begin{aligned} x_k(\delta) &= x_k^0 - \delta c_{kj} \quad (k \varepsilon Z) \\ x_j(\delta) &= \delta \\ x_i(\delta) &= 0 \quad (i \notin Z,\ i \neq j). \end{aligned}\right\} \tag{3.3}$$

$x(\delta)$ is chosen to satisfy $Ax(\delta) = b$, for by (3.1) and (3.2),

$$Ax(\delta) = \sum_{k\varepsilon Z} (x_k^0 - \delta c_{kj}) a^k + \delta a^j = \sum_{k\varepsilon Z} x_k^0 a^k = b.$$

Also, every component of $x(\delta)$ is non-negative for $0 \leq \delta < \delta_1$, where

$$\delta_1 = \underset{k}{\text{Min}} \left(\frac{x_k^0}{c_{kj}} \right) \tag{3.4}$$

and the minimum is taken over all $k \in Z$ with $c_{kj} > 0$. Since such a k exists, namely $\hat{k}$, and since every $x_k^0 > 0$, $0 < \delta_1 < \infty$. So in the range $0 \leq \delta < \delta_1$, $\underset{\sim}{x}(\delta) \in M$. Setting $\underset{\sim}{x}(\delta_1) = \underset{\sim}{x}^1$, we see not only that $\underset{\sim}{x}^1 \in M$, but also that $\underset{\sim}{x}^1$ is a (possibly degenerate) vertex of M. For the latter remark, first observe that $\underset{\sim}{x}^1$ has at most m non-zero components, because $x_i^1 = 0$ when $i \notin Z$ and $i \neq j$, and $x_\ell^1 = 0$ if $k = \ell$ is an index for which the minimum in (3.4) is attained. Therefore, $x_k^1 \neq 0$ is possible only if $k \in Z$ and $k \neq \ell$, or $k = j$. Next observe that the vectors $\underset{\sim}{a}^k$, $k \in Z$ and $k \neq \ell$, and $\underset{\sim}{a}^j$ are linearly independent. For suppose contrarily that they are dependent. Then there are numbers, d_k, $k \in Z$ and $k \neq \ell$, and d_j, not all zero, such that $\sum_{k\in Z, k\neq\ell} d_k \underset{\sim}{a}^k + d_j \underset{\sim}{a}^j = \underset{\sim}{0}$. This implies $d_j \neq 0$, because the $\underset{\sim}{a}^k$, $k \in Z$, are linearly independent. Without loss of generality, $d_j = 1$. Now apply (3.2) to get

$$\underset{\sim}{0} = \sum_{k\in Z, k\neq\ell} d_k \underset{\sim}{a}^k + \underset{\sim}{a}^j = c_{\ell j} \underset{\sim}{a}^\ell + \sum_{k\in Z, k\neq\ell} (d_k + c_{kj}) \underset{\sim}{a}^k.$$

Since the vectors $\underset{\sim}{a}^k$, $k \in Z$, are linearly independent, all coefficients vanish; in particular, $c_{\ell j} = 0$. But this is a contradiction because ℓ is an index where the minimum in (3.4) is attained, so $c_{\ell j} > 0$. Thus, the vectors $\underset{\sim}{a}^k$, $k \in Z$ and $k \neq \ell$, and $\underset{\sim}{a}^j$ are linearly independent, and $\underset{\sim}{x}^1$ is a vertex for which these vectors form a basis. The index set Z' belonging to $\underset{\sim}{x}^1$ is formed from the index set Z belonging to $\underset{\sim}{x}^0$ by dropping ℓ and adding j.

The new basis $\underset{\sim}{a}^k$, $k \in Z'$, can be used, as in (3.2), to represent the $\underset{\sim}{a}^i$, $i = 1,\ldots,n$: $\underset{\sim}{a}^i = \sum_{k\in Z'} c'_{ki}\underset{\sim}{a}^k$. We want to express the c'_{ki} in terms of the c_{ki}. Because $c_{\ell j} > 0$, it follows from (3.2) that, for $i = j$,

$$\underset{\sim}{a}^\ell = \frac{1}{c_{\ell j}}\left(\underset{\sim}{a}^j - \sum_{k\in Z, k\neq \ell} c_{kj}\underset{\sim}{a}^k\right)$$

and this implies that

$$\underset{\sim}{a}^i = \frac{c_{\ell i}}{c_{\ell j}}\underset{\sim}{a}^j + \sum_{k\in Z, k\neq\ell}\left(c_{ki} - \frac{c_{\ell i}c_{kj}}{c_{\ell j}}\right)\underset{\sim}{a}^k .$$

Consequently,

$$\left.\begin{array}{l} c'_{j\ell} = \dfrac{1}{c_{\ell j}}, \qquad c'_{k\ell} = -\dfrac{c_{kj}}{c_{\ell j}} \quad (k \in Z,\ k \neq \ell), \\ \left.\begin{array}{l} c'_{ji} = \dfrac{c_{\ell i}}{c_{\ell j}}, \\ c'_{ki} = c_{ki} - \dfrac{c_{\ell i}c_{kj}}{c_{\ell j}} \quad (k \in Z,\ k \neq \ell) \end{array}\right\} (i \neq \ell). \end{array}\right\} \qquad (3.5)$$

These conversion formulas for the c_{ki} are reminiscent of the Gauss-Jordan algorithm for solving a sytem of linear equations or inverting a matrix. Actually, the Gauss-Jordan algorithm can be derived as a consequence of the exchange process just described (E. Stiefel, 1960).

The treatment in Stiefel regards a vertex exchange as a change of variables. Thus let $\underset{\sim}{x}^0$ be a vertex of M with basis $\underset{\sim}{a}^k$, $k \in Z$. According to (3.1), $\sum_{k\in Z}\underset{\sim}{a}^k x_k^0 = \underset{\sim}{b}$, and for the remaining components, $x_i^0 = 0$, $i \notin Z$. Next let $\underset{\sim}{x}$ be an arbitrary point of M, so that $\sum_{i=1}^n x_i\underset{\sim}{a}^i = \underset{\sim}{b}$ and

$\underset{\sim}{x} \geq 0$.

Apply (3.2) to get

$$\sum_{k\varepsilon Z} \underset{\sim}{a}^k x_k^0 = \underset{\sim}{b} = \sum_{i=1}^n x_i \sum_{k\varepsilon Z} c_{ki}\underset{\sim}{a}^k = \sum_{k\varepsilon Z} \underset{\sim}{a}^k \sum_{i=1}^n c_{ki}x_i.$$

Since the vectors $\underset{\sim}{a}^k$ are linearly independent, this forces $x_k^0 = \sum_{i=1}^n c_{ki}x_i$, $k \varepsilon Z$.

Because $c_{ki} = \delta_{ki}$ for $k, i \varepsilon Z$, it follows that

$$x_k = x_k^0 - \sum_{i\not\varepsilon Z} c_{ki}x_i \quad (k \varepsilon Z). \tag{3.6}$$

The system of equations (3.6) may be interpreted as a solution of $\underset{\sim}{A}\underset{\sim}{x} = \underset{\sim}{b}$ in the variables x_k, $k \varepsilon Z$. Because the square submatrix of $\underset{\sim}{A}$ corresponding to the x_k, $k \varepsilon Z$, is non-singular, such a solution is possible. In a vertex exchange, going from vertex $\underset{\sim}{x}^0$ to vertex $\underset{\sim}{x}^1$ with basis $\underset{\sim}{a}^k$, $k \varepsilon Z'$, changes (3.6) to

$$x_k = x_k^1 - \sum_{i\not\varepsilon Z'} c'_{ki}x_i \quad (k \varepsilon Z') \tag{3.6a}$$

where the c'_{ki} are defined by (3.5). Conversely, we can obtain conversion formulas (3.5) as follows. From the equations (3.6), choose $x_\ell = x_\ell^0 - \sum_{i\not\varepsilon Z} c_{\ell i}x_i$ and solve this equation for some x_j with $j \not\varepsilon Z$ and $c_{\ell j} > 0$. Substitute this solution in the remaining equations, and reduce the system to (3.6a).

In the case of a degenerate vertex $\underset{\sim}{x}^0$, the above considerations for a regular vertex remain valid with certain modifications. It may happen that $x_k^0 = 0$, even for some $k \varepsilon Z$. If, in (3.4), $c_{kj} > 0$ only for those $k \varepsilon Z$

for which $x_k^0 > 0$, it will still be true that $\delta_1 > 0$, and the exchange process leads to a new vertex $\underset{\sim}{x}^1$.

If, on the other hand, there are indices $k \in Z$ with $c_{kj} > 0$ and $x_k^0 = 0$, we have instead that $\delta_1 = 0$ and $\underset{\sim}{x}(\delta_1) = \underset{\sim}{x}^0$. The exchange process keeps us at vertex $\underset{\sim}{x}^0$, but changes the basis to another associated to this vertex (as we observed in §2, several bases may be associated with a vertex which is degenerate).

Example:

$$\underset{\sim}{A} = \begin{pmatrix} 2 & 4 & -1 & 1 & 0 & 0 \\ -3 & 2 & -2 & 0 & 1 & 0 \\ 0 & -1 & -3 & 0 & 0 & 1 \end{pmatrix} \qquad \underset{\sim}{b} = \begin{pmatrix} 9 \\ 4 \\ 5 \end{pmatrix}$$

$$\underset{\sim}{x}^0 = (0,0,0,9,4,5)'.$$

$Z = \{4,5,6\}$. The second column of $\underset{\sim}{A}$ contains a positive matrix element, so we let $j = 2$. Then

$$\delta_1 = \mathrm{Min}\left(\frac{9}{4}, \frac{4}{2}\right) = 2,$$

so $\ell = 5$. By (3.3), the new vertex is $\underset{\sim}{x}^1 = (0,2,0,1,0,7)'$ and $Z' = \{2,4,6\}$.

3.2. The Simplex Method

The simplex method runs through several stages. At each stage, there is an exchange of vertices by the process just described, and after this exchange, the value of $Q(\underset{\sim}{x})$ has been reduced. As before, the matrix $\underset{\sim}{A}$ has rank m, $\underset{\sim}{x}^0$ is a vertex of M, $\underset{\sim}{a}^k$, $k \in Z$, is a basis at $\underset{\sim}{x}^0$, and the numbers c_{ki} are defined by (3.2). We set

$$Q^0 = Q(\underset{\sim}{x}^0) = \underset{\sim}{p}'\underset{\sim}{x}^0 = \sum_{k\varepsilon Z} p_k x_k^0,$$
$$t_i = \sum_{k\varepsilon Z} c_{ki} p_k \quad (i = 1,\ldots,n). \tag{3.7}$$

For arbitrary $\underset{\sim}{x} \varepsilon M$, we have by (3.6)

$$Q(\underset{\sim}{x}) = \sum_{k\varepsilon Z} p_k x_k + \sum_{i\not\varepsilon Z} p_i x_i =$$
$$= \sum_{k\varepsilon Z} p_k x_k^0 + \sum_{i\not\varepsilon Z} (p_i - \sum_{k\varepsilon Z} p_k c_{ki}) x_i,$$

so by (3.7),

$$Q(\underset{\sim}{x}) = Q^0 - \sum_{i\not\varepsilon Z} (t_i - p_i) x_i. \tag{3.8}$$

<u>Theorem 1</u>: Let $\underset{\sim}{x}^0$ be a regular vertex. Let $\hat{k} \varepsilon Z$ and $j \not\varepsilon Z$ be a pair of indices with $t_j > p_j$ and $c_{\hat{k}j} > 0$. Then the result of vertex exchange is a vertex $\underset{\sim}{x}^1$ with $Q(\underset{\sim}{x}^1) < Q^0 = Q(\underset{\sim}{x}^0)$.

<u>Proof</u>: For vertex $\underset{\sim}{x}^1$, we have $x_i^1 = 0$ for $i \not\varepsilon Z$ and $i \neq j$, and $x_j^1 = \delta_1 > 0$. By (3.8), $Q(\underset{\sim}{x}^1) = Q^0 - \delta_1(t_j - p_j) < Q^0$.

<u>Theorem 2</u>: Let $\underset{\sim}{x}^0$ be a vertex (regular or degenerate). Let there be a $j \not\varepsilon Z$ with $t_j > p_j$ and $c_{kj} \leq 0$ for all $k \varepsilon Z$. Then the optimization problem has no solution.

<u>Proof</u>: The vector $\underset{\sim}{x}(\delta)$ defined by (3.3) belongs to M for all $\delta > 0$. By (3.8), $Q(\underset{\sim}{x}(\delta)) = Q^0 - \delta(t_j - p_j)$. Therefore $Q(\underset{\sim}{x})$ is not bounded below on M.

Theorems 1 and 2 make evident that as δ increases from 0 through positive values, the path $\underset{\sim}{x}(\delta)$ defined by

(3.3) leaves $\underset{\sim}{x}^0$ along an edge of M in a direction where Q has smaller values than at $\underset{\sim}{x}^0$. In the case of theorem 2, this edge is infinitely long. In the case of theorem 1, when $\delta = \delta_1$, we have arrived at a new vertex, $\underset{\sim}{x}^1$. Let us suppose for a moment that M contains only regular vertices, and that we have found one of these, $\underset{\sim}{x}^0$. By repeated application of theorem 1, we obtain vertices $\underset{\sim}{x}^0, \underset{\sim}{x}^1, \underset{\sim}{x}^2, \ldots$ such that $Q(\underset{\sim}{x}^0) > Q(\underset{\sim}{x}^1) > Q(\underset{\sim}{x}^2) > \ldots$. In this process, no vertex can appear twice. Since M has only finitely many vertices, by §2, theorem 2, the process must end after finitely many stages, in one of the following two cases.

(1) There is an index $j \notin Z$ with $t_j > p_j$ and $c_{kj} \leq 0$ for all $k \in Z$. By theorem 2, the optimization problem has no solution.

(2) $t_j \leq p_j$ for all $j \notin Z$. Then the optimization problem is solved, by the following theorem.

<u>Theorem 3</u>: If $\underset{\sim}{x}^0$ is a vertex, possibly degenerate, such that $t_j \leq p_j$ for all $j \notin Z$, then $\underset{\sim}{x}^0$ is a minimal point.

<u>Proof</u>: Let $\underset{\sim}{x} = (x_1, \ldots, x_n)'$ be an arbitrary point of M. Because $\underset{\sim}{x} \geq 0$ and $t_j - p_j \leq 0$, $j \notin Z$, (3.8) implies that

$$Q(\underset{\sim}{x}) = Q^0 - \sum_{j \notin Z} (t_j - p_j) x_j \geq Q^0.$$

3.3. <u>Degenerate Vertices</u>

A degenerate vertex is characterized by having fewer than m positive components. If $\underset{\sim}{x}^0$ is such a degenerate vertex, and if the vectors $\underset{\sim}{a}^k$, $k \in Z$, form a basis at $\underset{\sim}{x}^0$, we first can find the c_{ki} by (3.2), and next the t_i by (3.7).

The following cases can arise.

1. $t_j \leq p_j$ for all $j \notin Z$. By theorem 3, $\underset{\sim}{x}^0$ is a minimal point.

2. There exists an index $j \notin Z$ such that $t_j > p_j$ and $c_{kj} \leq 0$ for all $k \in Z$. By theorem 2, the optimization problem has no solution.

3. There exist indices $j \notin Z$ with $t_j > p_j$, and for each such index j there exists an index $\hat{k} \in Z$ such that $c_{\hat{k}j} > 0$. For each such index j, we can define a δ_1 by (3.4). Since $\underset{\sim}{x}^0$ is a degenerate vertex, it is possible that a $\delta_1 = 0$.

3.1. For one of these indices j, $\delta_1 > 0$. Then a vertex exchange, described above, results in a different vertex $\underset{\sim}{x}^1$, which may also be degenerate, but for which $Q(\underset{\sim}{x}^1) < Q(\underset{\sim}{x}^0)$.

3.2. For all of these indices j, $\delta_1 = 0$. Then a vertex exchange, for any of these j, results in a change of basis at the same vertex $\underset{\sim}{x}^0$. Q is not reduced by the process.

Should case 3.2 arise several times in succession, we will have remained at vertex $\underset{\sim}{x}^0$ and merely changed bases each time. In particular, it may happen that a basis reappears after several stages. If the computation is carried forward, there will be a cyclic repetition of these stages. In practice, however, we may rely on the fact that such loops are extremely rare. To date, there are only a few examples, where such loops arise, in the literature, and these do not stem from practical applications of linear optimization, but were constructed to demonstrate their existence, cf. Gass, 1964, p. 119 ff. Should a degenerate vertex appear in a practical application of the simplex method, one should simply carry on the computations.

Nevertheless, it is desirable to construct a closed theory for the simplex method. To this end, we will show how one additional rule completes the process, so that loops cannot occur and a minimal point or its nonexistence is determined in finitely many steps.

If no degenerate vertices appear, the index $k \in Z$ which minimizes the quotient x_k^0/c_{kj} in (3.4) is uniquely determined. For if $\ell, \ell' \in Z$ were distinct indices with $\delta_1 = x_\ell^0/c_{\ell j} = x_{\ell'}^0/c_{\ell' j}$, then $x_\ell(\delta_1) = x_{\ell'}(\delta_1) = 0$. But then $\underset{\sim}{x}^1 = \underset{\sim}{x}(\delta_1)$ would have fewer than m positive components, and be degenerate. If, on the other hand, the index ℓ for which the minimum of (3.4) is attained, is always uniquely determined, and if there is a regular vertex at which to start the simplex method, then no subsequent vertex will be degenerate. So the case of a degenerate vertex is characterized by the non-uniqueness of the index ℓ which yields the minimum in (3.4) and the column vector $\underset{\sim}{a}^\ell$ that is dropped from the basis in a vertex exchange.

Our additional rule for avoiding loops will have the effect of determining a unique choice of index ℓ, even when there appear to be several choices available.

In order to formulate the additional rule, we need

the concept of a lexicographic ordering of vectors.

Definition: A vector $\underset{\sim}{v}$ with N components $v_1,\ldots,v_N$ is lexicographically positive ($\underset{\sim}{v} \succ \underset{\sim}{0}$), if $\underset{\sim}{v} \neq \underset{\sim}{0}$ and the first non-vanishing component is positive, i.e. if there is an index p, $1 \leq p \leq N$, such that $v_j = 0$ for $j < p$ and $v_p > 0$.

A vector $\underset{\sim}{v}$ is lexicographically greater than a vector $\underset{\sim}{u}$ ($\underset{\sim}{v} \succ \underset{\sim}{u}$), if $\underset{\sim}{v}-\underset{\sim}{u} \succ \underset{\sim}{0}$.

It is easily checked that the relation $\succ$ has all the requisite properties for an order relation:

1. If $\underset{\sim}{v} \succ \underset{\sim}{u}$, and $\underset{\sim}{u} \succ \underset{\sim}{w}$, then $\underset{\sim}{v} \succ \underset{\sim}{w}$.
2. If $\underset{\sim}{v} \succ \underset{\sim}{u}$, then $\underset{\sim}{v}+\underset{\sim}{w} \succ \underset{\sim}{u}+\underset{\sim}{w}$ for all $\underset{\sim}{w} \in R^N$.
3. If $\underset{\sim}{v} \succ \underset{\sim}{u}$ and $c > 0$, then $c\underset{\sim}{v} \succ c\underset{\sim}{u}$.

Given any two vectors $\underset{\sim}{u}$ and $\underset{\sim}{v}$, either $\underset{\sim}{u} \succ \underset{\sim}{v}$ or $\underset{\sim}{u} = \underset{\sim}{v}$ or $\underset{\sim}{v} \succ \underset{\sim}{u}$.

Let there be given a linear optimization problem (2.1), (2.2) where the matrix $\underset{\sim}{A}$ has rank m. Let $\underset{\sim}{x}^s = \underset{\sim}{x}^{start}$ be a known vertex of the set M of feasible vectors; this vertex will serve as the initial vertex to start the simplex method. By renumbering the indices, if necessary, we can arrange to have the column vectors $\underset{\sim}{a}^1,\ldots,\underset{\sim}{a}^m$ form the basis at $\underset{\sim}{x}^s$, where the simplex method initiates. The index set for the initial vector is then $Z = \{1,2,\ldots,m\}$. We now proceed under the rules of §3.1 and §3.2 and of the following.

Additional Rule: In the situation where the simplex method has produced a vertex $\underset{\sim}{x}^0$ (with basis vectors $\underset{\sim}{a}^k$,

$k \in Z$, and numbers c_{ki} and t_i determined by (3.2) and (3.7)) for which there is an index $j \notin Z$ with $t_j > p_j$ and with $k \in Z$ such that $c_{kj} > 0$, do the following.

For each such k, define a vector with $m+1$ components by

$$\underset{\sim}{w} = \left(\frac{x_k^0}{c_{kj}}, \frac{c_{k1}}{c_{kj}}, \frac{c_{k2}}{c_{kj}}, \ldots, \frac{c_{km}}{c_{kj}} \right)'. \tag{3.9}$$

Order these vectors lexicographically and choose the smallest. It will have index $k = \ell$, say.

Use the index ℓ to carry out the vertex exchange, as described in §3.1.

This additional rule amounts to the following. If there are several indices $k \in Z$ where x_k^0/c_{kj} is minimal in (3.4), select those for which c_{k1}/c_{kj} is minimal; if there are several of these, select those for which c_{k2}/c_{kj} is minimal; etc. This determines a unique index ℓ. For if $\ell \neq \ell'$ are indices defining equal vectors $\underset{\sim}{w}$ in (3.9), then the corresponding two rows in the square matrix of c_{ki}'s, $k \in Z$, $i = 1,\ldots,m$, are proportional, and this matrix is singular. In fact it is non-singular because it represents a change of basis: the linearly independent vectors $\underset{\sim}{a}^1,\ldots,\underset{\sim}{a}^m$ (the basis at $\underset{\sim}{x}^s$) are represented in terms of the vectors $\underset{\sim}{a}^k$, $k \in Z$, by $\underset{\sim}{a}^i = \sum_{k \in Z} c_{ki}\underset{\sim}{a}^k$ $(i = 1,\ldots,m)$.

For each vertex $\underset{\sim}{x}^0$ appearing in an application of the simplex method, define a vector $\underset{\sim}{v}^0$, with $m+1$ components, by

$$\underset{\sim}{v}^0 = (Q(\underset{\sim}{x}^0), t_1, \ldots, t_m)', \tag{3.10}$$

where the t_i are defined as in (3.7).

Theorem 4: In a vertex exchange carried out in accordance with the additional rule, the vector $\underset{\sim}{v}^0$ is replaced by a lexicographically smaller one.

Proof: It follows from (3.8) that $Q(\underset{\sim}{x}^0)$ is replaced by

$$Q(\underset{\sim}{x}^1) = Q(\underset{\sim}{x}^0) - \frac{x_\ell^0}{c_{\ell j}}(t_j - p_j);$$

$t_i = \sum_{k\varepsilon Z} c_{ki}p_k$ is replaced by $t_i' = \sum_{k\varepsilon Z'} c_{ki}'p_k$ $(i = 1,\ldots,m)$, where the c_{ki}' are defined as in (3.5). Now if $i \neq \ell$, then

$$\begin{aligned} t_i' &= \sum_{k\varepsilon Z, k\neq\ell}\left(c_{ki} - \frac{c_{\ell i}c_{kj}}{c_{\ell j}}\right)p_k + \frac{c_{\ell i}}{c_{\ell j}}p_j = \\ &= \sum_{k\varepsilon Z} c_{ki}p_k - \frac{c_{\ell i}}{c_{\ell j}}\sum_{k\varepsilon Z} c_{kj}p_k + \frac{c_{\ell i}}{c_{\ell j}}p_j = \\ &= t_i - \frac{c_{\ell i}}{c_{\ell j}}(t_j - p_j), \end{aligned}$$

while if $i = \ell$, then

$$t_\ell' = -\sum_{k\varepsilon Z, k\neq\ell}\frac{c_{kj}}{c_{\ell j}}p_k + \frac{1}{c_{\ell j}}p_j = t_\ell - \frac{c_{\ell\ell}}{c_{\ell j}}(t_j - p_j),$$

since $\ell \varepsilon Z$ implies $t_\ell = p_\ell$ and since $c_{\ell\ell} = 1$. The vertex exchange trades $\underset{\sim}{v}^0$ for the vector $\underset{\sim}{v}^1 = \underset{\sim}{v}^0 - \underset{\sim}{w}(t_j - p_j)$ where $\underset{\sim}{w}$ is the vector defined by (3.9) with $k = \ell$, i.e.,

$$\underset{\sim}{w} = \frac{1}{c_{\ell j}}(x_\ell^0, c_{\ell 1},\ldots,c_{\ell m})'.$$

By the rules of procedure, $t_j - p_j > 0$ and $c_{\ell j} > 0$.

It remains to show that the vector $(x_\ell^0, c_{\ell 1}, \ldots, c_{\ell m})'$ is lexicographically positive. By induction on the number of stages already completed, we will show that every vector of the form $\underset{\sim}{u}^k = (x_k^0, c_{k1}, \ldots, c_{km})'$, $k \in Z$, is lexicographically positive.

1. The vectors $\underset{\sim}{u}^k$ belonging to the initial vector $\underset{\sim}{x}^s$ are lexicographically positive because $x_k^s \geq 0$ and $c_{ki} = \delta_{ki}$ for $k, i = 1,\ldots,m$.

2. Let $\underset{\sim}{x}^0$ be a vertex occurring at some stage of the simplex process. By the induction hypothesis, all of the vectors $\underset{\sim}{u}^k$, $k \in Z$, formed with $\underset{\sim}{x}^0$ are lexicographically positive. After a vertex exchange, we obtain new vectors $\underset{\sim}{u}'^k$, $k \in Z'$, whose components we compute with the aid of (3.5). We find that, for $k \in Z'$ and $k \neq j$,

$$\underset{\sim}{u}'^k = (x_k^0 - \frac{x_\ell^0}{c_{\ell j}} c_{kj}, c_{k1} - \frac{c_{\ell 1} c_{kj}}{c_{\ell j}}, \ldots, c_{km} - \frac{c_{\ell m} c_{kj}}{c_{\ell j}})' =$$

$$= \underset{\sim}{u}^k - \frac{c_{kj}}{c_{\ell j}} \underset{\sim}{u}^\ell$$

and for $k = j$,

$$\underset{\sim}{u}'^j = (\frac{x_\ell^0}{c_{\ell j}}, \frac{c_{\ell 1}}{c_{\ell j}}, \ldots, \frac{c_{\ell m}}{c_{\ell j}})' = \frac{1}{c_{\ell j}} \underset{\sim}{u}^\ell .$$

By the induction hypothesis and because $c_{\ell j} > 0$, it is clear that $\underset{\sim}{u}'^j$ is lexicographically positive; similarly for $\underset{\sim}{u}'^k$, $k \neq j$, if $c_{kj} \leq 0$. But if $c_{kj} > 0$, k is an index calling the additional rule into play, according to which ℓ is to be chosen so that

$$\frac{\underset{\sim}{u}^k}{c_{kj}} - \frac{\underset{\sim}{u}^\ell}{c_{\ell j}} \succ 0$$

for every such index k (note that $\ell \notin Z'$). So even in this case, $\underset{\sim}{u}'^k \succ 0$.

From theorem 4 it follows easily that the additional rule precludes the appearance of loops. Every point $\underset{\sim}{x}^0$ and choice of basis at the point together determine uniquely a vector $\underset{\sim}{v}^0$. At every stage of the simplex process, $\underset{\sim}{v}^0$ is replaced by a lexicographically smaller vector. Therefore no basis can appear twice.

When we discuss duality in linear optimization problems in §5, we need the following result (the converse of theorem 3).

<u>Theorem 5</u>: If the vertex $\underset{\sim}{x}^0$ is a minimal point, then there exists a basis at $\underset{\sim}{x}^0$ for which $t_i \leq p_i$, $i = 1,\ldots,n$.

<u>Proof</u>: Suppose $\underset{\sim}{x}^0$ is regular. Then the basis at $\underset{\sim}{x}^0$ is uniquely determined. Theorems 1 and 2 imply that $t_i \leq p_i$ for $i = 1,\ldots,n$ (for $i \in Z$, $t_i = p_i$).

Suppose $\underset{\sim}{x}^0$ is degenerate. By theorem 4 and the additional rule, we can find a basis $\underset{\sim}{a}^k$, $k \in Z$, at $\underset{\sim}{x}^0$ for which the vector $\underset{\sim}{v}^0$ is lexicographically smallest. The numbers t_i defined by this basis satisfy $t_i \leq p_i$; otherwise $\underset{\sim}{x}^0$ would not be a minimal point (theorem 2) or $\underset{\sim}{v}^0$ would not be the lexicographically smallest vector (additional rule).

Degenerate vertices appear rather frequently in applications of the simplex process to practical cases. Loops, as already noted, have never appeared, in spite of the extremely large number of applied problems which have been solved. If the index ℓ which is used in a vertex exchange is not uniquely determined by (3.4), then we must choose among the available indices. This can be done by using the additional rule, or just as well by using a simpler rule, namely taking the smallest of the available indices. If the computation is being done by hand, we can survey the course of computation steadily. If a loop should somehow arise, we can deviate from our chosen rule for determining index ℓ, and thereby escape the loop. But if the simplex process is programmed, and the computation is done by machine, we have no discretion to deviate from the chosen rule. In this case, totally different problems also come up. Degenerate vertices may be recognized by one or more zero components x_k, $k \in Z$. Because of rounding error, numbers are rarely exactly zero, just of small absolute value. In a large-scale application it may be very difficult to program in a decision function which can distinguish between an absolutely small number that should be a zero and one that should not. A further discussion of these problems would take us beyond the frame of the present exposition.

3.4. Determination of the Initial Vertex

The description of the simplex method in §3.2 presupposed knowledge of an initial vertex $\underset{\sim}{v}^0$, from which the vertices $\underset{\sim}{v}^1$, $\underset{\sim}{v}^2$,... were constructed in successive stages, until the process ended either at a minimal point, or in the conclusion that the problem had no solution. In some applications, a vertex $\underset{\sim}{v}^0$ will become known while setting up the problem. In other cases, no vertex will become known and once the problem is set up, it may not even be apparent that any feasible points at all exist. We need a method for constructing an initial vertex $\tilde{x}^0$, in case one exists.

Consider the case where the optimization problem is presented as in (1.1a) - (1.3a). The constraints are $\tilde{A}\tilde{x} \leq \tilde{b}$ and $\tilde{x} \geq \tilde{0}$, with $\tilde{b} \geq \tilde{0}$, i.e. a vector with non-negative components. By introducing a slack variable vector $\tilde{y}$, the constraints are changed to $\tilde{A}\tilde{x}+\tilde{y} = \tilde{b}$, $\tilde{x} \geq \tilde{0}$, $\tilde{y} \geq \tilde{0}$. The column vectors belonging to $\tilde{y}$ in the expanded matrix are exactly the m unit vectors of R^m, and therefore a basis. So $\tilde{x} = \tilde{0}$, $\tilde{y} = \tilde{b}$ is a vertex.

Now suppose we are given a linear optimization problem of type (2.1), (2.2). The constraints are $\tilde{A}\tilde{x} = \tilde{b}$ and $\tilde{x} \geq \tilde{0}$. The rank of $\tilde{A}$ is m. Without loss of generality, we may assume that $\tilde{b} \geq \tilde{0}$, for if necessary we can multiply some of the equations of the system $\tilde{A}\tilde{x} = \tilde{b}$ by -1. Instead of directly attacking the problem with the given constraints and objective function $Q(\tilde{x}) = \tilde{p}'\tilde{x}$, we first attempt to solve the problem

$$y_1 + y_2 + \ldots + y_m = \text{Min!}, \quad \tilde{A}\tilde{x} + \tilde{y} = \tilde{b}, \quad \tilde{x} \geq \tilde{0}, \quad \tilde{y} \geq \tilde{0}. \tag{3.11}$$

For this problem, we know an initial vertex, namely $\tilde{x} = \tilde{0}$, $\tilde{y} = \tilde{b} \geq \tilde{0}$, as above. For this problem, the objective function is bounded below, by 0, and so there exists a solution $\tilde{x}^* \geq \tilde{0}$, $\tilde{y}^* \geq \tilde{0}$ (cf theorem 16, §5.6). If $\tilde{y}^* = \tilde{0}$, $\tilde{x}^*$ is a vertex for the original problem (2.1), (2.2). It may happen that the minimal point for problem (3.11) given by $\tilde{x}^*$, $\tilde{y}^*$ is degenerate, and that the corresponding basis contains column vectors belonging to components of $\underset{\sim}{y}^*$. Nevertheless, the column vectors of $\underset{\sim}{A}$ belonging to positive components of $\underset{\sim}{x}^*$ are linearly independent, and can be completed to a basis with other column vectors of $\underset{\sim}{A}$.

On the other hand, if the solution to (3.11) is not $\underset{\sim}{y}^* = \underset{\sim}{0}$, the initial problem (2.1), (2.2) has no feasible points. For every such feasible point $\underset{\sim}{x}$, after completion by $\underset{\sim}{y} = \underset{\sim}{0}$, yields a solution of problem (3.11) with value 0 for the objective function $y_1 + \ldots + y_m$.

§4. Algorithmic Implementation of the Simplex Method

By applying the rules of procedure developed in §3, we can implement the simplex method numerically. We can do this most expediently by using a computation tableau. Such a tableau is described extensively in §4.1 and §4.2. It is actually so simple, and so nearly self-evident, that the reader easily can assimilate the operation of the simplex method by considering the numerical examples in §4.3, and the rule for a vertex exchange in §4.2.

4.1. Description of the Tableau

For each stage of the simplex process, we assemble a tableau whose fields contain all the requisite data, namely

① The indices k belonging to index set Z.

② The indices i not belonging to index set Z.

③ The numbers c_{ki} (cf (3.2)) with $k \in Z$ and $i \notin Z$.

(For $k \in Z$ and $i \in Z$, $c_{ki} = \delta_{ki}$, which is 0 or 1; in the tableau they are superfluous.)

(4) The components x_k with $k \in Z$. (for $i \notin Z$, $x_i = 0$.)

(5) The numbers $t_i - p_i$ (cf (3.7)) for $i \notin Z$. For brevity of notation, set

$$d_i = t_i - p_i.$$

If $i \in Z$, we have $d_i = 0$. These d_i are dropped.

(6) The respective value of the objective function Q.

(7) For purposes of a sum test, we define and enter

$$\sigma_i = 1 - \sum_{k\in Z} c_{ki} - d_i \qquad (i \notin Z)$$

so that the column sums over fields (3), (5), and (7) are all 1.

(8) Correspondingly, define and enter

$$\sigma = 1 - \sum_{k\in Z} x_k - Q. \tag{4.2}$$

(9) This field is reserved for the x_k/c_{kj}, whose minimum is to be found in (3.4).

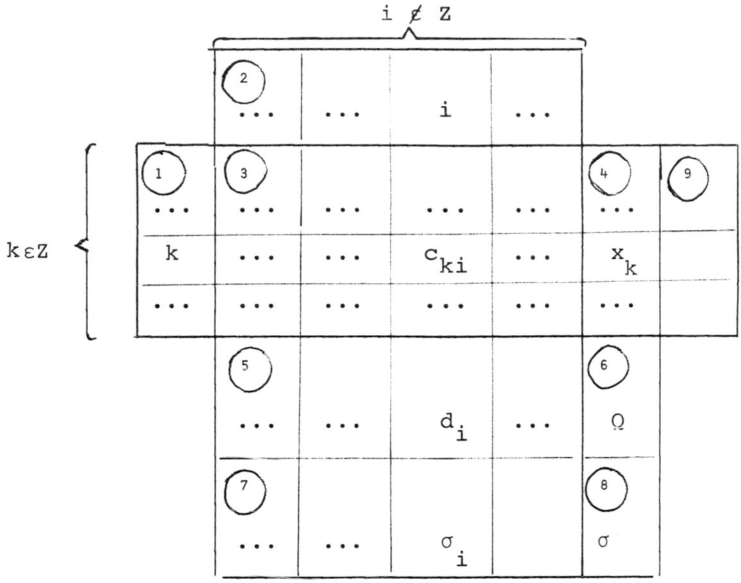

(This tableau, with 3 rows and 4 columns for the c_{ki}, would be suited for the case $m = 3$ and $n = 3+4 = 7$.)

The entries in fields (1) and (2) need not be the indices $k \in Z$ and $i \notin Z$. Instead it is frequently more expedient to use the notation for the respective variables, particularly when these are not described uniformly by $x_1,\ldots,x_n$.

To even begin, we must know how to fill in the tableau at the initial stage of the simplex process. We first treat the case where the optimization problem is originally of type (1.1a) - (1.3a), with constraints $\underset{\sim}{A}\underset{\sim}{x} \leq \underset{\sim}{b}$ and $\underset{\sim}{x} \geq \underset{\sim}{0}$, where $\underset{\sim}{b} \geq \underset{\sim}{0}$. As we saw in §3.4, this is the case where an initial vertex for the simplex process is found easily. Let the objective function be $Q(\underset{\sim}{x}) = \underset{\sim}{p}'\underset{\sim}{x}$, and let our objective be the minimum of $Q(\underset{\sim}{x})$, as in §3 (and not the maximum, as in (1.1a)). The introduction of slack variables changes the constraints to $\underset{\sim}{A}\underset{\sim}{x}+\underset{\sim}{y} = \underset{\sim}{b}$, $\underset{\sim}{x} \geq \underset{\sim}{0}$, and $\underset{\sim}{y} \geq \underset{\sim}{0}$. Choose the notation and indices of the components of $\underset{\sim}{x}$ and $\underset{\sim}{y}$ so that $\underset{\sim}{x} = (x_1,\ldots,x_q)'$ and $\underset{\sim}{y} = (x_{q+1},\ldots,x_n)'$, where $q = n-m$. As in (1.7), the vector $\underset{\sim}{p}$ in the objective function is completed by the m components $p_{q+1} = 0,\ldots,p_n = 0$.

By §3.4, $\underset{\sim}{x} = \underset{\sim}{0}$, $\underset{\sim}{y} = \underset{\sim}{b}$ provides an initial vertex for the simplex process. The corresponding basis consists of the m unit vectors

$$\underset{\sim}{a}^{q+1} = \begin{pmatrix} 1 \\ 0 \\ \\ \cdots \\ 0 \end{pmatrix}, \quad \underset{\sim}{a}^{q+2} = \begin{pmatrix} 0 \\ 1 \\ \\ \cdots \\ 0 \end{pmatrix}, \ldots, \underset{\sim}{a}^{n} = \begin{pmatrix} 0 \\ 0 \\ \\ \cdots \\ 1 \end{pmatrix}.$$

The index set Z turns out to be $Z = \{q+1, q+2,\ldots,n\}$.

For the column vectors $\underset{\sim}{a}^1,\ldots,\underset{\sim}{a}^q$ of matrix $\underset{\sim}{A}$ we have the representation (3.2)

$$\underset{\sim}{a}^i = \sum_{k=1}^{m} a_{ki}\underset{\sim}{a}^{q+k} \qquad (i = 1,\ldots,q).$$

So in field ③, for the c_{ki} we simply enter the elements a_{ki}, $k = 1,\ldots,m$; $i = 1,\ldots,q$, of matrix $\underset{\sim}{A}$.

Since $\underset{\sim}{y} = \underset{\sim}{b}$, we enter in field ④ the components $b_1,\ldots,b_m$ of vector $\underset{\sim}{b}$.

In field ⑤ we enter the d_i, and here $d_i = \sum_{k=q+1}^{n} c_{ki}p_k - p_i = -p_i$ $(i = 1,\ldots,q)$ because $p_{q+1} = \cdots = p_n = 0$; in field ⑥, the value of the objective function, namely $Q(\underset{\sim}{x}) = \sum_{k=q+1}^{n} p_k x_k = \underset{\sim}{0}$. The entries for the remaining fields require no elucidation. The tableau for the initial step thus has the following form.

	(2) 1	2	...	q		
(1) q+1	(3) a_{11}	a_{12}	...	a_{1q}	(4) b_1	(9)
...	...	...	...	...		
n	a_{m1}	a_{m2}	...	b_m		
	(5) $-p_1$	$-p_2$	...	$-p_q$	(6) 0	
	(7) σ_1	σ_2	...	σ_q	(8) σ	

Later, we will discuss how to fill in the initial talbeau when the problem is not of type (1.1a) - (13a) with constraints $\underset{\sim}{A}\underset{\sim}{x} \leq \underset{\sim}{b}$, $\underset{\sim}{x} \geq \underset{\sim}{0}$ where $\underset{\sim}{b} \geq \underset{\sim}{0}$, but of type (2.1), (2.2) with constraints $\underset{\sim}{A}\underset{\sim}{x} = \underset{\sim}{b}$, $\underset{\sim}{x} \geq \underset{\sim}{0}$. A problem of type (1.1a) - (1.3a) not satisfying $\underset{\sim}{b} \geq \underset{\sim}{0}$ can be reduced to the case $\underset{\sim}{b} \geq \underset{\sim}{0}$ by the introduction of slack variables.

4.2. Implementation of an Exchange Step

To every stage of the simplex process we can attach a tableau of the type described. We will now show how to derive the successor from a given tableau.

We begin by searching field (5) for a positive $d_i = t_i - p_i$, $i \notin Z$. If there are none, then all the $d_i \leq 0$ and by §3, theorem 3, the optimization problem is solved. If some $d_j > 0$, we mark the column belonging to index $j \notin Z$, and search this column for a positive c_{kj}. If all $c_{kj} \leq 0$, $k \in Z$, the optimization problem has no solution, by §3, theorem 2. If there exist positive c_{kj}, then for each such index k we form the quotient x_k/c_{kj} and enter the result in the corresponding row in field (9) (some rows in field (9) will remain empty, in general). Search the quotients x_k/c_{kj} for the smallest, say $x_\ell/c_{\ell j}$ (thus determining δ_1 in (3.4)). Now mark the row with index $\ell \in Z$. The (positive) number $c_{\ell j}$ in the intersection of the marked column and marked row plays a special role in transforming the tableau and is called the <u>pivot</u> or <u>pivot element</u>. The marked row and column are also called the <u>pivot row</u> and <u>pivot column</u>. By §3, the vector $\underset{\sim}{a}^\ell$ is to be removed from the basis and exchanged for the vector $\underset{\sim}{a}^j$. In the new tableau, we replace the index ℓ in field (1) by the index j, and the index j in field (2) by the index ℓ, and retain all other indices. The c_{ki} in field (3) of the old tableau are replaced by the c'_{ki} of (3.5) in the new tableau.

(3.3) with $\delta = \delta_1 = x_\ell/c_{\ell j}$ tells how to replace the x_k in field (4). The numbers d'_i which replace the d_i in field (5) may be found as follows.

For $i \neq \ell$,

$$\begin{aligned} d'_i = \sum_{k\in Z'} c'_{ki}p_k - p_i &= \sum_{\substack{k\in Z\\ k\neq\ell}} \left(c_{ki} - \frac{c_{\ell i}c_{kj}}{c_{\ell j}}\right)p_k + \frac{c_{\ell i}}{c_{\ell j}}\, p_j - p_i \\ &= \sum_{k\in Z} c_{ki}p_k - c_{\ell i}p_\ell - \frac{c_{\ell i}}{c_{\ell j}} \sum_{k\in Z} c_{kj}p_k + c_{\ell i}p_\ell + \frac{c_{\ell i}}{c_{\ell j}}\, p_j - p_i \\ &= d_i - \frac{c_{\ell i}}{c_{\ell j}}\, d_j, \end{aligned}$$

and for $i = \ell$

$$d'_\ell = \sum_{k\varepsilon Z'} c'_{k\ell} p_k - p_\ell = -\sum_{\substack{k\varepsilon Z\\ k\neq\ell}} \frac{c_{kj}}{c_{\ell j}} p_k + \frac{1}{c_{\ell j}} p_j - p_\ell$$

$$= -\sum_{k\varepsilon Z} \frac{c_{kj}}{c_{\ell j}} p_k + \frac{1}{c_{\ell j}} p_j = -\frac{d_j}{c_{\ell j}} .$$

Furthermore, Q in field ⑥ is replaced by $Q - \frac{x_\ell d_j}{c_{\ell j}}$, in view of (3.8). The last row, namely fields ⑦ and ⑧, is modified exactly like the rows above.

Summary:

I. Selection of the pivot element:

1. Search field ③ for a $d_j > 0$. j determines the pivot column.

2. Enter quotients x_k/c_{kj} in field ⑨ for all indices $k \varepsilon Z$ with $c_{kj} > 0$.

3. Search field ⑨ for the smallest number appearing. This determines the pivot row.

II. Transforming fields ③ through ⑧:

1. The pivot $c_{\ell j}$ is replaced by $1/c_{\ell j}$.

2. Each remaining new entry in the pivot row is a multiple of the old by $(1/c_{\ell j})$; each remaining new entry in the pivot column is a multiple of the old by $(-1/c_{\ell j})$.

3. All remaining numbers are to be replaced by the rectangle rule:

$$\begin{array}{rccc} & \text{Pivot column} & & \\ & \downarrow & & \\ \text{Pivot row} \longrightarrow & a & \cdots & b \\ & \vdots & & \vdots \\ & c & \cdots & d \end{array}$$

d is to be replaced by $d - (bc)/a$.

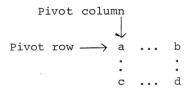

					*			
		...	i	...	j	...		
	...	...	...	...	...	...	...	...
	k	...	c_{ki}	...	c_{kj}	...	x_k	$\frac{x_k}{c_{kj}}$
	...	...	...	...	...	...	...	...
*	ℓ	...	$c_{\ell i}$	...	$c_{\ell j}$	...	x_ℓ	$\frac{x_\ell}{c_{\ell j}}$
	...	...	...	...	...	...	...	...
		...	d_i	...	d_j	...	Q	
		...	σ_i	...	σ_j	...	σ	

Old tableau

	...	i	...	ℓ	...		
...	...	...	...	...	...	...	
k	...	$c_{ki} - \frac{c_{\ell i} c_{kj}}{c_{\ell j}}$	...	$-\frac{c_{kj}}{c_{\ell j}}$	...	$x_k - \frac{x_\ell c_{kj}}{c_{\ell j}}$	
...	...	...	...	...	...	...	
j	...	$\frac{c_{\ell i}}{c_{\ell j}}$	...	$\frac{1}{c_{\ell j}}$	...	$\frac{x_\ell}{c_{\ell j}}$	
...	...	...	...	...	...	...	
	...	$d_i - \frac{c_{\ell i} d_j}{c_{\ell j}}$	...	$-\frac{d_j}{c_{\ell j}}$	...	$Q - \frac{x_\ell d_j}{c_{\ell j}}$	
	...	$\sigma_i - \frac{c_{\ell i}\sigma_j}{c_{\ell j}}$	...	$-\frac{\sigma_j}{c_{\ell j}}$	...	$\sigma - \frac{x_\ell \sigma_j}{c_{\ell j}}$	

New tableau

In practice, the last replacement is undertaken by subtracting a multiple of the already transformed pivot row from the row undergoing transformation (with the exception of the number in the pivot column). The factor used is exactly the number in the old tableau which is in the intersection of the pivot column and the row undergoing the transformation. Similarly, one can add a multiple of the transformed pivot column to the column undergoing transformation.

Once fields (1) through (8) of the new tableau have been filled in this way, all the data is assenbled for the next step of the simplex method. Before proceding it would be worthwhile to run a sum test to assure that no errors have crept into the data, especially when the computation is done by hand or by hand calculator.

The Sum Test:

In filling in the tableau for the initial step of the simplex method, we compute the σ_i and σ by (4.1) and (4.2). The column sums over fields (3), (5), and (7) as well as over fields (4), (6), and (8) are then all 1. In the transition to a new tableau, we modify the last row, i.e., fields (7) and (8), in exactly the same way as the rows above (except the pivot row) and then check that all the column sums are still 1, after the modifications. For if

$$\sum_{k \varepsilon Z} c_{ki} + d_i + \sigma_i = 1 \quad (i \not\in Z), \qquad \sum_{k \varepsilon Z} x_k + Q + \sigma = 1,$$

then also

$$-\sum_{\substack{k\varepsilon Z\\ k\neq \ell}} \frac{c_{kj}}{c_{\ell j}} + \frac{1}{c_{\ell j}} - \frac{d_j}{c_{\ell j}} - \frac{j}{c_{\ell j}} =$$

$$= -\frac{1}{c_{\ell j}}\left(\sum_{k\varepsilon Z} c_{kj} - c_{\ell j} - 1 + d_j + \sigma_j\right) = 1,$$

$$\sum_{\substack{k\varepsilon Z\\ k\neq \ell}} \left(c_{ki} - \frac{c_{\ell i}c_{kj}}{c_{\ell j}}\right) + \frac{c_{\ell i}}{c_{\ell j}} + d_i - \frac{c_{\ell i}d_j}{c_{\ell j}} + \sigma_i - \frac{c_{\ell i}\sigma_j}{c_{\ell j}}$$

$$= \sum_{k\varepsilon Z} c_{ki} - c_{\ell i} + d_i + \sigma_i - \frac{c_{\ell i}}{c_{\ell j}}\left(\sum_{k\varepsilon Z} c_{kj} - c_{\ell j} - 1 + d_j + \sigma_j\right) = 1$$

$$(i \notin Z, \quad i \neq j).$$

A similar result holds for the last column, consisting of fields (4), (6), and (8).

4.3. Example

Executing the simplex method with tableaux is simple and natural, as we will demonstrate with the sheep and cattle raising example -- number 2 of §1. As before, the constraints are

$$\begin{array}{rcrl} x_1 & & \leq & 50 \\ & x_2 & \leq & 200 \qquad x_1 \geq 0 \\ x_1 & + \; 0.2x_2 & \leq & 72 \qquad x_2 \geq 0 \\ 150x_1 & + \; 25x_2 & \leq & 10000 \end{array}$$

We want to find the minimum of the objective function, $Q(x_1,x_2) = -250x_1 - 45x_2$, subject to these constraints. (In the following tableaux the pivot element has been framed for emphasis.)

		*			
		1	2		
*	3	[1]	0	50	50
	4	0	1	200	--
	5	1	0.2	72	72
	6	150	25	10000	66.67
		250	45	0	
		-401	-70.2	-10321	

			*		
		3	2		
	1	1	0	50	--
	4	0	1	200	200
	5	-1	0.2	22	110
*	6	-150	[25]	2500	100
		-250	45	-12500	
		401	-70.2	9729	

		*			
		3	6		
	1	1	0	50	50
	4	6	-0.04	100	16.67
*	5	[0.2]	-0.008	2	10
	2	-6	0.04	100	--
		20	-1.8	-17000	
		-20.2	2.808	16749	

	5	6		
1	-5	0.04	40	
4	-30	0.2	40	
3	5	-0.04	10	
2	30	-0.2	160	
	-100	-1	-17200	
	101	2	16951	

Solution: $x_1 = 40$, $x_2 = 160$, $Q = -17200$.

A further example, which even contains a degenerate vertex, will show the application of the simplex process once again. The accompanying illustration, Fig. 4.1, provides a visual representation. The linear optimization problem is

$$\begin{aligned} x_1 \qquad\qquad\quad &\le 2, &\quad x_1 \ge 0 \\ x_1 + x_2 + 2x_3 &\le 4, &\quad x_2 \ge 0 \\ 3x_2 + 4x_3 &\le 6, &\quad x_3 \ge 0 \\ Q = x_1 + 2x_2 + 4x_3 &= \text{Max!} & \end{aligned}$$

Introducing slack variables y_1, y_2, y_3, we can write the constraints as equations

$$\begin{aligned} x_1 \qquad\qquad\quad + y_1 &= 2, &\quad y_1 \ge 0 \\ x_1 + x_2 + 2x_3 + y_2 &= 4, &\quad y_2 \ge 0. \\ 3x_2 + 4x_3 + y_3 &= 6, &\quad y_3 \ge 0. \end{aligned}$$

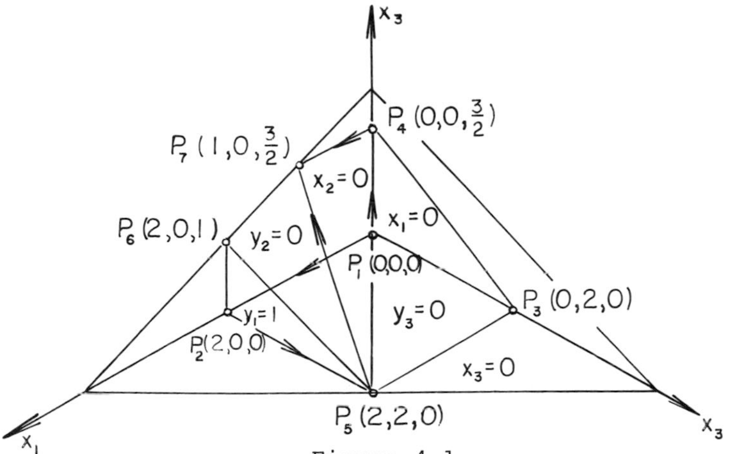

Figure 4.1

The set of feasible points, as shown in Fig. 4.1, is the polyhedron bounded by the six planes $x_1 = 0$, $x_2 = 0$, $x_3 = 0$, $y_1 = 0$, $y_2 = 0$, $y_3 = 0$. It has seven vertices P_j, $j = 1,\ldots,7$. The initial tableau at vertex P_1 is

		x_1	x_2	x_3 *			
	y_1	1	0	0	2	-	2
	y_2	1	1	2	4	2	4
*	y_3	0	3	**4**	6	3/2	-
		1	2	4	0	$P(_1)$	
		-2	-5	-9	-11		

Here we have used the alternative notation mentioned in 4.1, replacing the indices $i \notin Z$ and $k \in Z$ by the variables in fields ① and ②. Also note that we are solving the minimum problem $-Q = \text{Min!}$. Since the row of d_i's contains only positive numbers, we can choose any of the three columns as the pivot column. If we decide on the x_3-column as pivot column, the y_3-row becomes the pivot row, and we compute the new tableau

	x_1 *	x_2	y_3		
y_1	1	0	0	2	2
* y_2	1	-1/2	-1/2	1	1
y_3	0	3/4	1/4	3/2	-
	1	-1	-1	-6	(P_4)
	-2	7/4	9/4	5/2	

We have $x_1 = x_2 = 0$, $x_3 = 3/2$ (vertex P_4). The next exchange step is uniquely determined, and leads to the tableau

	y_2	x_2	y_3		
y_1	-1	1/2	1/2	1	-
x_1	1	-1/2	-1/2	1	-
x_3	0	3/4	1/4	3/2	-
	-1	-1/2	-1/2	-7	(P_7)
	2	3/4	5/4	9/2	

Now each $d_i < 0$, and we find ourselves at vertex P_7, where $-Q$ attains its minimum, and therefore Q its maximum value of 7. The edge-path $P_1P_4P_7$ is emphasized with arrows in Fig. 4.1.

Returning to the initial tableau, let us now choose a different column for the exchange, namely the x_1-column. We then obtain the tableau

			*			
		y_1	x_2	x_3		
	x_1	1	0	0	2	-
*	y_2	-1	1	2	2	2
	y_3	0	3	4	6	2
		-1	2	4	-2	(P_2)
		2	-5	-9	-7	

and vertex P_2, determined by $y_1 = x_2 = x_3 = 0$. The next exchange will pivot on the x_2-column. The y_2-row, as well as the y_3-row, can serve as pivot row. If we choose the y_2-row, we obtain the tableau

		*				
		y_1	y_2	x_3		
	x_1	1	0	0	2	2
	x_2	-1	1	2	2	-
*	y_3	3	-3	-2	0	0
		1	-2	0	-6	(P_5)
		-3	5	1	3	

and the degenerate vertex, P_5.

Now the exchange of y_1 for y_3 is uniquely determined. We obtain the new tableau

	y_3	y_2	x_3 *		
x_1	-1/3	1	2/3	2	3
* x_2	1/3	0	4/3	2	3/2
y_1	1/3	-1	-2/3	0	-
	-1/3	-1	2/3	-6	(P_5)
	1	2	-1	3	

and remain at vertex P_5. The next exchange is uniquely determined. We obtain the tableau

	y_3	y_2	x_2		
x_1	-1/2	1	-1/2	1	-
x_3	1/4	0	3/4	3/2	-
y_1	1/2	-1	1/2	1	-
	-1/2	-1	-1/2	-7	(P_7)
	5/4	2	3/4	9/2	

Again, for this second way to the solution, the resulting edge-path $P_1P_2P_5P_7$ has been indicated in Fig. 4.1 with arrows.

4.4. The Simplex Method with Equalities as Constraints

We still must discuss the procedure in the case where the optimization problem is not originally of the type (1.1a) - (1.3a) with $\underset{\sim}{b} \geq \underset{\sim}{0}$, but of the type (2.1), (2.2). By §3.4, we may assume that $\underset{\sim}{b} \geq \underset{\sim}{0}$. The first step is to solve the optimization problem (3.11). Thus the tableau for the initial stage is exactly the same as in the previously treated case, except that there are now n columns for the c_{ki}, d_i, and σ_i. In field (2), $i \notin Z$, we enter the indices $1,2,\ldots,n$, and in field (1), $k \in Z$, the indices $n+1,\ldots,n+m$. Since the objective function is now $y_1 + \ldots + y_m$ $(= x_{n+1} + \ldots + x_{n+m}$, if we set $x_{n+k} = y_k)$, we enter in field (5) the numbers

$$d_i = \sum_{k=n+1}^{n+m} c_{ki} = \sum_{k=1}^{m} a_{ki},$$

and in field (6),

$$Q = \sum_{k=n+1}^{n+m} x_k = \sum_{k=1}^{m} b_k \quad (\geq 0).$$

Beginning with this initial tableau, we apply the algorithm to solve problem (3.11), which, as previously noted, has a solution. If, for this solution, $Q > 0$, we see that problem (2.1), (2.2) which we actually wish to solve, has no feasible points. However, if we find a solution of (3.11) with $Q = 0$, i.e. with $x_{n+1} = \ldots = x_{n+m} = 0$, we also will have found an initial vertex for treating problem (2.1), (2.2).

As a rule, the terminal tableau for problem (3.11) will have the indices $i = n+1,\ldots,n+m$ among the $i \notin Z$

(because $x_i = 0$). In this case, the terminal tableau for problem (3.11) can be used immediately as the initial tableau for problem (2.1), (2.2); we strike the excess columns for the indices $i = n+1,\ldots,n+m$, recompute the $d_i = \sum_{k\varepsilon Z} c_{ki}p_k - p_i$ $(i \notin Z)$, as well as $Q = \sum_{k\varepsilon Z} p_k x_k$, and also the σ_i, $i \notin Z$, and σ.

Should the terminal tableau in problem (3.11) have several of the indices $n+1,\ldots,n+m$ among the $k \varepsilon Z$, the minimum for problem (3.11) will be attained at a degenerate vertex, because $x_k = 0$ for these indices. Then we have to remove these indices from Z with several more exchange steps; i.e. find a basis for this degenerate vertex which contains only column vectors $\underset{\sim}{a}^k$ from matrix $\underset{\sim}{A}$ $(k = 1,\ldots,n)$.

For these exchange steps, we cannot find the pivot element in the usual manner of the simplex process (find a positive d_j and a positive $c_{\ell j}$). Instead, we consider those rows whose indices $k \varepsilon Z$ are $\geq n+1$. If there is such a row, say the row with index ℓ, with a $c_{\ell j} \neq 0$ (positive or negative), having $j \leq n$, we carry out the exchange using $c_{\ell j}$ as the pivot. Since $x_l = 0$, $x_l/c_{\ell j} = 0$ also, and the x_k are unchanged as a result of this exchange step; thus we remain at the (degenerate) vertex, which we had found to be the solution for problem (3.11). We repeat this process as long as there still exists a row of this kind, with an index $\ell \varepsilon Z$, $\ell > n+1$, and a $c_{\ell j} \neq 0$, $j \leq n$. With each of these steps, one index $k \geq n+1$ is removed from the index set Z.

There exist two possibilities:

1. It is possible to remove all indices $k \geq n+1$ from Z. Then we have an initial tableau for a treatment of problem (2.1), (2.2).

2. There remain one or more indices $k \geq n+1$ in Z, and the respective rows of the tableau contain only numbers $c_{kj} = 0$ for $j \leq n$. According to the definition of c_{ki} in (3.2), this means that all the column vectors $\underset{\sim}{a}^i$ of matrix $\underset{\sim}{A}$ may be represented as a linear combination of fewer than m of these column vectors. The rank of matrix $\underset{\sim}{A}$ is then less than m. Again there are two possibilities.

2.1. The equations $\underset{\sim}{A}\underset{\sim}{x} = \underset{\sim}{b}$ are inconsistent. This case cannot arise, because the solution of problem (3.11) with $x_{n+1} = \ldots = x_{n+m} = 0$, is a solution of the system of linear equations $\underset{\sim}{A}\underset{\sim}{x} = \underset{\sim}{b}$.

2.2. Several of the equations in $\underset{\sim}{A}\underset{\sim}{x} = \underset{\sim}{b}$ are dependent on the remainder, and consequently dispensable. It is clear that we can take the terminal tableau for problem (3.11), which brought us to case 2, and make it into the initial tableau for problem (2.1), (2.2), by dropping those rows with index $k \geq n+1$ which have only zeros in columns with index $j \leq n$; as before, we also drop those columns with index $i \geq n+1$.

The algorithmic implementation of the simplex process has now been described completely, and this process is valid even when we do not know initially whether the matrix $\underset{\sim}{A}$ in $\underset{\sim}{A}\underset{\sim}{x} = \underset{\sim}{b}$ (2.2) has the full rank m.

Suppose we are faced with a combination of the two

types of optimization problems we have discussed, so that some of the constraints are written as $A_1x \le b^1$ with $b^1 \ge 0$, and the rest as $A_2x = b^2$ with $b^2 \ge 0$. In an actual computation, we can reduce the computational effort required, by also combining the two processes for determining the initial vertex. We can introduce a slack variable vector y^1 and an additional vector y^2 and write the constraints in the form

$$A_1x + y^1 = b^1, \quad A_2x + y^2 = b^2,$$
$$x \ge 0, \quad y^1 \ge 0, \quad y^2 \ge 0.$$

As an objective function, we first use the sum of the components of y^2, and as an initial vertex, we use $x = 0$, $y^1 = b^1$, $y^2 = b^2$. Once we have found a solution with $y^2 = 0$, we have an initial vertex for the original problem which has objective function $Q(x)$.

Let us add one further constraint to the sheep- and cattle-raising problem (Example 2 of §1). If we consider the fertilizer which this agricultural enterprise needs to cultivate its fields, then we must add an inequality to the original constraints, for example

$$10x_1 + x_2 \ge 550.$$

We now compute with a different objective function, namely $Q(x) = -250x_1 - 55x_2$. (With the previous objective function, the new constraint would have had no effect on the outcome.) We first introduce a slack variable. This converts the new constraint into $10x_1 + x_2 - x_3 = 550$, $x_3 \ge 0$. Let the slack variables for the original constraints be x_4, x_5, x_6, x_7. Now, to find the initial vertex we do not need to introduce new variables in all the constraints. It suffices to introduce one variable, $x_8 \geq 0$, in the new constraint, so that we have $10x_1 + x_2 - x_3 + x_8 = 550$, and now to minimize the objective function $\tilde{Q} = x_8$. The tableaux are:

	*				
	1	2	3		
* 4	1	0	0	50	50
5	0	1	0	200	-
6	1	0.2	0	72	72
7	150	25	0	10000	66.67
8	10	1	-1	550	55
	10	1	-1	550	
	-171	-27.2	3	-11421	

		*			
	4	2	3		
1	1	0	0	50	-
5	0	1	0	200	200
6	-1	0.2	0	22	110
7	-150	25	0	2500	100
* 8	-10	1	-1	50	50
	-10	1	-1	50	
	171	-27.2	3	-2871	

	4	8	3		
1	1	0	0	50	
5	10	-1	1	150	
6	1	-0.2	0.2	12	
7	100	-25	25	1250	
2	-10	1	-1	50	
	0	-1	0	0	
	-101	27.2	-24.2	-1511	

Thus the first part of the problem is solved, and we have found an initial vertex. The column with index 8 can be dropped. We have to recompute the last two rows. The final tableaux are:

		*			
		4	3		
	1	1	0	50	50
	5	10	1	150	15
*	6	1	0.2	12	12
	7	100	25	1250	12.5
	2	-10	-1	50	-
		300	55	-15250	
		-401	-79.2	13739	

	6	3		
1	-1	-0.2	38	
5	-10	-1	30	
4	1	0.2	12	
7	-100	5	50	
2	10	1	170	
	-300	-5	-18850	
	401	1	18551	

Solution: $x_1 = 38$, $x_2 = 170$, $Q = -18850$.

4.5. The Later Addition of a Variable

Occasionally the following situation occurs. We have solved the linear optimization problem

$$\left.\begin{array}{l} \underset{\sim}{p}'\underset{\sim}{x} = \sum\limits_{i=1}^{n} p_i x_i \; - \; \text{Min!}, \\[2ex] \sum\limits_{i=1}^{n} \underset{\sim}{a}^i x_i = \underset{\sim}{b}, \quad x_i \geq 0 \quad (i = 1,\ldots,n) \end{array}\right\} \qquad (4.3)$$

with the simplex method; and here we assume that this problem has a solution and that we have the terminal tableau of the simplex process at hand, where all the $d_i \leq 0$, $i \notin Z$. We are now to solve an expanded problem. To the data of problem (4.3), namely $\underset{\sim}{p} \in R^n$, $\underset{\sim}{b} \in R^m$, $\underset{\sim}{a}^i \in R^m$, has been added the vector $\underset{\sim}{a}^{n+1} \in R^m$ and the real number p_{n+1}, and the problem now reads

$$\left.\begin{aligned} &\sum_{i=1}^{n+1} p_i x_i = \text{Min!}, \\ &\sum_{i=1}^{n+1} \underset{\sim}{a}^i x_i = \underset{\sim}{b},\ x_i \geq 0 \quad (i = 1,\ldots,n+1). \end{aligned}\right\} \qquad (4.4)$$

This situation will arise in §10.2, during the discussion of a method of convex optimization.

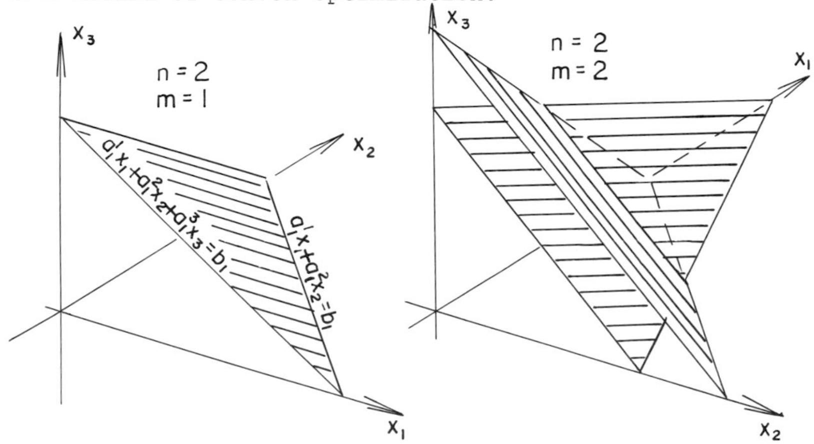

Figure 4.2

Figure 4.2 illustrates the problem for the case $n = 2$ with $m = 1$ and $m = 2$. We will now show how a great deal of computation may be avoided, if the treatment of problem (4.4) with the simplex process is begun at the terminal tableau of problem (4.3). Let $\bar{x} = (\bar{x}_1,\ldots,\bar{x}_n)'$ be the solution of problem (4.3), as determined by the simplex process, i.e. a vertex of the set M of feasible points. Clearly, $(\bar{x}_1,\ldots,\bar{x}_n, 0)'$ is a vertex for problem (4.4), and the solution of (4.4) by the simplex process can commence there. We keep the basis in the terminal tableau of (4.3), that is, the vectors $\underset{\sim}{a}^k$, $k \in Z$, and to the vectors $\underset{\sim}{a}^i$, $i \notin Z$, we add $\underset{\sim}{a}^{n+1}$. We have to add a new column to the terminal tableau of problem (4.3), and this is to be filled in with the numbers $c_{k,n+1}$ and d_{n+1}, which we find as follows.

We have

$$\underset{\sim}{a}^{n+1} = \sum_{k\varepsilon Z} c_{k,n+1}\underset{\sim}{a}^{k}, \tag{4.5}$$

so that the $c_{k,n+1}$ are the solution of a system of linear equations with a non-singular m-by-m matrix. Then we obtain the d_{n+1} from

$$d_{n+1} = \sum_{k\varepsilon Z} c_{k,n+1}p_k - p_{n+1}.$$

If $d_{n+1} \leq 0$, then $(\bar{x}_1,\ldots,\bar{x}_n, 0)'$ is a solution of problem (4.4). However, if $d_{n+1} > 0$, several exchange steps still remain to be carried out.

The solution of the system of linear equations (4.5) has a particularly simple form if the vectors $\underset{\sim}{a}^i$, $i = 1,\ldots,n$, include the m unit vectors of R^m, say the vectors $\underset{\sim}{a}^1,\ldots,\underset{\sim}{a}^m$ with $a_{\ell i} = \delta_{\ell i}$, ℓ, $i = 1,\ldots,m$. This case occurs when one works with slack variables, and later in an application in §10.2. We then have

$$\underset{\sim}{a}^{n+1} = \sum_{i=1}^{m} a_{i,n+1}\underset{\sim}{a}^{i}.$$

Using $\underset{\sim}{a}^i = \sum_{k\varepsilon Z} c_{ki}\underset{\sim}{a}^k$, it follows that

$$c_{k,n+1} = \sum_{i=1}^{m} c_{ki}a_{i,n+1} \quad (k \varepsilon Z). \tag{4.6}$$

The inverse of the matrix of the system of equations (4.5) is then explicitly presented. In an application of (4.6), one should note that, if the indices $i = 1,\ldots,m$

include some $i \varepsilon Z$, then $c_{ki} = \delta_{ki}$. The remaining c_{ki}, with $i \not\in Z$, can be taken from the terminal tableau of problem (4.3).

4.6. The Simplex Method for Variables without Sign Constraints

Occasionally there occur linear optimization problems in which no sign constraints are prescribed for some of the variables. In the discussion of duality in §5.1, the vector $\underset{\sim}{w}$ considered in problem D^1 has components which are not subject to any positivity constraints. Also, in the reduction of linear discrete Tchebychev approximation problems to linear optimization problems, described in §16.1, the variables can be of arbitrary sign. In both cases, the constraints are in the form of inequalities. If slack variables are introduced, so that the constraints are written as equations, then we obtain problems in which some of the variables may be of arbitrary sign, and the rest (namely the slack variables) are restricted in sign.

We now want to show how to treat such problems with the simplex method. Let the problem be of the following type:

$$Q(\underset{\sim}{x}) = \underset{\sim}{p}'\underset{\sim}{x} = \text{Min!}, \quad \underset{\sim}{A}\underset{\sim}{x} = \sum_{i=1}^{n} \underset{\sim}{a}^i x_i = \underset{\sim}{b}, \quad x_i \geq 0 \quad (i = 1,\ldots,q). \tag{4.7}$$

Here we have $q < n$, and the variables $x_{q+1},\ldots,x_n$ are unrestricted in sign. Let $\underset{\sim}{A}$ be an m-by-n matrix with $m < n$, and of rank m.

One way to treat such problems is by expressing the variables $x_{q+1},\ldots,x_n$ as the difference of two positively constrained variables:

$$x_i = x_i^+ - x_i^-,\ x_i^+ \geq 0,\ x_i^- \geq 0 \quad (i = q+1,\ldots,n). \tag{4.8}$$

The constraints $\underset{\sim}{A}\underset{\sim}{x} = \underset{\sim}{b}$ then read

$$\sum_{i=1}^{q} \underset{\sim}{a}^i x_i + \sum_{i=q+1}^{n} \underset{\sim}{a}^i x_i^+ + \sum_{i=q+1}^{n} (-\underset{\sim}{a}^i) x_i^- = \underset{\sim}{b}. \tag{4.9}$$

The reformed problem contains only positively constrained variables and thus can be treated with the simplex method, as previously described. The case where the column vectors for x_i^+ and x_i^- simultaneously belong to the basis cannot occur. For these are the linearly dependent vectors $\underset{\sim}{a}^i$ and $-\underset{\sim}{a}^i$. Of the two variables x_i^+ and x_i^-, only one can be positive at a time.

A second way to treat problem (4.7) consists of a suitable modification of the simplex process. We dispense with the splitting (4.8), set up the simplex tableau, as described in §4.1, but keep to the following rules in executing an exchange step.

1. If, among the variables x_i with $i \notin Z$, there is one of the positively unrestricted variables $x_{q+1},\ldots,x_n$, say x_j, and if $d_j \neq 0$, the corresponding column can be chosen as the pivot column. Since $j \notin Z$, the variable x_j has value 0. (3.8) then implies that if $d_j = t_j - p_j > 0$, in field ⑤ of the simplex tableau, $Q(\underset{\sim}{x})$ can be decreased by increasing x_j. If $d_j < 0$, $Q(\underset{\sim}{x})$ can be decreased by decreasing x_j. Naturally, we can also choose any column

as pivot column if it corresponds to a positively restricted variable x_j, $j \notin Z$, with $d_j > 0$.

2. In §3.1, we can now permit the δ in formula (3.3) to assume negative values for the case of an unrestricted variable x_j. After the appropriate modifications, (3.3) and (3.4) indicate the manner of determining the pivot row:

(a) $d_j > 0$: Form the quotients x_k/c_{kj} for all $k \in Z$ with $k \leq q$ and $c_{kj} > 0$. Find the smallest of these numbers. The corresponding row serves as the pivot row.

(b) $d_j < 0$: Form the quotients x_k/c_{kj} for all $k \in Z$ with $k \leq q$ and $c_{kj} < 0$. The number of smallest magnitude determines the pivot row.

Rules (a) and (b) say that for the δ_1 in (3.4) we choose the number δ of greatest possible magnitude for which all the restricted components of vector $\underset{\sim}{x}(\delta)$ in (3.3) are still non-negative.

In this manner, then, we alter the rules in I. of the Summary of the simplex method in §4.2. The rules in II. for modifying the simplex tableau remain unchanged.

After the appropriate changes in Theorems 2 and 3 of §3.2 are made, we see that the simplex process ends when one of the following two cases occurs.

A. $d_j \leq 0$ for all positively restricted x_j with $j \notin Z$, and $d_j = 0$ for all unrestricted x_j, $j \notin Z$. Then we have a minimal solution of the optimization problem.

B. For every x_j with $j \notin Z$ and $d_j > 0$, all

$c_{kj} \leq 0$ for $k \in Z$ and $k \leq q$. For every unrestricted x_j with $j \notin Z$ and $d_j < 0$, $c_{kj} \geq 0$ for all $k \in Z$ and $k \leq q$. Then the objective function is not bounded within the framework of the constraints.

For purposes of conceptual clarification, note that the points $\tilde{x}$ which arise in this modified simplex process are not necessarily vertices of the set M of feasible points. This happens when there are unrestricted variables among the x_j with $j \notin Z$. For then one can let these components of $\tilde{x}$ increase in a positive direction and decrease in a negative direction, without leaving M, so that $\tilde{x}$ can be expressed as a strictly convex combination of two distinct points of M. All theorems needed to justify the simplex process remain valid after the appropriate modifications. The process used to determine the initial vertex for the simplex method, described in §3.4 and §4.4, can be applied unaltered.

For this modified simplex process, the computational effort is less than for the version first described. Without the splitting of x_i into $x_i^+ - x_i^-$, there are no additional columns which have to be carried in the simplex tableaux. But primarily, where the modified process changes the sign of an unrestricted variable in one exchange step, the first version requires two steps.

4.7. Special Forms of the Simplex Method

The problem of the numerical treatment of linear optimization problems is solved in principle by the simplex process as described and established in this, and the preceding, section. There exist several special forms, and further developments, of the simplex method which are presented and briefly described in the following.

A. The revised simplex method. This version of the simplex method is particularly suited for a machine treatment of large scale problems. The basic problem is of the same type as before:

$$Q(\underset{\sim}{x}) = \underset{\sim}{p}'\underset{\sim}{x} = \text{Min!}, \quad \underset{\sim}{A}\underset{\sim}{x} = \underset{\sim}{b}, \quad \underset{\sim}{x} \geq \underset{\sim}{0}.$$

At any stage of the simplex process, we have a basis of linearly independent column vectors $\underset{\sim}{a}^k$, $k \in Z$. These are collected into an m-by-m matrix $\tilde{\underset{\sim}{A}}$. This matrix is non-singular. If we know the inverse matrix

$$\tilde{\underset{\sim}{A}}^{-1} = (\alpha_{kj})_{k\in Z;\ j = 1,\dots,m}$$

we can easily compute all the numbers required to carry out an exchange step.

By (3.2) the m-by-n matrix of the numbers c_{ki}, $k \in Z$, $i = 1,\dots,n$, is

$$\underset{\sim}{C} = \tilde{\underset{\sim}{A}}^{-1}\underset{\sim}{A}. \tag{4.10}$$

By (3.1), we can write the vector $\tilde{\underset{\sim}{x}}^0 \in R^m$ with components x_k^0, $k \in Z$, as

$$\tilde{\underset{\sim}{x}}^0 = \tilde{\underset{\sim}{A}}^{-1}\underset{\sim}{b}. \tag{4.11}$$

Finally, letting $\tilde{\underset{\sim}{p}} \in R^m$ be the vector with components p_k, $k \in Z$, we have by (3.7) that

$$\underset{\sim}{t}' = (t_i) = \tilde{\underset{\sim}{p}}'\underset{\sim}{C} = \tilde{\underset{\sim}{p}}'\tilde{\underset{\sim}{A}}^{-1}\underset{\sim}{A}. \tag{4.12}$$

If we arrange the simplex process so that, at every step, both the matrix $\tilde{A}^{-1}$ and the vectors $\tilde{x}^0$ and $\tilde{p}'\tilde{A}^{-1}$ are known, then we can proceed in the following manner.

We compute the components t_i, $i \notin Z$, of the vector t by (4.12). If we find a component $t_j > p_j$, we can carry out the exchange step by Theorem 1 of §3.2. We do not need to compute the complete matrix C (4.10), but only the column for index j. With this column and the vector $\tilde{x}^0$, we can determine the index $\ell \in Z$ by (3.4), and thus the vector a^ℓ which is to be exchanged for a^j. To compute the inverse matrix $\tilde{A}^{-1}$ for the next step, we can use the conversion formulas (3.5).

If the number of variables, n, in a linear optimization problem is much greater than the number of equations, m, the revised simplex method requires far fewer computations than the standard form. Recomputing the complete matrix C, as for the standard form, requires some $(n - m)m$ multiplications, while recomputing the matrix $\tilde{A}^{-1}$ requires only some m^2 multiplications, as does the computation of one column of C.

Further computational advantages follow from a further development of the revised simplex method, the so called product form. Then the matrix $\tilde{A}^{-1}$ is no longer presented explicitly, but rather is computed as a product of simpler matrices. For this and further details of the revised method, we refer to Gass, 1964, Ch. 6.1.

B. <u>The dual simplex method</u>. In the standard simplex

process (§3) we determine a series of points $\underset{\sim}{x}^t$, each of which is feasible, and the last of which is optimal if a solution exists. Then the numbers d_j in the simplex tableau are all ≤ 0. In this variant we determine instead a series of points $\underset{\sim}{x}^t$, which are not necessarily feasible, but each of which is "optimal" in the sense that every $d_j \leq 0$ in the simplex tableau, and the last of which is feasible if a solution exists.

This process will not be described here in detail, but see Gass, 1964, Ch. 9.2. Actually, one may as well begin with the dual problem (to be described in §5 below) to a linear optimization problem, solve the dual by the usual simplex process, and then apply the rule given at the end of §5.1 to determine the solution of the original problem. This means recommends itself, among other times, when it is easier to determine an initial vertex for the dual problem than for the original.

C. Integer linear optimization. In actual problems which lead to linear optimization problems it is frequently the case that the variables can only assume integer values, e.g. Example 2 of §1.1, which dealt with numbers of cows and sheep. If we use the simplex process as described so far to solve such problems, we generally obtain solution vectors, not all of whose components are integral. A less than satisfactory way of obtaining an integer solution is to round off the non-integral components of the solution vector. In general, the vector obtained in this way is not feasible or not optimal in the subset of feasible vectors with integer components.

There are several modifications of the simplex process for solving such integer linear optimization problems. Gomory, 1963, suggests a process which begins by using the simplex method to find a solution which is non-integral in general; after such a one has been found, there is a stepwise addition of constraints which reduces the range of feasible points; after finitely many such steps, one obtains a range for which the optimal solution is integral (or the conclusion that there are no feasible vectors with integer components.)

In a further development of this process, called the all integer method and described in Gass, 1964, Ch. 9.3, this reduction of the range of feasible points through the introduction of additional constraints, is already undertaken in the course of the exchange steps of the simplex process. In that way all of the values appearing in the execution of the simplex process turn out to be integral.

4.8. Transportation Problems and their Solution by the Simplex Method

One example of a transportation problem was already given in §1.2. The general case of such a problem is the following. There are M (≥ 1) supply depots, $S_1,\dots,S_M$, and N (≥ 1) demand depots, $R_1,\dots,R_N$. A commodity (sugar, in the given example) is present in supply depot S_j in the amount of s_j units, $j = 1,\dots,M$. At demand depot R_k, there is a demand for r_k units, $k = 1,\dots,N$. The total supply is presumed equal to the total demand, so that

$$\sum_{j=1}^{M} s_j = \sum_{k=1}^{N} r_k = C. \tag{4.13}$$

The movement of one unit of the commodity from S_j to R_k entails a cost of p_{jk}. If x_{jk} units are transported from S_j to R_k, $j = 1,\ldots,M$, $k = 1,\ldots,N$, the total cost will be

$$Q = \sum_{j=1}^{M} \sum_{k=1}^{N} p_{jk} x_{jk}. \tag{4.14}$$

The numbers x_{jk} are to be determined so as to minimize Q under the constraints

$$\left.\begin{aligned} \sum_{k=1}^{N} x_{jk} &= s_j && (j = 1,\ldots,M), \\ \sum_{j=1}^{M} x_{jk} &= r_k && (k = 1,\ldots,N), \\ x_{jk} &\geq 0 && (\text{all } j,k). \end{aligned}\right\} \tag{4.15}$$

Incidentally, the case where (4.13) is invalid can easily be reduced to the given case. For example, if the total demand is less than the total supply, we add a fictitious demand depot, with transportation costs $p_{jk} = 0$, to absorb the excess supply. We assume that all of the numbers r_k and s_j are positive. If one of them were equal to zero, we could reduce the problem to one with a smaller M or N.

Let us formulate the transportation problem in the language of matrices and vectors. Let

$$\underset{\sim}{x} = (x_{11},\ldots,x_{1N},x_{21},\ldots,x_{MN})',$$

$$\underset{\sim}{p} = (p_{11},\ldots,p_{1N},p_{21},\ldots,p_{MN})',$$

$$\underset{\sim}{\bar{b}} = (s_1,\ldots,s_M,r_1,\ldots,r_N)'$$

and let $\underset{\sim}{\bar{A}}$ be the (N+M)-by-(NM) matrix of form (1.12). Then the problem reads

$$Q(\underset{\sim}{x}) = \underset{\sim}{p}'\underset{\sim}{x} = \text{Min!}, \quad \underset{\sim}{\bar{A}}\underset{\sim}{x} = \underset{\sim}{\bar{b}}, \quad \underset{\sim}{x} \geq \underset{\sim}{0}. \tag{4.16}$$

The column vector $\underset{\sim}{\bar{a}}^{jk}$ of matrix $\underset{\sim}{\bar{A}}$ has a 1 in the jth row and in the $(M+k)^{th}$ row, and is otherwise zero. Matrix $\underset{\sim}{\bar{A}}$ has rank N+M-1, by the remarks of §2.3. Its rank is therefore one less than the number of rows. The considerations undertaken in §2.3 show that any matrix formed from $\underset{\sim}{\bar{A}}$ by deleting a row has the full rank, N+M-1. When we apply the simplex process from now on, we delete the last row of $\underset{\sim}{\bar{A}}$ and also the last component of $\underset{\sim}{\bar{b}}$. In this way we obtain an (N+M-1)-by-(NM) matrix $\underset{\sim}{A}$ of rank N+M-1, and a vector $\underset{\sim}{b} \in R^{N+M-1}$.

One can easily show that the transportation problem (4.16) always has a solution if all the s_j and all the r_k, and hence C, are positive. The vector given by $x_{jk} = s_j r_k/C$, $j = 1,\ldots,M$, $k = 1,\ldots,N$, is feasible. In addition, the set of feasible points is bounded (since $0 \leq x_{jk} \leq \text{Min}(s_j,r_k)$), and therefore a polyhedron. By Theorem 6, §2.2, the objective function Q attains its minimum at a vertex of this polyhedron.

In principle, it is possible to treat such a transportation problem with the simplex method of the form previously described. However, this entails the use of tableaux of a prohibitive size, namely of the order of the size of matrix A. The process about to be described uses tableaux of a size on the order of M-by-N, containing all the requisite data for an exchange step; in content it is identical to the simplex process.

If the simplex process in standard form were used, we would work with the constraints $\underset{\sim}{A}\underset{\sim}{x} = \underset{\sim}{b}$ (instead of $\bar{\underset{\sim}{A}}\underset{\sim}{x} = \bar{\underset{\sim}{b}}$) and execute a number of exchange steps. As a basis we would always have a system of N+M-1 column vectors $\underset{\sim}{a}^{jk}$ from matrix $\underset{\sim}{A}$. An exchange step would consist of removing one of these vectors from the basis, and replacing it with some other vector.

The form of the simplex process which is tailored to the transportation problem requires several concepts and results from graph theory, both for its description and its verification. A _graph_ (König, 1936) consists of a set of _nodes_ which are connected one to another by line segments, called _edges_. A node may have no, one, or several out-going edges; an edge always connects two nodes (alternatively, these nodes are _incident_ with the edge). In general, the two endpoints of an edge may coincide; the edge then degenerates to a _loop_. However, this case will not occur here.

The supply depots, S_j, and demand depots, R_j, of the transportation problem will now be symbolized by points and assigned to nodes, as in Figure 4.3.

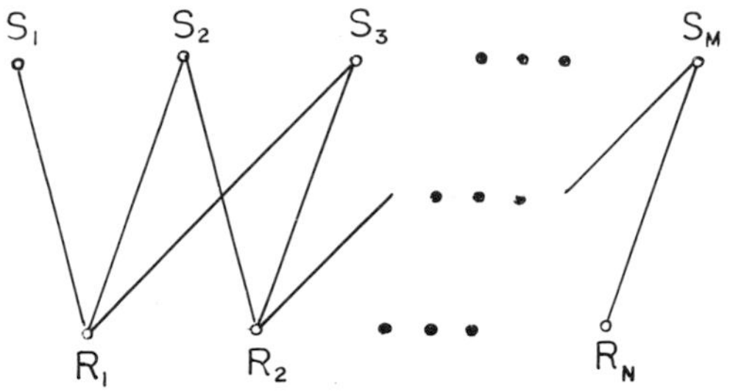

Figure 4.3

To a subset V of column vectors $\underset{\sim}{a}^{jk}$ of $\underset{\sim}{A}$, we can assign a graph G by connecting S_j to R_k with an edge iff $a^{jk} \varepsilon V$. In this way, we obtain a <u>bipartite</u> graph (i.e., the set of nodes consists of two classes, and edges only connect nodes in different classes). An edge connecting S_j and R_k will henceforth be denoted by α_{jk}.

An <u>edge-path</u> is defined as an alternating sequence of nodes and edges (e.g., S_1, α_{11}, R_1, α_{31}, S_3, α_{32}, R_2 in Fig. 4.3), where each edge is incident with the bracketing nodes in the sequence, and where no edge appears more than once. When the initial node and the terminal node of the edge-path are the same, we have a <u>closed</u> edge-path. And a graph is <u>connected</u> if there exists an edge-path from any node of the graph to any other node.

In Fig. 4.4, S_1, α_{11}, R_1, α_{21}, S_2, α_{23}, R_3, α_{13}, S_1 is a closed edge-path. The graph illustrated is connected.

<u>Theorem 1</u>. A subset V of column vectors of $\underset{\sim}{A}$ is linearly dependent iff the associated graph contains a closed edge-path.

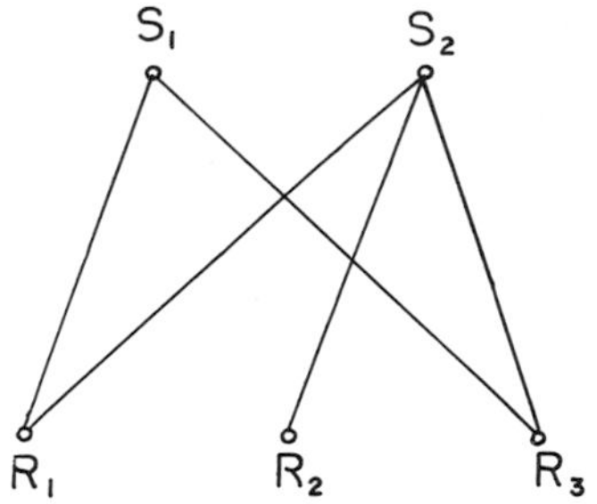

Figure 4.4

Proof: I. Suppose the graph G contains a closed edge-path, say

$$R_{k_1}, \alpha_{j_1k_1}, S_{j_1}, \alpha_{j_1k_2}, R_{k_2}, \alpha_{j_2k_2}, S_{j_2}, \ldots, \alpha_{j_tk_1}, R_{k_1}.$$

If we take note of the remark regarding the components of $\bar{\underset{\sim}{a}}^{jk}$ made below (4.16), it follows at once that

$$\underset{\sim}{a}^{j_1k_1} - \underset{\sim}{a}^{j_1k_2} + \underset{\sim}{a}^{j_2k_2} - + \ldots - \underset{\sim}{a}^{j_tk_1} = \underset{\sim}{0}.$$

II. Let the vectors $\underset{\sim}{a}^{jk} \varepsilon V$ be linearly dependent. Then there exists a non-empty subset $V' \subset V$ such that

$$\sum_{\underset{\sim}{a}^{jk} \varepsilon V'} \gamma_{jk} \underset{\sim}{a}^{jk} = \underset{\sim}{0}, \tag{4.17}$$

where all $\gamma_{jk} \neq 0$. This equation is also valid for the vectors $\bar{\underset{\sim}{a}}^{jk}$, for the last row of matrix $\bar{\underset{\sim}{A}}$, which was dropped to make $\underset{\sim}{A}$, is a linear combination of the rows of $\underset{\sim}{A}$. By the above remark on the components of $\bar{\underset{\sim}{a}}^{jk}$, every index which appears at all in (4.17) must appear at least twice. The set of edges α_{jk} with $\underset{\sim}{a}^{jk} \varepsilon V'$ and nodes S_j and R_k with indices j and k appearing in (4.17) then form a subgraph G' of G, in which every node is incident with at least two edges. Therefore G', and hence G, contains a closed edge-path. For if we depart from some node of G' along an edge, and continue on through G', being careful never to pass along any edge more than once, eventually we must arrive at a node our path crossed previously, and then we have a closed edge-path.

Theorem 2: If G is the graph associated with a subset V of N+M-1 linearly independent column vectors of matrix A, then G is connected.

Remark. A basis which appears in the simplex process is such a set V.

Proof: By Theorem 1, G contains no closed edge-paths. G has N+M-1 edges and exactly N+M nodes. By induction on n we will show the following. A graph G with n edges and at most n+1 nodes which contains no closed edge-paths is connected.

n = 1: The graph G consists of one edge and its two endpoints (two nodes), and thus is connected. (We even could have begun the induction with the trivial case n = 0.)

n > 1: Since there are no closed edge-paths, there must exist a node which is incident with only one edge, the one connecting the node to the rest of the graph. If we remove this node and this edge, we are left with a graph G', which has n-1 edges and at most n nodes and no closed edge-paths. The induction hypothesis implies that G' is connected, and therefore G is also.

With these preparations behind us, we can proceed to

describe the simplex process for the transportation problem. The numbers p_{jk} are arranged in a tableau, P, with M rows and N columns. Similarly, at every step of the process, the numbers x_{jk} are entered in a tableau, X, along with the numbers s_j and r_k.

P:

p_{11}	p_{12}	...	p_{1N}
...	...	...	...
p_{M1}	p_{M2}	...	p_{MN}

X:

s \ r	r_1	r_2	...	r_N
s_1	x_{11}	x_{12}	...	x_{1N}
...	...	...	...	...
s_M	x_{M1}	x_{M2}	...	x_{MN}

Our first task is to find an intial vertex for the simplex process. To do this, we can use the "north west corner" rule. In determining numbers x_{jk} which satisfy all the constraints, we begin in the "north west corner" of the X-tableau, i.e. the upper left, and set $x_{11} = \operatorname{Min}(r_1, s_1)$. So if $s_1 \geq r_1$, the total demand at depot R_1 is met with supplies transported from S_1; and if $s_1 \leq r_1$, the total supply at S_1 is shipped to R_1. In case $s_1 > r_1$, we set $x_{j1} = 0$ for all $j \geq 2$, and set $x_{12} = \operatorname{Min}(s_1 - r_1, r_2)$; then either the whole of the supply remaining at S_1 is shipped to R_2 or the total demand at R_2 is met from S_1. Correspondingly, if $s_1 < r_1$, we set $x_{1k} = 0$ for all $k \geq 2$ and set $x_{21} = \operatorname{Min}(r_1 - s_1, s_2)$. We continue in this way. If the choice of x_{jk} implies that the demand at depot R_k is met and that depot S_j has a remaining supply of $s_j' > 0$, then we choose

$x_{j,k+1} = \text{Min}(r_{k+1}, s'_j)$; if the choice of x_{jk} implies that the supply at depot S_j has been exhausted and that R_k has an as yet unmet demand of $r'_k > 0$, then we choose $x_{j+1,k} = \text{Min}(r'_k, s_{j+1})$. All x_{jk} not determined by this rule, we set equal to 0 (but conveniently, do not enter these zeros in the X-tableau).

Example

X:

s \ r	1	6	3	4
5	1	4		
3		2	1	
6			2	4

If the case does not occur, where the demand at a depot before the last is filled by exhausting the supply at another depot, as in the example, N+M-1 entries in the X-tableau will be positive. The first of these is x_{11} and the last is x_{MN} (total demand equals total supply). Except for x_{MN}, every positive number in the tableau has a positive neighbor to the right or below. We see that the number of positive entries x_{jk} is precisely N+M-1 by considering the special case where only the first column and the last row contain positive x_{jk}. In this case, the number is N+M-1. In every other case, we obtain the same number.

Once we have obtained N+M-1 positive x_{jk} in this way, we choose the corresponding vectors $\tilde{a}^{jk}$ as the basis vectors. In the graph G associated to the set V of these vectors, the nodes S_j and R_k correspond to the rows and columns of the X-tableau, and the edges α_{jk} correspond to the positive x_{jk}. An edge-path in graph G corresponds to a zigzag path in the X-tableau, which runs alternately horizontally and vertically from one positive x_{jk} to another positive x_{jk}, without repetitions.

By design, a tableau filled in by the north west corner rule admits no zigzag paths which return to a row or column once crossed. Hence the graph G contains no closed edge-paths, and so, by theorem 1, the vectors of V are linearly independent.

If, in this construction, the case does occur, where the demand at depot R_k is filled by exhausting the supply at depot S_j, then we carry on the construction with R_{k+1} and S_{j+1}, first adding either $\underset{\sim}{a}^{j+1,k}$ or $\underset{\sim}{a}^{j,k+1}$ to the basis and setting $x_{j+1,k} = 0$ or $x_{j,k+1} = 0$, respectively. The vertex thus obtained is degenerate.

Again, and in summary, the prescriptions of the north west corner rule are these.

For $t = 1,2,\ldots,N+M-1$ determine numbers j_t and k_t (with $j_t+k_t = t+1$), and also σ_t and ρ_t by the rule

$$j_1 = k_1 = 1, \quad \sigma_1 = s_1, \quad \rho_1 = r_1,$$

$$\left.\begin{array}{l} \left.\begin{array}{ll} j_{t+1} = j_t+1, & \sigma_{t+1} = s_{j_{t+1}} \\ k_{t+1} = k_t, & \rho_{t+1} = \rho_t-\sigma_t \end{array}\right\} \text{ if } \sigma_t \le \rho_t \\ \\ \left.\begin{array}{ll} j_{t+1} = j_t, & \sigma_{t+1} = \sigma_t-\rho_t \\ k_{t+1} = k_t+1, & \rho_{t+1} = r_{k_{t+1}} \end{array}\right\} \text{ if } \sigma_t > \rho_t \end{array}\right\} \quad (t = 1,2,\ldots,N+M-2). \tag{4.18}$$

Set $x_{j_t k_t} = \mathrm{Min}(\sigma_t, \rho_t)$ $(t = 1,2,\ldots,N+M-1)$, and set all remaining $x_{jk} = 0$. Add the vectors $\underset{\sim}{a}^{j_t k_t}$ $(t = 1,2,\ldots,N+M-1)$ to the basis.

Next we have to show how to execute an exchange step. To begin, we need the numbers defined by (3.7) which we accordingly denote by t_{jk}. If the vector $\underset{\sim}{a}^{jk}$ belongs to the basis, $t_{jk} = p_{jk}$. Of the vectors $\underset{\sim}{a}^{jk}$ not belonging to the basis, we consider only those for inclusion in the basis, for which $t_{jk} > p_{jk}$. The vector $\underset{\sim}{t}$ with components t_{jk} can be determined from (4.12). We have $\underset{\sim}{t}' = \tilde{\underset{\sim}{p}}'\tilde{\underset{\sim}{A}}^{-1}\underset{\sim}{A}$. Here $\tilde{\underset{\sim}{A}}$ is the square submatrix of $\underset{\sim}{A}$ made up of basis vectors, and $\tilde{\underset{\sim}{p}}$ is the vector of those p_{jk} belonging to basis vectors $\underset{\sim}{a}^{jk}$. If we set

$$\tilde{\underset{\sim}{p}}'\tilde{\underset{\sim}{A}}^{-1} = \underset{\sim}{u}' = (u_1,\ldots,u_M,\ v_1,\ldots,v_{N-1}),$$

we obtain $\underset{\sim}{u}'$ as a solution of the system of linear equations $\underset{\sim}{u}'\tilde{\underset{\sim}{A}} = \tilde{\underset{\sim}{p}}'$. In expanded form, this reads

$$\left.\begin{array}{lll} u_j+v_k & = p_{jk} & (k \le N-1) \\ u_j & = p_{jk} & (k = N) \end{array}\right\} \quad (j,k \text{ with } \underset{\sim}{a}^{jk} \in V). \tag{4.19}$$

By adding $v_N = 0$, we can extend $\underset{\sim}{u}$ to a vector $\bar{\underset{\sim}{u}} \in R^{N+M}$. The u_j and v_k can be computed in the following manner. From the P-tableau we select those p_{jk} which belong to basis vectors $\underset{\sim}{a}^{jk}$, and enter them in a new tableau, the T-tableau, as indicated for the above example with $M = 3$ and $N = 4$.

P:

3	2	5	7
1	4	1	0
0	2	2	3

T:

u \ v	$v_1 = 3$	$v_2 = 2$	$v_3 = -1$	$v_4 = 0$
$u_1 = 0$	3	2		
$u_2 = 2$		4	1	
$u_3 = 3$			2	3

(4.20)

From the T-tableau we can compute the u_j and v_k recursively, using (4.19). The rationale is that the graph associated to basis V contains no closed edge-paths. Once the vector $\underset{\sim}{u}$ has been determined, we have $\underset{\sim}{t}' = \underset{\sim}{u}'\underset{\sim}{A}$, i.e.,

$$t_{jk} = u_j + v_k \quad (j = 1,\ldots,M;\ k = 1,\ldots,N). \qquad (4.21)$$

Those t_{jk} with $\underset{\sim}{a}^{jk} \in V$ satisfy $t_{jk} = p_{jk}$ and are already contained in the tableau (4.20); the free squares are filled with t_{jk} computed from (4.21):

T:

u \ v	3	2	-1	0
0	3	2	-1	0
2	5	4	1	2
3	6	5	2	3

If all the $t_{jk} \leq p_{jk}$, theorem 3 of §3.2 implies that the solution to the problem is at hand. However if there are any $t_{jk} > p_{jk}$, then the corresponding $\underset{\sim}{a}^{jk} \notin V$ are the candidates for an exchange with a basis vector. In the example, these are $\underset{\sim}{a}^{21}$, $\underset{\sim}{a}^{24}$, $\underset{\sim}{a}^{31}$, and $\underset{\sim}{a}^{32}$. Let us choose one of these vectors, say $\underset{\sim}{a}^{\hat{j}\hat{k}}$. The basis vector $\underset{\sim}{a}^{jk}$ which will be exchanged for $\underset{\sim}{a}^{\hat{j}\hat{k}}$ is determined in the following manner. We define the vector $\underset{\sim}{x}(\delta)$ as in (3.3), i.e., $x_{\hat{j}\hat{k}}(\delta) = \delta$, and for the remaining components, $x_{jk}(\delta) = 0$ if $\underset{\sim}{a}^{jk} \notin V$ and if $\underset{\sim}{a}^{jk} \in V$, $x_{jk}(\delta)$ is defined so as to satisfy the constraints. For this it is not necessary to compute all the numbers c_{ki}. Instead, we add $\underset{\sim}{a}^{\hat{j}\hat{k}}$ to the vectors $\underset{\sim}{a}^{jk} \in V$, obtaining a system of $N+M$ linearly dependent vectors. The graph associated with this system contains a closed edge-path, by theorem 1. Since the graph associated with V contains no closed edge-path (and is a subgraph), this closed edge-path must contain the edge $\alpha_{\hat{j}\hat{k}}$. Corresponding to the closed edge-path, in the X-tableau, is an alternately horizontal and vertical, closed zigzag path. So for those edges α_{jk} of the graph which lie in the closed edge-path, we take the corresponding corners x_{jk} of the zigzag path, and alternately define

$$x_{jk}(\delta) = x_{jk} - \delta \quad \text{and} \quad x_{jk}(\delta) = x_{jk} + \delta$$

For the rest, we simply let $x_{jk}(\delta) = x_{jk}$. Then all of the constraints $\underset{\sim}{A}\underset{\sim}{x}(\delta) = \underset{\sim}{b}$ are satisfied for arbitrary δ. In the example, $\underset{\sim}{a}^{\hat{j}\hat{k}} = a^{31}$:

X:

s \ r	1	6	3	4
5	$1 - \delta$	$4 + \delta$		
3		$2 - \delta$	$1 + \delta$	
6	δ		$2 - \delta$	4

By (3.4), we now must find the largest $\delta \geq 0$ for which all the positivity constraints $x_{jk}(\delta) \geq 0$ are satisfied. In the example, this $\delta = 1$. For $\delta = 1$, we have $x_{11}(\delta) = 0$, so the vector $\tilde{a}^{31}$ should be exchanged for $\tilde{a}^{11}$. We obtain a new X-tableau:

X:

s \ r	1	6	3	4
5		5		
3		1	2	
6	1		1	4

This process corresponds exactly to the simplex process described in §3. We obtain a new basis, therefore, of linearly independent vectors, and can repeat this exchange process until we have arrived at a minimal solution. We have arrived when all of the numbers in the T-tableau are less than or equal to the numbers in the P-tableau. Should degenerate vertices appear, all the previous considerations, for the general form of the simplex process, apply. In particular, one could formulate an additional rule to avoid loops, but in practice, we can do without. Finally, we include for completeness the remaining exchange steps for the example.

P:

3	2	5	7
1	4	1	0
0	2	2	3

X: (Q=36)

s \ r	1	6	3	4
5	1−δ	4+δ		
3		2−δ	1+δ	
6	δ		2−δ	4

δ = 1

T:

u \ v	3	2	−1	0
0	3	2	−1	0
2	5	4	1	2
3	(6)	5	2	3

X: (Q=30)

s \ r	1	6	3	4
5		5		
3		1−δ	2+δ	
6	1	δ	1−δ	4

δ = 1

T:

u \ v	−3	2	−1	0
0	−3	2	−1	0
2	−1	4	1	2
3	0	(5)	2	3

X: (Q=27)

s \ r	1	6	3	4
5		5		
3		0−δ	3	δ
6	1	1+δ		4−δ

δ = 0 (degenerate vertex)

T:

u \ v	−3	−1	−4	0
3	0	2	−1	3
5	2	4	1	(5)
3	0	2	−1	3

X: (Q=27)

s \ r	1	6	3	4
5		5		
3			3−δ	0+δ
6	1	1	δ	4−δ

T:

u \ v	-3	-1	1	0
3	0	2	4	3
0	-3	-1	1	0
3	0	2	(4)	3

$\delta = 3$

X: (Q=21)

s \ r	1	6	3	4
5		5		
3				3
6	1	1	3	1

T:

u \ v	-3	-1	-1	0
3	0	2	2	3
0	-3	-1	-1	0
3	0	2	2	3

Solution

§5. Dual Linear Optimization Problems

The following duality theorems are of great theoretical interest. In applications, dual problems arise in game theory (§18), and in particular formulations of a question, e.g., in statics (§5.4). Because of the bracketing theorem (5.7), dual problems are of significance for the numerical treatment of optimization problems.

5.1. Duality with Constraints in the Form of Equations

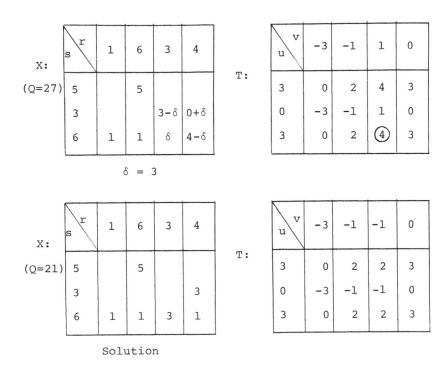

With a minimum problem of linear optimization we can associate a dual maximum problem. The most important theorems on the relationship between the two problems will be deduced here from the theory of the simplex method. Later (in §5.6) we will present a different approach to duality.

As before, let there be given vectors $\underset{\sim}{b} \varepsilon R^m$, $\underset{\sim}{p} \varepsilon R^n$ and a real m-by-n matrix $\underset{\sim}{A}$, where $m < n$ and $\underset{\sim}{A}$ has rank m. Let us consider the following two problems.

D^0: Find an $\underset{\sim}{x} \varepsilon R^n$ with

$$\underset{\sim}{A}\underset{\sim}{x} = \underset{\sim}{b}, \quad \underset{\sim}{x} \geq \underset{\sim}{0}, \quad Q(\underset{\sim}{x}) = \underset{\sim}{p}'\underset{\sim}{x} = \text{Min!} \tag{5.1}$$

D^1: Find a $\underset{\sim}{w} \varepsilon R^m$ with

$$\underset{\sim}{A}'\underset{\sim}{w} \leq \underset{\sim}{p}, \quad G(\underset{\sim}{w}) = \underset{\sim}{w}'\underset{\sim}{b} = \text{Max!} \tag{5.2}$$

Note that $\underset{\sim}{w}$ in D^1 is subject to no positivity constraints. D^1 is called the <u>first dual problem of</u> $\underline{D^0}$. We also say briefly that D^0 and D^1 are dual to each other. A feasible vector which yields the minimum, respectively maximum, for one of these problems will be called simply the optimal solution.

<u>Theorem 1</u>: If $\underset{\sim}{x}$ is a feasible vector for D^0 and if $\underset{\sim}{w}$ is a feasible vector for D^1, then

$$Q(\underset{\sim}{x}) \geq G(\underset{\sim}{w}).$$

<u>Proof</u>: Since $\underset{\sim}{x} \geq \underset{\sim}{0}$ and $\underset{\sim}{p}' \geq \underset{\sim}{w}'\underset{\sim}{A}$, we have

$$Q(\underset{\sim}{x}) = \underset{\sim}{p}'\underset{\sim}{x} \geq \underset{\sim}{w}'\underset{\sim}{A}\underset{\sim}{x} = \underset{\sim}{w}'\underset{\sim}{b} = G(\underset{\sim}{w}). \tag{5.3}$$

We will now make use of an n-by-(2m+n) matrix $\underset{\sim}{S}$ and a vector $\underset{\sim}{r} \varepsilon R^{2m+n}$, which are defined by

$$\underset{\sim}{S} = (-\underset{\sim}{A}'|\underset{\sim}{A}'| - \underset{\sim}{E}_n)$$
$$\underset{\sim}{r}' = (-\underset{\sim}{b}'|\underset{\sim}{b}'|\underset{\sim}{0}_n')$$

where the notation is almost obvious, and where $\underset{\sim}{E}_n$ is the n-by-n identity matrix and $\underset{\sim}{0}_n$ the zero vector of R^n.

<u>Theorem 2</u>: D^0 has a finite optimal solution iff D^1 does. The extreme values for the two problems are equal (if they exist).

<u>Proof</u>: I. Let $\underset{\sim}{x}^0$ be a finite minimal solution for D^0. By theorems 6 and 7 of §2, we may assume that $\underset{\sim}{x}^0$ is a vertex, say $\underset{\sim}{x}^0 = (x_1^0,\ldots,x_m^0, 0,\ldots,0)'$. The column vectors $\underset{\sim}{a}^1,\ldots,\underset{\sim}{a}^m$ constitute the basis at $\underset{\sim}{x}^0$. The numbers c_{ki} and t_i defined by (3.2) and (3.7) are collected into a matrix $\underset{\sim}{C}$ and a vector $\underset{\sim}{t}$,

$$\underset{\sim}{C} = (c_{ki})_{\substack{k=1,\ldots,m,\\ i=1,\ldots,n,}} \qquad \underset{\sim}{t} = (t_1,\ldots,t_n)'.$$

By theorem 5, §3, we may suppose that

$$t_i \le p_i \quad (i = 1,\ldots,n). \tag{5.4}$$

Setting $\tilde{\underset{\sim}{A}} = (\underset{\sim}{a}^1|\cdots|\underset{\sim}{a}^m)$, $\tilde{\underset{\sim}{x}}^0 = (x_1^0,\ldots,x_m^0)'$, $\tilde{\underset{\sim}{p}} = (p_1,\ldots,p_m)'$, we have $\tilde{\underset{\sim}{A}}\tilde{\underset{\sim}{x}}^0 = \underset{\sim}{b}$ and $\tilde{\underset{\sim}{A}}\underset{\sim}{C} = \underset{\sim}{A}$, so that

$$\tilde{\underset{\sim}{x}}^0 = \tilde{\underset{\sim}{A}}^{-1}\underset{\sim}{b}, \quad \underset{\sim}{C} = \tilde{\underset{\sim}{A}}^{-1}\underset{\sim}{A}. \tag{5.5}$$

Furthermore, by (3.7) and (5.4),

$$\tilde{\underset{\sim}{p}}'\underset{\sim}{C} = \underset{\sim}{t}' \le \underset{\sim}{p}'. \tag{5.6}$$

We set $\underset{\sim}{w}^0 = (\tilde{\underset{\sim}{A}}^{-1})'\tilde{\underset{\sim}{p}}$ and show that $\underset{\sim}{w}^0$ is a solution for problem D^1. By (5.5) and (5.6), $\underset{\sim}{w}^{0'}\underset{\sim}{A} = \tilde{\underset{\sim}{p}}'\tilde{\underset{\sim}{A}}^{-1}\underset{\sim}{A} = \tilde{\underset{\sim}{p}}'\underset{\sim}{C} \le \underset{\sim}{p}'$, so that $\underset{\sim}{w}^0$ satisfies the constraints of D^1. Also by (5.5),

$$G(\underset{\sim}{w}^0) = \underset{\sim}{w}^{0'}\underset{\sim}{b} = \tilde{\underset{\sim}{p}}'\tilde{\underset{\sim}{A}}^{-1}\underset{\sim}{b} = \tilde{\underset{\sim}{p}}'\tilde{\underset{\sim}{x}}^0 = Q(\underset{\sim}{x}^0).$$

Now we apply theorem 1 to conclude that $G(\underset{\sim}{w}^0)$ is the maximal value of $G(\underset{\sim}{w})$, subject to the constraints of D^1, and agrees with the minimal value of D^0.

II. Suppose D^1 has a finite optimal solution. Let $\underset{\sim}{w} = \underset{\sim}{w}^1-\underset{\sim}{w}^2$ with $\underset{\sim}{w}^1 \ge \underset{\sim}{0}$, $\underset{\sim}{w}^2 \ge \underset{\sim}{0}$ and $\underset{\sim}{w}^3 = \underset{\sim}{p}-\underset{\sim}{A}'\underset{\sim}{w}$; then D^1 is equivalent to the problem

$$\underset{\sim}{A}'(-\underset{\sim}{w}^1+\underset{\sim}{w}^2)-\underset{\sim}{w}^3 = -\underset{\sim}{p}, \quad \underset{\sim}{w}^1 \ge \underset{\sim}{0}, \quad \underset{\sim}{w}^2 \ge \underset{\sim}{0}, \quad \underset{\sim}{w}^3 \ge \underset{\sim}{0}$$

$$\underset{\sim}{b}'(-\underset{\sim}{w}^1 + \underset{\sim}{w}^2) = \text{Min!}$$

Letting $\underset{\sim}{v}'=(\underset{\sim}{w}^{1'}|\underset{\sim}{w}^{2'}|\underset{\sim}{w}^{3'})$ and $\underset{\sim}{S}$ and $\underset{\sim}{r}$ be as above, we obtain the following problem.

$\tilde{D}^1$: Find a $\underset{\sim}{v} \in R^{2m+n}$ such that

$$\underset{\sim}{S}\underset{\sim}{v} = -\underset{\sim}{p}, \quad \underset{\sim}{v} \ge \underset{\sim}{0}, \quad \underset{\sim}{r}'\underset{\sim}{v} = \text{Min!}$$

$\tilde{D}^1$ is equivalent to D^1; since D^1 has a finite maximal solution by assumption, $\tilde{D}^1$ has a finite minimal solution. On the other hand, $\tilde{D}^1$ is precisely of form D^0. By part I of this proof, therefore, the first dual problem of $\tilde{D}^1$, namely $\tilde{D}^2$, also has a finite maximal solution, and this maximal value is equal to the minimal value of $\tilde{D}^1$, and therefore, to the negative of the maximal value of D^1.

The first dual problem of $\tilde{D}^1$ is

$\tilde{D}^2$: Find an $\underset{\sim}{x} \in R^n$ with $\underset{\sim}{S}'\underset{\sim}{x} \leq \underset{\sim}{r}$, $\underset{\sim}{x}'(-\underset{\sim}{p}) = \text{Max!}$ By definition of $\underset{\sim}{S}$ and $\underset{\sim}{r}$, the constraints of $\tilde{D}^2$ mean that $-\underset{\sim}{A}\underset{\sim}{x} \leq -\underset{\sim}{b}$, $\underset{\sim}{A}\underset{\sim}{x} \leq \underset{\sim}{b}$, $-\underset{\sim}{x} \leq \underset{\sim}{0}$. Therefore, $\tilde{D}^2$ is equivalent to problem D^0: Find an $\underset{\sim}{x} \in R^n$ with $\underset{\sim}{A}\underset{\sim}{x} = \underset{\sim}{b}$, $\underset{\sim}{x} \geq \underset{\sim}{0}$, $\underset{\sim}{p}'\underset{\sim}{x} = \text{Min!}$. By the previous remarks, the minimal value of D^0 is equal to the maximal value of D^1.

<u>Theorem 3</u>: If the function Q is not bounded below on the set of feasible vectors for D^0, then the set of feasible vectors for D^1 is empty. If the function G is not bounded above on the set of feasible vectors for D^1, then the set of feasible vectors for D^0 is empty.

<u>Proof</u>: If $\underset{\sim}{w}$ is a feasible vector for D^1, theorem 1 implies that $Q(\underset{\sim}{x}) \geq G(\underset{\sim}{w})$ for all feasible vectors $\underset{\sim}{x}$ of D^0. $Q(\underset{\sim}{x})$ is then bounded below. The second assertion is proven in the same way.

A further, immediate consequence of theorems 1 and 2 is

<u>Theorem 4</u>: A feasible vector $\underset{\sim}{x}^0$ of D^0 is a minimal solution of D^0 iff there exists a feasible vector $\underset{\sim}{w}^0$ of D^1 such that $\underset{\sim}{p}'\underset{\sim}{x}^0 = \underset{\sim}{b}'\underset{\sim}{w}^0$. Then $\underset{\sim}{w}^0$ is a maximal solution of D^1. The corresponding assertion holds if one starts at D^1 instead of D^0.

As before, the column vectors of $\underset{\sim}{A}$ are denoted by $\underset{\sim}{a}^i$, $i = 1,\ldots,n$.

<u>Theorem 5</u>: A feasible vector $\underset{\sim}{x}^0 = (x_1^0,\ldots,x_n^0)'$ for D^0 is a minimal solution of D^0 iff there exists a feasible vector $\underset{\sim}{w}^0$ for D^1 with the following properties.

$a^{k'}w^0 = p_k$ for every index k with $x_k^0 > 0$; $x_i^0 = 0$ for every index i with $a^{i'}w^0 < p_i$. Then w^0 is a maximal solution of D^1. The corresponding assertion holds if one starts at D^1 instead of D^0.

Proof: We will show that the conditions of the theorem are equivalent to the condition $p'x^0 = b'w^0$ of theorem 4. Since $Ax^0 = b$, these conditions can be written as $p'x^0 = x^{0'}A'w^0 = w^{0'}Ax^0$, i.e., $(p'-w^{0'}A)x^0 = \sum_{i=1}^{n}(p_i-a^{i'}w^0)x_i^0 = 0$. Since $p_i-a^{i'}w^0 \geq 0$, $x_i^0 \geq 0$, this sum is zero iff every summand is zero.

Theorem 6: If problems D^0 and D^1 both have feasible vectors, then both problems have optimal solutions.

Proof: Let M be the set of feasible vectors x for D^0 and let $\hat{w}$ be a feasible vector for D^1. By theorem 1, $Q(x) \geq G(\hat{w})$ for all $x \in M$. $Q(x)$ is bounded below on M. By theorem 3 of §2, there exists a vertex $\bar{x}$ of M. Beginning at $\bar{x}$, the simplex method finds a finite minimal solution x^0 of D^0; for the case that no solution exists, considered in theorem 3 of §2, cannot occur here because $Q(x)$ is bounded below on M. By theorem 2, there also exists a finite maximal solution w^0 of D^0.

Theorems 1 and 6 show how a two-sided bound on the extreme value Q^0 of D^0 is obtained from a pair of feasible vectors, $\hat{x}$ for D^0 and $\hat{w}$ for D^1. Knowing these vectors, we find that

$$G(\hat{w}) \leq Q^0 \leq Q(\hat{x}). \tag{5.7}$$

Example: Sheep and cattle raising, cf §1. By introducing slack variables, we write the problem as a minimal problem in form (5.1) with $m = 4$ and $n = 6$.

$$\underset{\sim}{A} = \begin{pmatrix} 1 & 0 & 1 & 0 & 0 & 0 \\ 0 & 1 & 0 & 1 & 0 & 0 \\ 1 & 0.2 & 0 & 0 & 1 & 0 \\ 150 & 25 & 0 & 0 & 0 & 1 \end{pmatrix}, \quad \underset{\sim}{b} = \begin{pmatrix} 50 \\ 200 \\ 72 \\ 10000 \end{pmatrix}$$

$$\underset{\sim}{p} = \begin{pmatrix} -250 \\ -45 \\ 0 \\ 0 \\ 0 \\ 0 \end{pmatrix}$$

$\hat{\underset{\sim}{x}} = (36, 180, 14, 20, 0, 100)'$ is a feasible vector for D^0 and $\hat{\underset{\sim}{w}} = (-50, -10, -50, -1)'$ is a feasible vector for D^1. Then

$$G(\hat{\underset{\sim}{w}}) = -18100 \leq Q^0 \leq -17100 = Q(\hat{\underset{\sim}{x}}).$$

The actual value is $Q^0 = -17{,}200$.

We now want to show how to obtain a numerical solution for the dual problem D^1, given that we have already treated D^0 with the simplex method, and obtained a solution $\underset{\sim}{x}^0$ which is a vertex of the set M of feasible points for D^0. From theorem 5 we know that a solution $\underset{\sim}{w}^0$ of D^1 is determined by the system of linear equations

$$\underset{\sim}{a}^{k'} \underset{\sim}{w}^0 = p_k \quad (k \in Z)$$

where the vectors $\underset{\sim}{a}^k$, $k \in Z$, constitute a basis at vertex $\underset{\sim}{x}^0$.

The solution for this system of equations has a particularly simple form when the vectors $\underset{\sim}{a}^i$, $i = 1,\ldots,n$, include the m unit vectors of R^m; say $\underset{\sim}{a}^1,\ldots,\underset{\sim}{a}^m$, with $a_{i\ell} = \delta_{i\ell}$, $i,\ell = 1,\ldots,m$. Since $\sum_{k\varepsilon Z} c_{ki}\underset{\sim}{a}^k = \underset{\sim}{a}^i$ $(i = 1,\ldots,m)$, the c_{ki} are the elements of the inverse of the matrix of the above system of equations, and thus $w_i^0 = \sum_{k\varepsilon Z} c_{ki}p_k$ $(i = 1,\ldots,m)$. By (3.7), the w_i^0 are equal to the t_i defined there. The numbers d_i are taken from the terminal tableau for problem D^0; by §4.1, $d_i = t_i - p_i$. Therefore,

$$w_i^0 = d_i + p_i,$$

and the solution of the dual problem can immediately be read off of the simplex tableau. Later (in §10.2 and §18.6) this remark will prove useful.

5.2. Symmetric Dual Problems with Inequalities as Constraints

Let $\underset{\sim}{A}$ now be an m-by q matrix, $\underset{\sim}{b} \in R^m$, and $\underset{\sim}{p} \in R^q$:

$$\underset{\sim}{A} = \begin{pmatrix} a_{11} & \cdots & a_{1q} \\ \cdots & \cdots & \cdots \\ a_{m1} & \cdots & a_{mq} \end{pmatrix}, \quad \underset{\sim}{b} = \begin{pmatrix} b_1 \\ \cdots \\ b_m \end{pmatrix}, \quad \underset{\sim}{p} = \begin{pmatrix} p_1 \\ \cdots \\ p_q \end{pmatrix}.$$

We again formulate two problems.

$\hat{D}^0$: Find an $\underset{\sim}{x} \in R^q$ such that

$$\underset{\sim}{A}\underset{\sim}{x} \geq \underset{\sim}{b}, \quad \underset{\sim}{x} \geq \underset{\sim}{0}, \quad Q(\underset{\sim}{x}) = \underset{\sim}{p}'\underset{\sim}{x} = \text{Min!}$$

$\hat{D}^1$: Find a $\underset{\sim}{w} \in R^m$ such that

$$\underset{\sim}{A}'\underset{\sim}{w} \leq \underset{\sim}{p}, \quad \underset{\sim}{w} \geq \underset{\sim}{0}, \quad G(\underset{\sim}{w}) = \underset{\sim}{w}'\underset{\sim}{b} = \text{Max!}$$

Remark: Since the constraints are in the form of inequalities, we do not need any conditions on m and q (such as $m < n$ in §5.1) nor on the rank of $\underset{\sim}{A}$.

Theorem 7: $\hat{D}^0$ and $\hat{D}^1$ are dual to each other in the sense that theorems 1 through 4 and 6 of §5.1 are valid (with the obvious transfer of notation).

Proof: After the introduction of a slack variable vector $\underset{\sim}{y} \in R^m$, D^0 is equivalent to the problem $\underset{\sim}{A}\underset{\sim}{x}-\underset{\sim}{y} = \underset{\sim}{b}$, $\underset{\sim}{x} \geq \underset{\sim}{0}$, $\underset{\sim}{y} \geq \underset{\sim}{0}$, $\underset{\sim}{p}'\underset{\sim}{x} = \text{Min!}$ This is a problem of type D^0 with the matrix $(\underset{\sim}{A}| - \underset{\sim}{E}_m)$ in place of $\underset{\sim}{A}$. This is a matrix whose row number m is less than its column number $m+q$, and whose rank is clearly equal to the row number m. By §5.1, the problem dual to this is $\underset{\sim}{A}'\underset{\sim}{w} \leq \underset{\sim}{p}$, $-\underset{\sim}{w} \leq \underset{\sim}{0}$, $\underset{\sim}{w}'\underset{\sim}{b} = \text{Max!}$, which is exactly $\hat{D}^1$.

It follows from the proof of theorem 7 that we have the following theorem in place of theorem 5.

Theorem 5a: A feasible vector $\underset{\sim}{x}^0 = (x_1^0,\ldots,x_q^0)'$ for $\hat{D}^0$ is a minimal solution for $\hat{D}^0$ iff there exists a feasible vector $\underset{\sim}{w}^0 = (w_1^0,\ldots,w_m^0)'$ for $\hat{D}^1$ with the following properties (where the $\underset{\sim}{a}^{i}{}'$, $i = 1,\ldots,m$, are row vectors of $\underset{\sim}{A}$).

$$\left.\begin{array}{rcl} x_k^0 > 0 & \text{implies} & a^{k'}w^0 = p_k \\ a^{k'}w^0 < p_k & \text{implies} & x_k^0 = 0 \end{array}\right\} \quad k = 1,\ldots,q$$

$$\left.\begin{array}{rcl} w_i^0 > 0 & \text{implies} & \tilde{a}^{t'}x^0 = b_i \\ \tilde{a}^{i'}x^0 > b_i & \text{implies} & w_i^0 = 0 \end{array}\right\} \quad i = 1,\ldots,m.$$

We might ask ourselves when a problem $\hat{D}^0$ is self-dual. Clearly, we must have $A = -A'$ (so matrix A must be square and skew-symmetric), $b = -p$, and $m = q$. Such self-dual problems, however, are of no great practical or theoretical importance in linear optimization (unlike self-adjoint problems in differential equations, for example.) Nevertheless, one should note that the important theorem 12 in §5.5 on skew-symmetric matrices is basically an assertion about self-dual problems.

5.3. Duality for Mixed Problems

We can combine the results of §5.1 and §5.2 and formulate a duality theorem for problems, some of whose constraints are equations and the rest of which are inequalities, and some of whose variables are subject to positivity constraints while the rest are not. Let there be given an m-by-n matrix

$$A = \begin{pmatrix} A_{11} & A_{12} \\ A_{21} & A_{22} \end{pmatrix} \begin{array}{l} m_1 \text{ rows} \\ m_2 \text{ rows} \end{array}$$

$$\begin{array}{cc} n_1 \text{ columns} & n_2 \text{ columns} \end{array}$$

where $m_1 + m_2 = m$ and $n_1 + n_2 = n$. Here, $m_1 < n$, and the matrix $(\underset{\sim}{A}_{11}|\underset{\sim}{A}_{12})$ has rank m_1. Let there also be given the vectors

$$\underset{\sim}{b} = \begin{pmatrix} \underset{\sim}{b}^1 \\ \underset{\sim}{b}^2 \end{pmatrix}, \quad \underset{\sim}{p} = \begin{pmatrix} \underset{\sim}{p}^1 \\ \underset{\sim}{p}^2 \end{pmatrix},$$

with $\underset{\sim}{b}^1 \in R^{m_1}$, $\underset{\sim}{b}^2 \in R^{m_2}$, $\underset{\sim}{p}^1 \in R^{n_1}$, $\underset{\sim}{p}^2 \in R^{n_2}$. The two problems which will prove to be dual are

$\tilde{D}^0$: Find $\underset{\sim}{x} = \begin{pmatrix} \underset{\sim}{x}^1 \\ \underset{\sim}{x}^2 \end{pmatrix}$ $(\underset{\sim}{x}^1 \in R^{n_1},\ \underset{\sim}{x}^2 \in R^{n_2})$ such that

$$\begin{aligned} &\underset{\sim}{A}_{11}\underset{\sim}{x}^1 + \underset{\sim}{A}_{12}\underset{\sim}{x}^2 = \underset{\sim}{b}^1, && \underset{\sim}{x}^1 \geq \underset{\sim}{0}, \\ &\underset{\sim}{A}_{21}\underset{\sim}{x}^1 + \underset{\sim}{A}_{22}\underset{\sim}{x}^2 \geq \underset{\sim}{b}^2, && \underset{\sim}{x}^2 \text{ unrestricted in sign} \\ &\underset{\sim}{p}'\underset{\sim}{x} = \underset{\sim}{p}^{1'}\underset{\sim}{x}^1 + \underset{\sim}{p}^{2'}\underset{\sim}{x}^2 = \text{Min!} \end{aligned}$$

$\tilde{D}^1$: Find $\underset{\sim}{w} = \begin{pmatrix} \underset{\sim}{w}^1 \\ \underset{\sim}{w}^2 \end{pmatrix}$ $(\underset{\sim}{w}^1 \in R^{m_1},\ \underset{\sim}{w}^2 \in R^{m_2})$ such that

$$\begin{aligned} &\underset{\sim}{A}'_{11}\underset{\sim}{w}^1 + \underset{\sim}{A}'_{21}\underset{\sim}{w}^2 \leq \underset{\sim}{p}^1, && \underset{\sim}{w}^1 \text{ unrestricted in sign} \\ &\underset{\sim}{A}'_{12}\underset{\sim}{w}^1 + \underset{\sim}{A}'_{22}\underset{\sim}{w}^2 = \underset{\sim}{p}^2, && \underset{\sim}{w}^2 \geq 0, \\ &\underset{\sim}{w}'\underset{\sim}{b} = \underset{\sim}{w}^{1'}\underset{\sim}{b}^1 + \underset{\sim}{w}^{2'}\underset{\sim}{b}^2 = \text{Max!} \end{aligned}$$

Theorem 8: $\tilde{D}^0$ and $\tilde{D}^1$ are dual to each other in the sense that theorems 1 through 4 and 6 of §5.1 are valid.

Proof: Represent the unrestricted vector $\underset{\sim}{x}^2$ as a difference, $\bar{\underset{\sim}{x}}^2 - \bar{\bar{\underset{\sim}{x}}}^2$, where $\bar{\underset{\sim}{x}}^2 \geq 0$ and $\bar{\bar{\underset{\sim}{x}}}^2 \geq 0$, and introduce a slack variable vector $\underset{\sim}{y}^2 \in R^{m_2}$ so as to transform $\tilde{D}^0$ into the equivalent problem

$$\underset{\sim}{A}_{11}\underset{\sim}{x}^1 + \underset{\sim}{A}_{12}\bar{\underset{\sim}{x}}^2 - \underset{\sim}{A}_{12}\bar{\bar{\underset{\sim}{x}}}^2 = \underset{\sim}{b}^1,$$

$$\underset{\sim}{A}_{21}\underset{\sim}{x}^1 + \underset{\sim}{A}_{22}\bar{\underset{\sim}{x}}^2 - \underset{\sim}{A}_{22}\bar{\bar{\underset{\sim}{x}}}^2 - \underset{\sim}{y}^2 = \underset{\sim}{b}^2,$$

$$\underset{\sim}{x}^1 \geq \underset{\sim}{0},\ \bar{\underset{\sim}{x}}^2 \geq \underset{\sim}{0},\ \bar{\bar{\underset{\sim}{x}}}^2 \geq \underset{\sim}{0},\ \underset{\sim}{y}^2 \geq \underset{\sim}{0},$$

$$\underset{\sim}{p}^{1'}\underset{\sim}{x}^1 + \underset{\sim}{p}^{2'}\underset{\sim}{x}^2 - \underset{\sim}{p}^{2'}\bar{\bar{\underset{\sim}{x}}}^2 = \text{Min!}$$

of type D^0. The conditions on the rank, row number, and column number demanded by §5.1 are satisfied by the matrix of this problem. The dual problem is

$$\begin{aligned} \underset{\sim}{A}_{11}'\underset{\sim}{w}^1 + \underset{\sim}{A}_{21}'\underset{\sim}{w}^2 &\leq \underset{\sim}{p}^1, \\ \underset{\sim}{A}_{12}'\underset{\sim}{w}^1 + \underset{\sim}{A}_{22}'\underset{\sim}{w}^2 &\leq \underset{\sim}{p}^2, \\ -\underset{\sim}{A}_{12}'\underset{\sim}{w}^1 - \underset{\sim}{A}_{22}'\underset{\sim}{w}^2 &\leq -\underset{\sim}{p}^2 \\ -\underset{\sim}{w}^2 &\leq \underset{\sim}{0}, \\ \underset{\sim}{w}^{1'}\underset{\sim}{b}^1 + \underset{\sim}{w}^{2'}\underset{\sim}{b}^2 &= \text{Max!}; \end{aligned}$$

and this is equivalent to $\tilde{D}^1$.

We do not attempt to formulate an analog of theorem 5. If $n_1 = n$, $m_1 = m$, and $n_2 = m_2 = 0$, that is, if the submatrices $\underset{\sim}{A}_{12}$, $\underset{\sim}{A}_{21}$, and $\underset{\sim}{A}_{22}$ do not even appear, then problems $\tilde{D}^0$ and $\tilde{D}^1$ reduce to D^0 and D^1. If $n_1 = n$, $m_2 = m$, and $n_2 = m_1 = 0$, so that $\underset{\sim}{A}_{11}$, $\underset{\sim}{A}_{12}$, and $\underset{\sim}{A}_{22}$ do not appear, we obtain $\hat{D}^0$ and $\hat{D}^1$.

There is a one-to-one assignment of the constraints of $\tilde{D}^0$ to the variables of $\tilde{D}^1$ and, conversely, of the variables of $\tilde{D}^0$ to the constraints of $\tilde{D}^1$. A component of $\underset{\sim}{w}^1$ (resp. $\underset{\sim}{w}^2$) is assigned to that constraint of $\tilde{D}^0$ which has the corresponding component of $\underset{\sim}{b}^1$ (resp. $\underset{\sim}{b}^2$) on the right-hand side. A similar assignment is made for

$\underset{\sim}{x}^1$, $\underset{\sim}{x}^2$ and $\underset{\sim}{p}^1$, $\underset{\sim}{p}^2$.

The form of $\tilde{D}^0$ and $\tilde{D}^1$ shows that the inequality constraints are assigned to positively constrained variables and the equality constraints are assigned to unrestricted variables.

5.4. Linear Optimization and Duality in Statics

(Due to W. Prager, 1962)

The example treated here results in a pair of dual optimization problems, each of which has physical significance. A rigid, weightless, four-sided plate is supported at its four corners.

The following, idealized assumptions are made. The supports are rigid. They may be subjected to an arbitrarily high load by tension (the plate is firmly connected to the supports, so that it cannot be lifted off). They may be subjected to loading by compression up to a creep limit, F_j, $j = 1,\ldots,4$. Thus the jth support remains rigid and unchanged in length while subject to a force P with $-\infty < P \leq F_j$. If P exceeds the creep limit F_j, the support collapses.

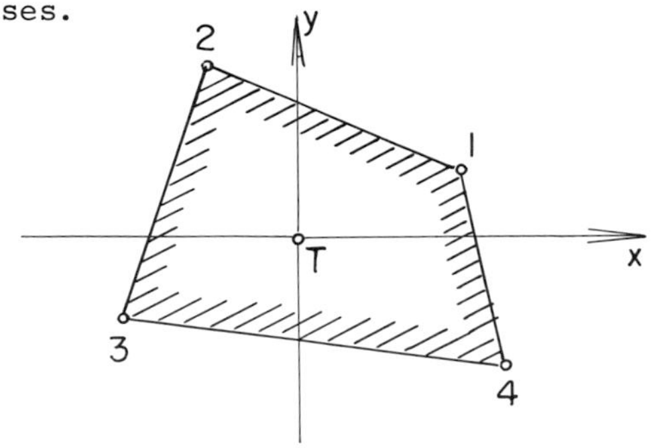

Figure 5.1. A plate supported at four places.

Our problem is to find the greatest load to which any point T of the plate may be subjected without causing a collapse of the supports. This maximum admissible load is called the limit load P^* at the point T, and naturally depends on the location of T.

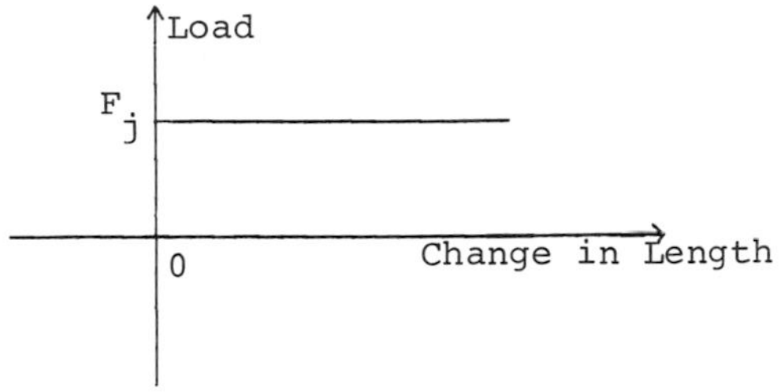

Figure 5.2. Behavior of a support under load

Let P_j, $j = 1,\ldots,4$, denote the force acting on the jth support. As the load P at point T increases from zero, the force $P_j = F_j$ at some corner is eventually reached.

If the forces at the other corners are still $P_j < F_j$, the supported plate will not yet collapse; for then we have the (statically determined) case of a loaded plate supported at three corners. Only when the force acting at a second corner exceeds the creep limit will a collapse result (which consists of a rotation about the axis connecting the two remaining corners).

We choose our coordinate system so that T is the origin and the corners of the quadrilateral have coordinates (ξ_j,η_j), $j = 1,\ldots,4$. If P is the load at the point T, we have the equilibrium constraints

$$P = \sum_{j=1}^{4} P_j, \tag{5.8}$$

$$\sum_{j=1}^{4} P_j \xi_j = 0, \quad \sum_{j=1}^{4} P_j \eta_j = 0. \tag{5.9}$$

We want to find the maximal value P^* of P for which there still exist P_j satisfying (5.8), (5.9), and

$$P_j \le F_j \quad (j = 1,\dots,4),$$

so that the creep limits are not exceeded. We are dealing, therefore, with a linear optimization problem in four variables P_j, without positivity constraints, and six constraints, two equalities and four inequalities. Using the notation of §5.3, we write the problem in form $\tilde{D}^1$, where

$$\underset{\sim}{w}^1 = \begin{pmatrix} P_1 \\ P_2 \\ P_3 \\ P_4 \end{pmatrix}, \quad \underset{\sim}{A}'_{11} = \begin{pmatrix} 1 & 0 & 0 & 0 \\ 0 & 1 & 0 & 0 \\ 0 & 0 & 1 & 0 \\ 0 & 0 & 0 & 1 \end{pmatrix}, \quad \underset{\sim}{A}'_{12} = \begin{pmatrix} \xi_1 & \xi_2 & \xi_3 & \xi_4 \\ \eta_1 & \eta_2 & \eta_3 & \eta_4 \end{pmatrix}$$

$$\underset{\sim}{p}^1 = \begin{pmatrix} F_1 \\ F_2 \\ F_3 \\ F_4 \end{pmatrix}, \quad \underset{\sim}{p}^2 = \begin{pmatrix} 0 \\ 0 \end{pmatrix}, \quad \underset{\sim}{b}^1 = \begin{pmatrix} 1 \\ 1 \\ 1 \\ 1 \end{pmatrix},$$

and $\underset{\sim}{w}^2$, $\underset{\sim}{A}'_{21}$, $\underset{\sim}{A}'_{22}$, and $\underset{\sim}{b}^2$ do not appear (so $m_2 = 0$, $n_1 = 4$, $n_2 = 2$, and $m_1 = 4$):

$$\underset{\sim}{A}'_{11}\underset{\sim}{w}^1 \le \underset{\sim}{p}^1, \quad \underset{\sim}{A}'_{12}\underset{\sim}{w}^1 = \underset{\sim}{p}^2, \quad P = \underset{\sim}{w}^{1'}\underset{\sim}{b}^1 = \text{Max!}.$$

The dual problem then is to find an $\underset{\sim}{x} = \begin{pmatrix} \underset{\sim}{x}^1 \\ \underset{\sim}{x}^2 \end{pmatrix}$

$(\underset{\sim}{x}^1 \in R^4, \underset{\sim}{x}^2 \in R^2)$ with $\underset{\sim}{A}_{11}\underset{\sim}{x}^1 + \underset{\sim}{A}_{12}\underset{\sim}{x}^2 = \underset{\sim}{b}^1$, $\underset{\sim}{x}^1 \geq \underset{\sim}{0}$, $\underset{\sim}{p}^{1'}\underset{\sim}{x}^1 + \underset{\sim}{p}^{2'}\underset{\sim}{x}^2 = \text{Min!}$.

Setting $\underset{\sim}{x}^1 = (v_1, v_2, v_3, v_4)'$, $\underset{\sim}{x}^2 = (\omega_x, \omega_y)'$, we obtain the problem

$$\left.\begin{aligned} v_j + \xi_j\omega_x + \eta_j\omega_y &= 1 \\ v_j &\geq 0 \end{aligned}\right\} (j = 1, \ldots, 4), \qquad \begin{aligned}(5.10)\\(5.11)\end{aligned}$$

$$\sum_{j=1}^{4} F_j v_j = \text{Min!} \qquad (5.12)$$

where v_j can be interpreted as the virtual deflection at the jth corner, ω_x as the virtual rotation about the axis $x = 0$, and ω_y as the virtual rotation about the axis $y = 0$. (5.10) is a consequence of the assumption that the plate is rigid, if the point T (the origin of the coordinate system) is subjected to a virtual deflection $v = 1$, in the direction of the applied load, and the corners are subjected to a virtual deflection v_j and also to virtual rotations ω_x and ω_y. $v_j \geq 0$ follows from the assumption that the supports may be subjected to arbitrarily high loads by tension without a change in length. A positive deflection $v_j > 0$ can only occur if the force acting at the jth corner is F_j, since the supports remain rigid under a smaller load. The virtual work (= force times virtual deflection) at the corners thus adds to $\sum_{j=1}^{4} F_j v_j$, while at the point T the virtual work is $Pv = P$, since $v = 1$. By the principle of virtual work,

$$P = \sum_{j=1}^{4} F_j v_j.$$

(5.12) requires us to find the smallest load $P = P^{**}$

for which a positive virtual deflection $v = 1$ is possible at the point T. For $P < P^{**}$ any such virtual deflection is impossible, and the system remains rigid.

Duality theorem 2 (for $\tilde{D}^0$ and $\tilde{D}^1$) yields the conclusion that $P^* = P^{**}$, and this we have just shown by physical arguments. In addition, the $\tilde{D}^0$-$\tilde{D}^1$ version of theorem 5 shows that, given a solution $P = P^* = P^{**}$ of the dual problems, we can have $v_j > 0$ only if $P_j = F_j$ (at the corner in question).

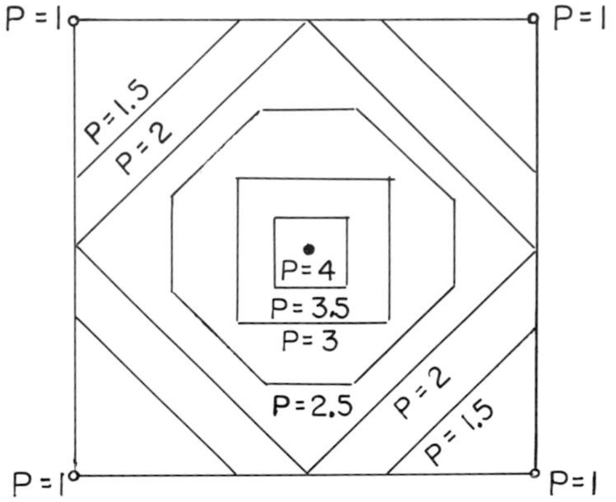

Figure 5.3. Square plate with creep limits $F_j = 1$

The case of a square plate with equal creep limits $F_j = 1$, $j = 1,\ldots,4$, at all four corners and the positivity constraints $P_j \geq 0$, $j = 1,\ldots,4$, was already treated in 1823 by Fourier. This is perhaps the first example of a linear optimization problem.

In the cited work of Prager, further examples (e.g., plastic designs of beams and frames) are treated.

5.5. Theorems of the Alternative for Systems of Linear Equations and Inequalities

In this section, we will prove several theorems from which we also can derive the duality properties of linear optimization problems, and which will be applied in the discussion of convex optimization in the following chapter.

This route to the duality theorems is at once short and elementary, independent of the previous sections, and free of any considerations of polyhedra or degeneracy.

As a starting point, we use the theorem of the alternative on the solubility of homogeneous and inhomogeneous systems of linear equations. The formulation is chosen with subsequent applications in mind. $\underset{\sim}{A}$ is to be an m-by-n matrix of arbitrary rank, and $\underset{\sim}{b}$ a vector in R^m. All quantities are once again real.

Theorem 9: Either the system of equations

$$\underset{\sim}{A}\underset{\sim}{x} = \underset{\sim}{b} \tag{5.13}$$

has a solution $\underset{\sim}{x} \in R^n$, or the system of equations

$$\underset{\sim}{A}'\underset{\sim}{y} = \underset{\sim}{0}, \quad \underset{\sim}{b}'\underset{\sim}{y} = 1 \tag{5.14}$$

has a solution $\underset{\sim}{y} \in R^m$.

Proof: I. (5.13) and (5.14) are not simultaneously solvable. For if $\underset{\sim}{x} \in R^n$ and $\underset{\sim}{y} \in R^m$ were solutions, then

$$0 = \underset{\sim}{x}'\underset{\sim}{A}'\underset{\sim}{y} = (\underset{\sim}{A}\underset{\sim}{x})'\underset{\sim}{y} = \underset{\sim}{b}'\underset{\sim}{y} = 1$$

II. If (5.13) has no solution, (5.14) is solvable.

For then $\underset{\sim}{b}$ is not a linear combination of the column vectors of $\underset{\sim}{A}$, which implies the following. If matrix $\underset{\sim}{A}$ has rank r, as does matrix $\underset{\sim}{A}'$ therefore, the (n+1)-by-m matrix $\begin{pmatrix} \underset{\sim}{A}' \\ \hline \underset{\sim}{b}' \end{pmatrix}$ has rank r+1. Furthermore, the (n+1)-by-(m+1) matrix

$$\left(\begin{array}{c|c} & 0 \\ \underset{\sim}{A}' & \cdots \\ & 0 \\ \hline \underset{\sim}{b}' & 1 \end{array}\right)$$

also has rank r+1. Since both matrices have the same rank, it follows from the theory of linear equations that (5.14) has a solution.

The next theorem of the alternative we prove will require a non-negative solution $\underset{\sim}{x}$ in (5.13). We make the following definition in order to phrase the proof simply, and also to clarify the significance of the theorem.

<u>Definition</u>: Let $\underset{\sim}{a}^1,\ldots,\underset{\sim}{a}^n$ be vectors in R^m. The <u>cone</u> generated by $\underset{\sim}{a}^1,\ldots,\underset{\sim}{a}^n$ is the set of all linear combinations $\sum_{i=1}^{n} \underset{\sim}{a}^i x_i$ with $x_i \geq 0$, $i = 1,\ldots,n$, and is denoted by $K(\underset{\sim}{a}^1,\ldots,\underset{\sim}{a}^n)$. See Figure 5.4.

<u>Theorem 10</u>: Either the system

$$\underset{\sim}{A}\underset{\sim}{x} = \underset{\sim}{b}, \quad \underset{\sim}{x} \geq \underset{\sim}{0} \tag{5.15}$$

has a solution $\underset{\sim}{x} \in R^n$, or the system

$$\underset{\sim}{A}'\underset{\sim}{y} \geq 0, \quad \underset{\sim}{b}'\underset{\sim}{y} < 0 \tag{5.16}$$

has a solution $\underset{\sim}{y} \in R^m$.

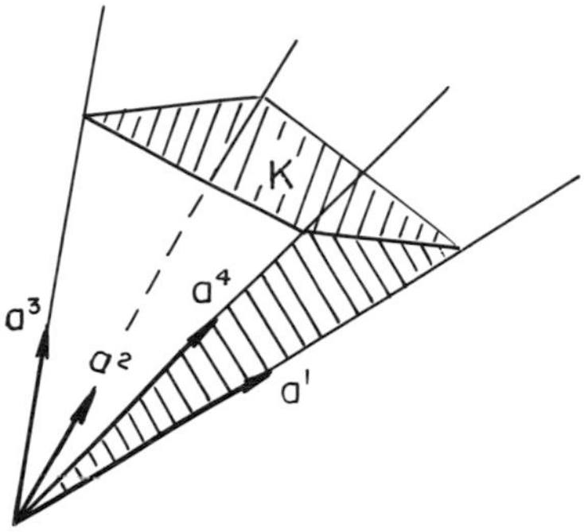

Figure 5.4. The cone $K(a^1, a^2, a^3, a^4)$

Proof: I. (5.15) and (5.16) are not simultaneously solvable. If $x \in R^n$ and $y \in R^m$ were solutions, then

$$0 > b'y = (Ax)'y = x'A'y \geq 0.$$

II. If $Ax = b$ has no solution whatsoever, then by theorem 9, there exists a $\hat{y}$ such that $A'\hat{y} = 0$, $b'y = 1$. Then $y = -\hat{y}$ is a solution of (5.16).

III. It remains to show that (5.16) is solvable whenever every solution x of $Ax = b$ has at least one negative component. The proof will be by induction on the column number n of matrix A.

$n = 1$: A contains only one column vector, a^1. Suppose $a^1 x_1 = b$ and $x_1 < 0$. Now $b \neq 0$, for otherwise $x_1 = 0$ would have been a solution of (5.15). Then $y = -b$ is a solution of (5.16) because

$$a^{1'} \underset{\sim}{y} = -a^{1'} \underset{\sim}{b} = -x_1 a^{1'} a^{1} \geq 0, \quad \underset{\sim}{b'} \underset{\sim}{y} = -\underset{\sim}{b'} \underset{\sim}{b} < 0.$$

Induction step: Suppose the conclusion of the theorem holds for column number $n-1$. We have to show, for column number n, that $\underset{\sim}{b} \notin K(a^{1}, \ldots, a^{n})$ implies that (5.16) has a solution $\underset{\sim}{y} \in R^{m}$. $\underset{\sim}{b} \notin K(a^{1}, \ldots, a^{n})$ implies first of all that $\underset{\sim}{b} \notin K(a^{1}, \ldots, a^{n-1})$. By the induction hypothesis, there exists a vector $\underset{\sim}{v} \in R^{m}$ such that $a^{i'} \underset{\sim}{v} \geq 0$, $i = 1, \ldots, n-1$, and $\underset{\sim}{b'} \underset{\sim}{v} < 0$. If, in addition, $a^{n'} \underset{\sim}{v} \geq 0$, we can set $\underset{\sim}{y} = \underset{\sim}{v}$. It remains to investigate the case $a^{n'} \underset{\sim}{v} < 0$. Define vectors $\hat{a}^{1}, \ldots, \hat{a}^{n-1}$, $\hat{\underset{\sim}{b}}$ by

$$\hat{\underset{\sim}{a}}^{i} = (a^{i'} \underset{\sim}{v}) a^{n} - (a^{n'} \underset{\sim}{v}) a^{i}, \quad \hat{\underset{\sim}{b}} = (\underset{\sim}{b'} \underset{\sim}{v}) a^{n} - (a^{n'} \underset{\sim}{v}) \underset{\sim}{b}.$$

The following two cases (a) and (b) are now possible.

(a) $\hat{\underset{\sim}{b}} \in K(a^{1}, \ldots, a^{n-1})$

Then there exist non-negative z_i, $i = 1, \ldots, n-1$, such that $\sum_{i=1}^{n-1} \hat{\underset{\sim}{a}}^{i} z_i = \hat{\underset{\sim}{b}}$.

It follows that

$$\underset{\sim}{b} = \sum_{i=1}^{n-1} a^{i} z_i - \frac{1}{\underset{\sim n \sim}{a'v}} \left[\sum_{i=1}^{n-1} (a^{i'} \underset{\sim}{v}) z_i - \underset{\sim}{b'} \underset{\sim}{v} \right] a^{n}.$$

Since $z_i \geq 0$, $a^{i'} \underset{\sim}{v} \geq 0$ $(i = 1, \ldots, n-1)$, $a^{n'} \underset{\sim}{v} < 0$, $\underset{\sim}{b'} \underset{\sim}{v} < 0$, we would have $\underset{\sim}{b} \in K(a^{1}, \ldots, a^{n})$. Thus this case cannot occur.

(b) $\hat{\underset{\sim}{b}} \notin K(\hat{\underset{\sim}{a}}^{1}, \ldots, a^{n-1})$

By the induction hypothesis there is a vector $\underset{\sim}{w} \in R^{m}$ with $\hat{\underset{\sim}{a}}^{i'} \underset{\sim}{w} \geq 0$, $i = 1, \ldots, n-1$, and $\hat{\underset{\sim}{b}}' \underset{\sim}{w} < 0$. Then

$y = (a^{n'}w)v - (a^{n'}v)w$ is a solution of (5.16) because $a^{i'}y = (a^{n'}w)(a^{i'}v) - (a^{n'}v)(a^{i'}w) = \hat{a}^{i'}w \geq 0$ $(i = 1,\ldots, n-1)$, $a^{n'}y = 0$, and $b'y = b'w < 0$.

Remark. In part III. of the above proof, we could just as easily have used the separation theorem for convex sets given in the Appendix. The cone $K(a^1,\ldots,a^n)$ is a convex, closed subset of R^m. Suppose $b \notin K(a^1,\ldots,a^n)$. Then the cone is a proper subset of R^m. Let it be the set B_1. For a point b not in the closed set B_1, there is an open ball $S = \{w \mid \|w-b\| < \eta\}$ ($\eta > 0$, and $\|x\| = (\sum_i x_i^2)^{1/2}$ is the usual euclidean norm), about b which contains no points of B_1. Indeed, the open set $B_2 = \{\alpha w \mid w \in S, \alpha > 0\}$ contains no points of B_1. By the separation theorem, there exists a non-zero vector $a \in R^m$ and a real β such that $a'u \leq \beta < a'v$ for $u \in B_1$ and $v \in B_2$. Since $0 \in B_1$, $\beta \geq 0$. Since $\alpha b \in B_2$ for all $\alpha > 0$, we cannot have $\beta > 0$. Therefore, $\beta = 0$. For $v = b$ we then have $a'b > 0$, and for $u = a^i$, $i = 1,\ldots,n$, we have $a'a^i \leq 0$. $y = -a$ is then the vector, whose existence had to be shown in part III. of the above proof.

We see that theorem 10 may be given the following formulation. Either b lies in the cone $K(a^1,\ldots,a^n)$ or there exists a hyperplane through the origin separating b from the cone.

We may deduce a result about skew-symmetric matrices from theorem 10 which is important in the treatment of duality. The following theorem serves as preparation. A is an arbitrary real m-by-n matrix once again.

Theorem 11: The systems $A'\underset{\sim}{y} \geq \underset{\sim}{0}$ and $A\underset{\sim}{x} = \underset{\sim}{0}, \underset{\sim}{x} \geq \underset{\sim}{0}$ have solutions $\underset{\sim}{\tilde{y}}$ and $\underset{\sim}{\tilde{x}}$ with $A'\underset{\sim}{\tilde{y}} + \underset{\sim}{\tilde{x}} > \underset{\sim}{0}$.

(The notation where a vector is $> \underset{\sim}{0}$ means, as usual, that all the components are positive.)

Proof: Let the a^i, $i = 1,\ldots,n$, be the column vectors of $\underset{\sim}{A}$. For $k = 1,\ldots,n$, we consider the systems

$$\sum_{\substack{i=1\\i\neq k}}^{n} \underset{\sim}{a}^i x_i = -\underset{\sim}{a}^k, \quad x_i \geq 0 \quad (i \neq k) \qquad (5.17)$$

and

$$\underset{\sim}{a}^{i'}\underset{\sim}{y} \geq \underset{\sim}{0} \quad (i = 1,\ldots,n; \ i \neq k), \quad \underset{\sim}{a}^{k'}\underset{\sim}{y} > \underset{\sim}{0} \quad (5.18)$$

For fixed k, exactly one of these systems has a solution, by theorem 10. If (5.17) is solvable, there exists a vector $\underset{\sim}{\tilde{x}}^k \varepsilon R^n$ with $A\underset{\sim}{\tilde{x}}^k = \underset{\sim}{0}$ and $\underset{\sim}{\tilde{x}}^k \geq \underset{\sim}{0}$, for which the component $x_k^k = 1$. If (5.18) is solvable, there exists a vector $\underset{\sim}{\tilde{y}}^k \varepsilon R^m$ with $A'\underset{\sim}{\tilde{y}}^k \geq \underset{\sim}{0}$, for which $\underset{\sim}{a}^{k'}\underset{\sim}{\tilde{y}}^k > \underset{\sim}{0}$. The indices k for which (5.17) is solvable define an index set Z_1, and those, for which (5.18) is solvable, an index set Z_2. Now $Z_1 \cup Z_2 = \{1,2,\ldots,n\}$. Setting $\underset{\sim}{\tilde{x}} = \sum_{k\varepsilon Z_1} \underset{\sim}{\tilde{x}}^k$, and $\underset{\sim}{\tilde{y}} = \sum_{k\varepsilon Z_2} \underset{\sim}{\tilde{y}}^k$, we have $A'\underset{\sim}{\tilde{y}} \geq \underset{\sim}{0}$, $A\underset{\sim}{\tilde{x}} = \underset{\sim}{0}$, $\underset{\sim}{\tilde{x}} \geq \underset{\sim}{0}$, and $A'\underset{\sim}{\tilde{y}} + \underset{\sim}{\tilde{x}} > \underset{\sim}{0}$.

Theorem 12: Let $\underset{\sim}{A}$ be a real, skew-symmetric n-by-n matrix. Then there exists a vector $\underset{\sim}{w} \varepsilon R^n$ such that

$$A\underset{\sim}{w} \geq \underset{\sim}{0}, \quad \underset{\sim}{w} \geq \underset{\sim}{0}, \quad A\underset{\sim}{w} + \underset{\sim}{w} > \underset{\sim}{0}.$$

Proof: A real, skew-symmetric matrix is characterized by the property $A' = -A$. The systems

$$\left(\frac{E_n}{A}\right) y \geq 0 \quad \text{and} \quad (E_n \mid -A)\begin{pmatrix} x \\ z \end{pmatrix} = 0, \begin{pmatrix} x \\ z \end{pmatrix} \geq 0$$

(where E_n is the n-dimensional identity matrix) have solutions $\tilde{y}$, $\tilde{x}$, $\tilde{z}$ (by theorem 11, with x there corresponding to $\begin{pmatrix} x \\ z \end{pmatrix}$ here) such that $\tilde{y} \geq 0$, $A\tilde{y} \geq 0$, $\tilde{x} - A\tilde{z} = 0$, $\tilde{x} \geq 0$, $\tilde{z} \geq 0$, $\tilde{y} + \tilde{x} > 0$, and $A\tilde{y} + \tilde{z} > 0$, so also $\tilde{y} + A\tilde{z} > 0$. Setting $w = \tilde{y} + \tilde{z}$, we obtain $Aw \geq 0$, $w \geq 0$, and $Aw + w > 0$.

5.6. Another Approach to the Treatment of Duality

In §5.2 we already showed that the problems

$$\hat{D}^0\text{: } Q(x) = p'x = \text{Min!}, \quad Ax \geq b, \quad x \geq 0;$$
$$\hat{D}^1\text{: } G(w) = b'w = \text{Max!}, \quad A'w \leq p, \quad w \geq 0$$

are dual to each other. Here A is an m-by-q matrix, x, $p \in R^q$, and b, $w \in R^m$. If x and y are feasible vectors for $\hat{D}^0$ and $\hat{D}^1$, then (cf. theorem 1)

$$w'b \leq w'Ax \leq p'x. \tag{5.19}$$

We want to deduce these duality results once again, and from theorem 12 (following A. J. Goldman, A. W. Tucker, 1956). The square matrix of order $m + q + 1$ given by

$$\begin{pmatrix} O_m & A & -b \\ -A' & O_q & p \\ b' & -p' & 0 \end{pmatrix}$$

is skew-symmetric. Here O_m and O_q are the square zero

matrices of order m and q, respectively. By theorem 12, there exists a vector

$$\begin{pmatrix} \tilde{w} \\ \tilde{x} \\ t \end{pmatrix} \in R^{m+q+1}$$

with

$$\tilde{w} \geq 0 \, (\tilde{w} \in R^m), \quad \tilde{x} \geq 0 \, (\tilde{x} \in R^q), \quad t \geq 0 \quad (t \text{ real}) \tag{5.20}$$

$$A\tilde{x} - bt \geq 0, \quad -A'\tilde{w} + pt \geq 0, \tag{5.21}$$

$$b'\tilde{w} - p'\tilde{x} \geq 0 \tag{5.22}$$

$$A\tilde{x} - bt + \tilde{w} > 0, \quad -A'\tilde{w} + pt + \tilde{x} > 0 \tag{5.23}$$

$$b'\tilde{w} - p'\tilde{x} + t > 0 \tag{5.24}$$

We must now distinguish the case $t > 0$ from the case $t = 0$.

<u>Theorem 13</u>. Let $t > 0$. Then there exist optimal solutions x^0 and w^0 of $\hat{D}^0$ and $\hat{D}^1$ such that

$$b'w^0 = p'x^0, \tag{5.25}$$

$$Ax^0 + w^0 > b, \quad A'w^0 - x^0 < p. \tag{5.26}$$

<u>Proof</u>: Set $x^0 = \frac{1}{t}\tilde{x}$, and $w^0 = \frac{1}{t}\tilde{w}$. It follows from (5.20) and (5.21) that x^0 and w^0 are feasible for $\hat{D}^0$ and $\hat{D}^1$, respectively. (5.25) and the optimality of w^0 and x^0 follow from (5.19) and (5.22). (5.26) follows from (5.23).

<u>Theorem 14</u>. Let $t = 0$. Then the following assertions hold.

(a) At least one of the problems $\hat{D}^0$ and $\hat{D}^1$ has no feasible vectors.

(b) If the set of feasible vectors for one of these two problems, D^0 and D^1, is not empty, then this set is not bounded, and neither is the objective function bounded on this set.

(c) Neither of the two problems has an optimal solution.

<u>Proof</u>: (a) Suppose $\underset{\sim}{x}^1$ and $\underset{\sim}{w}^1$ were feasible vectors for $\hat{D}^0$ and $\hat{D}^1$. By (5.24) and (5.21), with $t = 0$, we would have

$$\underset{\sim}{p}'\underset{\sim}{\tilde{x}} < \underset{\sim}{b}'\underset{\sim}{\tilde{w}} \leq (A\underset{\sim}{x}^1)'\underset{\sim}{\tilde{w}} = \underset{\sim}{x}^{1'}A'\underset{\sim}{\tilde{w}} \leq 0, \tag{5.27}$$

and on the other hand,

$$0 \leq \underset{\sim}{w}^{1'}A\underset{\sim}{\tilde{x}} = (A'\underset{\sim}{w}^1)'\underset{\sim}{\tilde{x}} \leq \underset{\sim}{p}'\underset{\sim}{\tilde{x}}.$$

(b) Suppose $\underset{\sim}{x}^1$ is a feasible vector for $\hat{D}^0$. Then the vector $\underset{\sim}{x}^1 + \lambda\underset{\sim}{\tilde{x}}$ is feasible for all $\lambda \geq 0$ because $A\underset{\sim}{x}^1 \geq \underset{\sim}{b}$ and $A\underset{\sim}{\tilde{x}} \geq \underset{\sim}{0}$. The objective function $\underset{\sim}{p}'(\underset{\sim}{x}^1+\lambda\underset{\sim}{\tilde{x}}) = \underset{\sim}{p}'\underset{\sim}{x}^1 + \lambda\underset{\sim}{p}'\underset{\sim}{\tilde{x}}$ is not bounded below for $\lambda \geq 0$, since by (5.27), $\underset{\sim}{p}'\underset{\sim}{\tilde{x}} < \underset{\sim}{0}$.

(c) follows from (b).

Theorems 13 and 14 immediately imply theorems 2 through 6 of §5.1, as carried over for problems $\hat{D}^0$ and $\hat{D}^1$. Indeed, from (5.26) we obtain the following conclusion, which exceeds that of theorem 5a in §5.2.

<u>Theorem 15</u>: If both of the problems $\hat{D}^0$ and $\hat{D}^1$ have feasible vectors, then there exists a pair of optimal

solutions $\underset{\sim}{x}^0$ and $\underset{\sim}{w}^0$ with the following properties. A component x_k^0 of $\underset{\sim}{x}^0$ is positive iff the constraint for index k, of the dual problem $\hat{D}^1$, is satisfied by $\underset{\sim}{w}^0$ as an equality. Correspondingly for $\underset{\sim}{w}^0$.

The following existence theorems also can be proven easily now.

Theorem 16: Problem $\hat{D}^0$ has an optimal solution iff the set of its feasible points is not empty, and the objective function $Q(\underset{\sim}{x})$ is bounded below on this set.

Proof: Necessity, of the given condition for the existence of an optimal solution, is trivial. Sufficiency follows because theorem 14(b) implies that $t \neq 0$, and $t > 0$ implies the existence of an optimal solution, by theorem 13.

Remark: This theorem could have been proved by reference to the simplex method, which delivers a solution under the given conditions.

5.7. Linear Optimization Problems with Infinitely many Constraints

New phenomena appear in the study of linear optimization problems with finitely many variables when a finite number of constraints is replaced by a continuum of the same. If the functions at hand are continuous, it suffices, of course, to consider countably infinitely many constraints, rather than a whole continuum.

Let us begin with a simple example. Suppose we are

to find two real variables, x_1 and x_2, such that, for a given fixed constant c, the following objective function assumes the smallest possible value:

$$Q(x_1,x_2) = x_1 + cx_2 = \text{Infimum}. \tag{5.28}$$

Here the prescribed constraints are

$$x_1 + x_2 t \geq t^{1/2} \quad \text{for} \quad 0 < t < 1, \tag{5.29}$$

where t is either any real number in the given interval (a continuum of constraints) or any rational number in the interval (countably infinitely many constraints). Geometrically, this means that we are to find all pairs (x_1,x_2) such that the graph of $x_1 + x_2t$, considered as a function of t, either lies completely above the parabola $t^{1/2}$, or at most is tangent to this parabola. See Figure 5.5a.

We now consider various cases, as determined by the value of the constant c.

1) $c = 0$. The graph of the function $x_1 + x_2t$ at the point $t = 0$ has ordinate x_1. The set of all these ordinates, for all lines lying above the parabola, has infimum zero although zero is not attained. Thus no minimum exists, and consequently, neither does a minimal solution.

2) $c = 1/4$. Then $Q = x_1 + (1/4)x_2 = \text{Min}$. Here Q is the ordinate of $x_1 + x_2t$ at the point $t = 1/4$. Thus Min. $Q = 1/2$, and the uniquely determined minimal solution is given by $x_1 = 1/4$, $x_2 = 1$. See Figure 5.5b.

3) $c = 1$. Then $Q = x_1 + x_2 = \text{Min}$. This means that Q is the ordinate of the function $x_1 + x_2t$ at the point

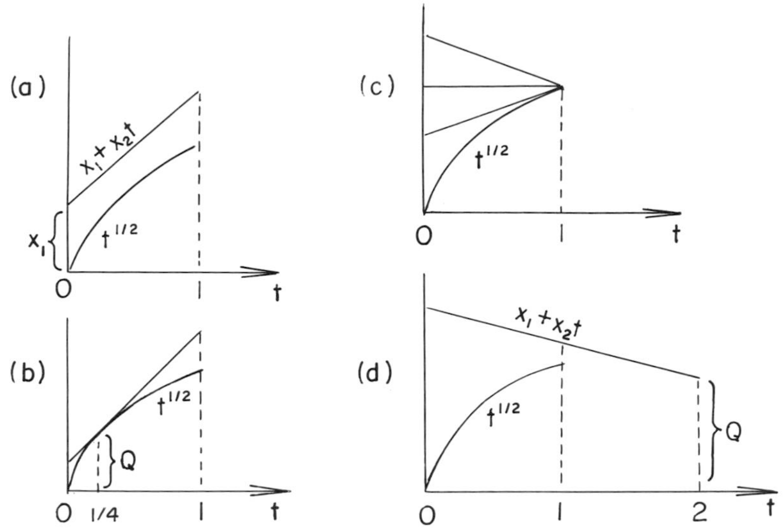

Figure 5.5. A continuum of constraints

$t = 1$. Figure 5.5c shows that Min. $Q = 1$, and that there are infinitely many solutions. Choose an arbitrary $x_2 \leq 1/2$, and set $x_1 = 1 - x_2$.

4) $c = 2$. Then $Q = x_1 + 2x_2 = \text{Min.}$ So Q is the ordinate of $x_1 + x_2 t$ at the point $t = 2$. If we allow x_1 and x_2 to take arbitrary real values, Figure 5.5d shows that the values of Q are not bounded below. The optimization problem has no minimum, no infimum, and no solution. However, if we add the constraints $x_1 \geq 0$ and $x_2 \geq 0$, then there again is a uniquely determined minimum: Min.$Q = 1$ for $x_1 = 1$ and $x_2 = 0$.

One example of an application of such a linear optimization problem with infinitely many constraints is given by the linear continuous Tchebychev approximation problem,

(17.3) in §17.2.

Now we shall introduce the problem more generally, as problem D^0, and construct a dual problem D^1.

Let B be a closed, bounded region in euclidean m-space, R^m, consisting of m-dimensional vectors $\underset{\sim}{t}$, and let $C(B)$ be the space of real-valued continuous functions $h(\underset{\sim}{t})$ defined on B.

Now let $f(\underset{\sim}{t}), v_1(\underset{\sim}{t}),\ldots,v_n(\underset{\sim}{t})$ be given, fixed functions belonging to $C(B)$, and let $\underset{\sim}{c} = (c_1,\ldots,c_n)'$ be a constant vector in R^n. Further, let A be the linear map from the vector space R^n to the function space $C(B)$, defined by assigning to the vector

$$\underset{\sim}{x} = (x_1,\ldots,x_n)' \in R^n$$

the function

$$A\underset{\sim}{x} = \sum_{j=1}^{n} v_j(\underset{\sim}{t})x_j. \tag{5.30}$$

Now we formulate problem D^0 of determining a vector $\underset{\sim}{x}$, where the set of feasible points is defined by

$$M^0\text{:}\quad \underset{\sim}{x} \in R^n,\ \underset{\sim}{x} \geq \underset{\sim}{0},\quad A\underset{\sim}{x} \geq f, \tag{5.31}$$

and the objective function by

$$Q = \underset{\sim}{c}'\underset{\sim}{x} = \sum_{j=1}^{n} c_j x_j = \text{Min. or perhaps Infimum.} \tag{5.32}$$

Here $A\underset{\sim}{x} \geq f$ is written out as

$$\sum_{j=1}^{n} v_j(\underset{\sim}{t})x_j \geq f(\underset{\sim}{t}) \quad \text{for all} \quad \underset{\sim}{t} \in B.$$

To formulate the dual problem, we work in the so-called "dual space" of $C(B)$, whose elements are the continuous linear functionals. (The following theorem remains valid when restricted to point functionals and their linear combinations. Then it would still be sufficient for the following example.)

Let the space of continuous linear functionals be denoted by F^*. All quantities are to be real. A real functional Φ assigns to every element $h \in C(B)$ a real number $\Phi(h)$. A functional Φ is non-negative, written $\Phi \geq \Theta$, iff $\Phi(h) \geq 0$ for every non-negative function $h(\underset{\sim}{t})$, i.e. $h(\underset{\sim}{t}) \geq 0$ for all $\underset{\sim}{t} \in B$. One example of a non-negative functional is the point functional $\Phi(h) = h(P)$, where P is a fixed point in B. Another is any linear combination of such point functionals, where the coefficients c_ν are non-negative and the points P_ν in B are fixed:

$$\Phi(h) = \sum_{\nu=1}^{N} c_\nu h(P_\nu).$$

Finally, as the limiting case, there is the integral with a non-negative weight function $g(\underset{\sim}{t})$, the existence of the integral being assumed,

$$\Phi(h) = \int_B g(\underset{\sim}{t}) h(\underset{\sim}{t}) d\underset{\sim}{t}.$$

Now we can introduce the operator A^*, adjoint to A, which assigns to every functional $\Phi \in F^*$ a vector $A^*\Phi$ in R^n by the rule

$$A^*\Phi = (\Phi(v_1),\ldots,\Phi(v_n))'. \tag{5.33}$$

Then the problem D^1 dual to D^0 is given by the set of feasible points (feasible functionals),

$$M^1\colon \quad \Phi \in F^*, \quad \Phi \geq \Theta, \quad \underset{\sim}{A}^*\Phi \leq \underset{\sim}{c}, \tag{5.34}$$

and the objective function,

$$\Phi(f) = \text{Max} \quad \text{or perhaps Supremum.} \tag{5.35}$$

The formulation of problems D^0 and D^1 evinces the same symmetry as did the finite problems in §5.6. So we have

<u>Theorem 17</u>: (Weak Duality Theorem), (Krabs, 1968). If both sets of feasible points M^0 and M^1 are not empty, then for arbitrary $\underset{\sim}{x} \in M^0$ and arbitrary $\Phi \in M^1$,

$$\Phi(f) \leq \underset{\sim}{c}'\underset{\sim}{x}. \tag{5.36}$$

An inequality such as (5.36) ordinarily plays a fundamental role in duality theory, for it implies that the range of the numbers $\Phi(f)$ is bounded above, while the range of the numbers $\underset{\sim}{c}'\underset{\sim}{x}$ is bounded below. Theorem 17 therefore has as an immediate consequence the existence of the two numbers

$$\alpha = \sup_{\Phi\varepsilon M^1} \Phi(f), \qquad \beta = \inf_{x\varepsilon M^0} \underset{\sim}{c}'\underset{\sim}{x} \tag{5.37}$$

together with the inequality

$$\alpha \leq \beta. \tag{5.38}$$

In general, α and β will not coincide, and if one can prove that $\alpha = \beta$, the assertion is called a "strong duality theorem".

For the case at hand, the weak duality theorem is proven easily. Indeed, let $\underset{\sim}{x} \in M^0$ and $\Phi \in M^1$ be chosen arbitrarily.

Since $\Phi \geq \Theta$, $f \leq A\underset{\sim}{x}$ implies that $\Phi(f) \leq \Phi(A\underset{\sim}{x})$. By the linearity of Φ, we have

$$\Phi(A\underset{\sim}{x}) = \Phi(\sum_j v_j(\underset{\sim}{t})x_j) = \sum_{j=1}^{n} \Phi(v_j(\underset{\sim}{t}))\cdot x_j = (A^*\Phi)'\underset{\sim}{x}$$

since $\underset{\sim}{x} \geq \underset{\sim}{0}$, (5.34) implies that

$$(A^*\Phi)'\underset{\sim}{x} \leq \underset{\sim}{c}'\underset{\sim}{x};$$

combining the two yields (5.36).

A simple numerical example. Problem D^0 is to find two real unknowns, x_1 and x_2, in the feasible set,

$$M^0\colon\ x_1t + x_2t^2 \geq -1+2t \quad \text{for} \quad 0 \leq t \leq 1, \quad x_j \geq 0, \quad j = 1,2,$$

which optimize the objective function

$$Q = x_1 + 2x_2 = \text{Min.}$$

Geometrically, M^0 means that, on the interval $B = [0,1]$, the parabola $x_1t + x_2t^2$ lies above the line $-1 + 2t$, or, at worst, is tangent to it. See Figure 5.6. An immediate computation shows that the region M^0 of feasible points (x_1,x_2) in the (x_1,x_2)-plane consists of the intersection of the positive, or first, quadrant with the half-plane $x_1 + x_2 \geq 1$. See Figure 5.7. The illustration also shows immediately that the objective function Q attains its minimum, $\beta = 1$, on this feasible set at the point $(x_1,x_2) = (1,0)$.

Let us note in passing that without the conditions $x_j \geq 0$, $j = 1,2$, the set of feasible points M^0 would be expanded by the two regions $\tilde{M}$ and $\tilde{\tilde{M}}$ shown in Figure 5.7. On this extended feasible set, the range of Q would not be

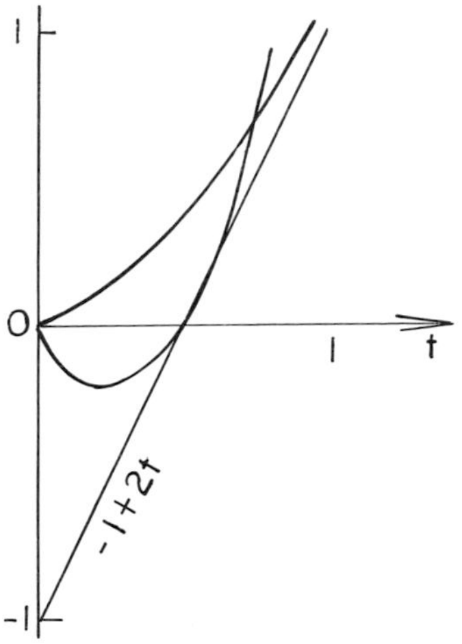

Figure 5.6. The feasible set of parabolas.

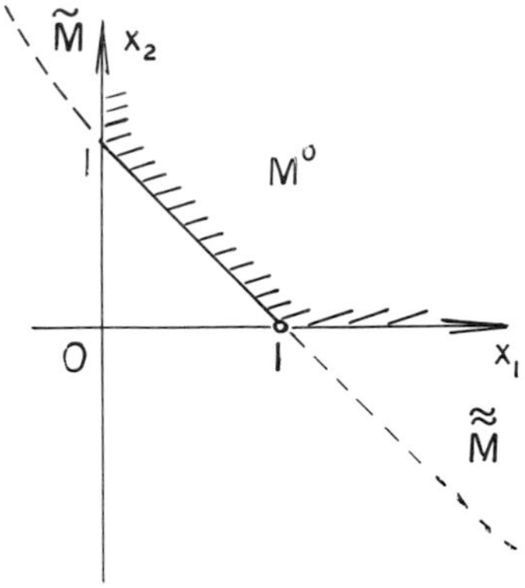

Figure 5.7. The feasible set M^0.

bounded below, and there would be no duality theorem.

The dual problem D^1 reads

$$M^1:\quad \Phi(t) \le c_1 = 1, \quad \Phi(t^2) \le c_2 = 2, \quad \Phi \ge \Theta,$$
$$\Phi(f) = \Phi(-1 + 2t) = \text{supremum}.$$

We try the simplest approach, with a point functional $\Phi(h) = h(t_0)$ for some fixed t_0 in B. Then we attempt to determine t_0 from the conditions

$$t_0 \le 1, \quad t_0^2 \le 2, \quad 0 \le t_0 \le 1, \quad -1+2t_0 = \text{Max}.$$

The maximum of $\alpha = 1$ is obtained at $t_0 = 1$, so in this case, $\alpha = \beta$, and we have strong duality.

The great significance of the duality theorem for numerical analysis stems from the fact that, whenever the feasible sets of the dual problems are not empty, we immediately obtain upper and lower bounds on the extreme values of the problems.

II. CONVEX OPTIMIZATION

Among the non-linear optimization problems, those of convex optimization are the ones which still have a number of properties in common with the linear problems; and because of the Kuhn-Tucker Theorem, they are susceptible to a more thoroughgoing theoretical treatment.

§6. Introduction

6.1. Non-linear Optimization Problems

In order to apply linear optimization to problems in applied fields, one is often forced to make strong idealizations. The mathematical model then reproduces the actual facts of the case only imprecisely, and as a rule, the solution of the idealized optimization problem is not the optimum for the actual underlying problem. One such idealization is the frequent assumption that the coefficients of the variables in the objective function and in the constraints are constant. If we drop this assumption, we obtain non-linear optimization problems, among others.

In example 1 of §1, we assumed that the net profit from the production of x_k units of a product was proportional to x_k, and therefore given by $p_k x_k$. In practice, both the influence of supply and demand, and the possible economies of large-scale production, will insure that p_k is not constant, but a function of $x_1,\ldots,x_q$. The net profit will be

$$Q(x_1,\ldots,x_q) = \sum_{k=1}^{q} p_k(x_1,\ldots,x_q)\cdot x_k.$$

Generally, one frequently gets a better grasp on reality with a non-linear, rather than a linear, problem.

In this chapter, we will treat non-linear optimization problems of the following type. Find an $\underset{\sim}{x} \in R^n$ such that

$$\left.\begin{array}{l} f_j(\underset{\sim}{x}) \le 0 \quad (j = 1,\ldots,m), \quad \underset{\sim}{x} \ge \underset{\sim}{0}, \\ F(\underset{\sim}{x}) = \text{Min!} \end{array}\right\} \qquad (6.1)$$

Here F and $f_1,\ldots,f_m$ are continuous, real-valued functions of $\underset{\sim}{x} \in R^n$; thus functions in the n variables $x_1,\ldots,x_n$. We again use the term (inequality) <u>constraints</u> for the inequalities $f_j(\underset{\sim}{x}) \le 0$, and the term <u>positivity</u> (or, sign) <u>constraint</u> for $\underset{\sim}{x} \ge \underset{\sim}{0}$, i.e. for $x_1 \ge 0,\ldots,x_n \ge 0$. The function $F(\underset{\sim}{x})$ is called the <u>objective function</u>. Occasionally we consider optimization problems where some or all of the positivity constraints are lacking.

Linear optimization problems are included in (6.1). To make the distinction precise, we introduce the concept of an affine linear function.

Definition: A real-valued function $\phi(\underset{\sim}{x})$ of the vector $\underset{\sim}{x} \in R^n$ is called affine linear iff

$$\phi(\alpha\underset{\sim}{x} + (1-\alpha)\underset{\sim}{y}) = \alpha\phi(\underset{\sim}{x}) + (1-\alpha)\phi(\underset{\sim}{y}),$$

where $\underset{\sim}{x},\underset{\sim}{y} \in R^n$ and α is an arbitrary real number.

Theorem 1: $\phi(\underset{\sim}{x})$ is affine linear iff

$$\phi(\underset{\sim}{x}) = \underset{\sim}{a}'\underset{\sim}{x} + \beta$$

for $\underset{\sim}{a} \in R^n$ and β real.

Proof: Trivially, $\underset{\sim}{a}'\underset{\sim}{x} + \beta$ is affine linear. Suppose $\phi(\underset{\sim}{x})$ is affine linear. Set $\psi(\underset{\sim}{x}) = \phi(\underset{\sim}{x}) - \phi(\underset{\sim}{0})$, and let α be real. Then we have

$$\psi(\alpha\underset{\sim}{x}) = \phi(\alpha\underset{\sim}{x}) - \phi(\underset{\sim}{0}) = \phi(\alpha\underset{\sim}{x} + (1-\alpha)\underset{\sim}{0}) - \phi(\underset{\sim}{0}) =$$

$$= \alpha\phi(\underset{\sim}{x}) + (1-\alpha)\phi(\underset{\sim}{0}) - \phi(\underset{\sim}{0}) = \alpha\psi(\underset{\sim}{x}).$$

For $\underset{\sim}{x},\underset{\sim}{y} \in R^n$, we have

$$\psi(\underset{\sim}{x}+\underset{\sim}{y}) = \phi(\underset{\sim}{x}+\underset{\sim}{y}) - \phi(\underset{\sim}{0}) = \phi(\tfrac{1}{2}\cdot 2\underset{\sim}{x} + \tfrac{1}{2}\cdot 2\underset{\sim}{y}) - \phi(\underset{\sim}{0}) =$$

$$= \tfrac{1}{2}\phi(2\underset{\sim}{x}) + \tfrac{1}{2}\phi(2\underset{\sim}{y}) - \phi(\underset{\sim}{0}) =$$

$$= \tfrac{1}{2}\psi(2\underset{\sim}{x}) + \tfrac{1}{2}\psi(2\underset{\sim}{y}) = \psi(\underset{\sim}{x}) + \psi(\underset{\sim}{y}).$$

It follows by induction that

$$\psi(\sum_{i=1}^{q} \alpha_i\underset{\sim}{x}^i) = \sum_{i=1}^{q} \alpha_i\psi(\underset{\sim}{x}^i).$$

Now choose $q = n$, for $\underset{\sim}{x}^i$, the n unit vectors of

R^n, for α_i, the components of some vector $\underset{\sim}{x}$, and let $\underset{\sim}{a}$ denote the vector with components $\psi(\underset{\sim}{x}^i)$, and the last formula becomes

$$\psi(\underset{\sim}{x}) = \underset{\sim}{a}'\underset{\sim}{x}.$$

Setting $\beta = \phi(\underset{\sim}{0})$, we finally obtain $\phi(\underset{\sim}{x}) = \underset{\sim}{a}'\underset{\sim}{x} + \beta$.

If at least one of the functions $F, f_1,\ldots,f_m$ in (6.1) is not affine linear, problem (6.1) is called nonlinear.

We now give an example with affine linear constraints and a quadratic objective function. The details may be found in Boot, 1964.

<u>Optimal Use of Milk in the Netherlands</u>

The milk available for domestic use can be made into four products and sold. These are butter, cheese, cottage cheese, and finally, the (essentially unchanged) milk itself. Of the milk available for processing in the course of a year, h_1 tons consists of fat, and h_2 tons of dried matter. The four milk products contain, per ton,

	Milk	Butter	Cheese	Cottage Cheese
Fat	d_{11}	d_{12}	d_{13}	d_{14}
Dried Matter . .	d_{21}	d_{22}	d_{23}	d_{24}

If there is to be an annual production of x_1 tons of milk, x_2 tons of butter, x_3 tons of cheese, and x_4 tons of cottage cheese, the constraints must be

$$x_k \geq 0 \quad (k = 1,\dots,4);$$
$$\sum_{k=1}^{4} d_{jk}x_k \leq h_j \quad (j = 1,2). \tag{6.2}$$

The prices of milk products may be set by the government; let p_k be the price (in gulden) per ton of the kth product, $k = 1,\dots,4$. The amount of each product consumed will depend on the price of that product. For the year 1960, we know an "equilibrium point": the prices in the year 1960 were $\bar{p}_1,\dots,\bar{p}_4$, and the quantities consumed, $\bar{x}_1,\dots,\bar{x}_4$. We take this known price-consumption relationship as the norm.

If z is some quantity, and $\bar{z}$ a normal value for z, we denote the relative deviation from the norm by $\hat{z}$, where

$$\hat{z} = \frac{z-\bar{z}}{\bar{z}}. \tag{6.3}$$

For small deviations in price from the norm, the effect on demand, that is, on consumption, can be expressed in the following, linearized form.

$$\hat{x}_1 = -\varepsilon_1\hat{p}_1, \qquad \hat{x}_2 = -\varepsilon_2\hat{p}_2,$$
$$\hat{x}_3 = -\varepsilon_3\hat{p}_3+\varepsilon_{34}\hat{p}_4, \quad \hat{x}_4 = \varepsilon_{43}\hat{p}_3-\varepsilon_4\hat{p}_4.$$

The price elasticities $\varepsilon_1,\dots,\varepsilon_4$ and the cross elasticities ε_{34} and ε_{43} are known, positive constants. They were determined empirically through a study of consumer behavior. The $\bar{x}_i$ and the $\bar{p}_i$, $i = 1,\dots,4$, are also known, so these equations can be solved for $x_1,\dots,x_4$, with the help of (6.3). We obtain equations of the form

$$x_j = -\sum_{k=1}^{4} a_{2+j,k}p_k + b_{2+j} \quad (j = 1,\ldots,4). \tag{6.4}$$

Making this substitution in the second condition of (6.2), we obtain a constraint of the form

$$\sum_{k=1}^{4} a_{jk}p_k \le b_j \quad (j = 1,2). \tag{6.5}$$

Since $x_k \ge 0$, $k = 1,\ldots,4$, it then follows from (6.4) that

$$\sum_{k=1}^{4} a_{jk}p_k \le b_j \quad (j = 3,\ldots,6). \tag{6.6}$$

The receipts from the sale of the products total $\sum_{k=1}^{4} p_k x_k$ gulden. Using (6.4) to eliminate the x_k, we obtain an expression, Q, for the receipts which is quadratic in $p_1,\ldots,p_4$. We want to maximize $Q(p_1,\ldots,p_4)$ within the framework of the constraints (6.5) and (6.6), by an appropriate choice of $p_1,\ldots,p_4$. However, prices cannot be dictated arbitrarily within the bounds of (6.5) and (6.6), for the interests of the consumer have to be considered. Milk, for example, must be sold at a reasonably popular price. Therefore we weight the relative deviations in price from the norm, for each of the four products, with weights $\eta_1,\ldots,\eta_4$. We also choose a constant, $K > 0$, and demand that $\sum_{k=1}^{4} \eta_k \hat{p}_k \le K$.

This leads to the "political constraint"

$$\sum_{k=1}^{4} a_{7k}p_k \le b_7.$$

Setting $\underset{\sim}{A} = (a_{jk})_{j = 1,\ldots,7;\ k = 1,\ldots,4}$, $\underset{\sim}{b} = (b_1,\ldots,b_7)'$, and $\underset{\sim}{p} = (p_1,\ldots,p_4)'$, we obtain the non-linear optimization problem of finding a $\underset{\sim}{p} \in R^4$ such that

$$\underset{\sim}{A}\underset{\sim}{p} \leq \underset{\sim}{b}, \quad \underset{\sim}{p} \geq \underset{\sim}{0}; \quad Q(\underset{\sim}{p}) = \sum_{i,k=1}^{4} c_{ik}p_i p_k + \sum_{i=1}^{4} c_i p_i = \text{Max!}$$

The phenomena which are possible for non-linear optimization problems of type (6.1) will be explained with the aid of Figure 6.1.

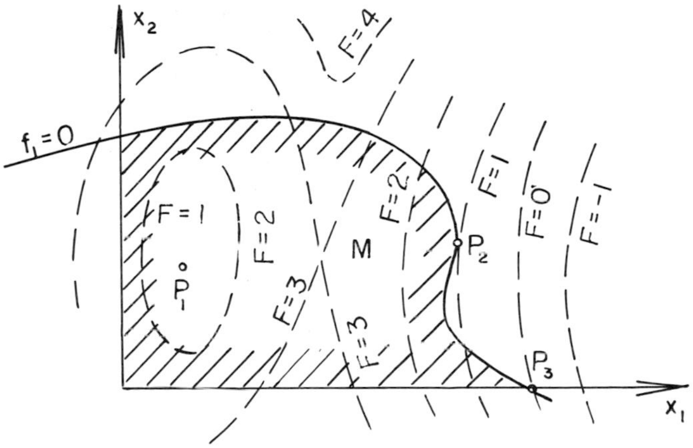

Figure 6.1. A non-linear optimization problem.

The illustration shows an optimization problem in two variables, x_1 and x_2, with one constraint, $f_1(x_1,x_2) \leq 0$. The axes, $x_1 = 0$ and $x_2 = 0$, as well as the curve $f_1 = 0$, form the boundary of the set M of those points which satisfy all of the constraints, $f_1(x_1,x_2) \leq 0$, $x_1 \geq 0$, and $x_2 \geq 0$. The dashed lines in the illustration are the level lines, F = constant, of the objective function $F(x_1,x_2)$. On M, the relative minima of $F(x_1,x_2)$ are at the points P_1, P_2, and P_3. The absolute minimum is at P_3. For the general case of problem (6.1) there will, as a rule, also be several relative minima, from which the absolute minimum is to be found. With large-scale problems, it can be very difficult to obtain any kind of overview of all the relative minima. A satisfying theoretical treatment of the general case of problem type (6.1) has not been discovered yet. If we restrict the problem, however, by requiring convexity properties of the objective function $F(\underset{\sim}{x})$ and of the functions $f_j(\underset{\sim}{x})$ appearing in the constraints, we can construct a theory which sensibly describes the behavior of the solutions of such restricted problems. It will turn out that, in this case, the objective function $F(\underset{\sim}{x})$ attains its absolute minimum relative to M on a convex point-set, if at all, and that there are no further relative minima. It is this theory of convex optimization which we will now develop.

6.2. Convex Functions

The concept of a convex point-set was already defined in §2.

Definition: Let B be a convex subset of R^n. A real-valued function $\phi(\underset{\sim}{x})$, defined for all $\underset{\sim}{x} \in B$, is called convex on B iff

$$\phi(\alpha\underset{\sim}{x} + (1-\alpha)\underset{\sim}{y}) \leq \alpha\phi(\underset{\sim}{x}) + (1-\alpha)\phi(\underset{\sim}{y}), \qquad (6.7)$$

for all $\underset{\sim}{x},\underset{\sim}{y} \in B$, and all α such that $0 < \alpha < 1$. $\phi(\underset{\sim}{x})$ is called strongly convex on B iff, whenever $\underset{\sim}{x} \neq \underset{\sim}{y}$, there is strict inequality ($<$) instead of inequality ($\leq$) in (6.7).

Figure 6.2 shows a function which is convex on $[0,1]$, but not strictly convex. Note too, that every affine linear function is, by definition, convex but not strictly convex.

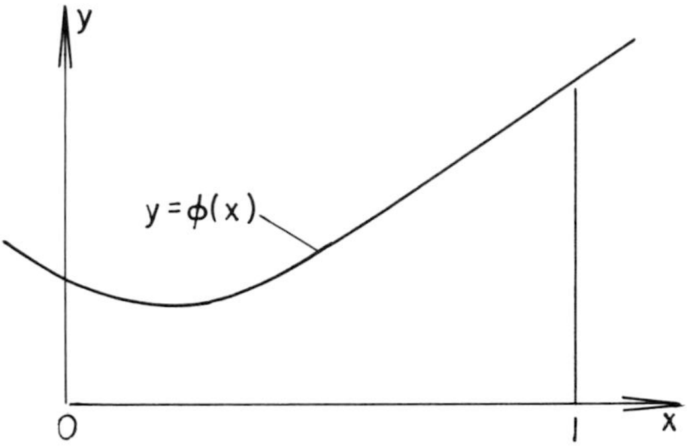

Figure 6.2.

We now show that, for a function $\phi(\underset{\sim}{x})$ defined on an open convex set, convexity implies continuity. Let $\underset{\sim}{x}^1,\ldots,$ $\underset{\sim}{x}^{n+1}$ be points in R^n; then $S(\underset{\sim}{x}^1,\ldots,\underset{\sim}{x}^{n+1})$ denotes the (n-dimensional) simplex which is the convex hull of these points, i.e., the set of all points which are a convex combination of $\underset{\sim}{x}^1,\ldots,\underset{\sim}{x}^{n+1}$. The euclidean norm in R^n is denoted by $||\underset{\sim}{x}|| = (\sum x_i^2)^{1/2}$.

<u>Lemma</u>. Let $\phi(\underset{\sim}{x})$ be a function defined and convex on a simplex

$$S = S(\underset{\sim}{x}^1,\ldots,\underset{\sim}{x}^{n+1}),$$

then $\phi(\underset{\sim}{x})$ is bounded on S with upper bound

$$\phi(\underset{\sim}{x}) \leq M = \max_{i=1,\ldots,n+1} \phi(\underset{\sim}{x}^i) \quad \text{for} \quad \underset{\sim}{x} \in S.$$

<u>Proof</u>: Let $\underset{\sim}{x} \in S$, so $\underset{\sim}{x} = \sum_{i=1}^{n+1} \alpha_i \underset{\sim}{x}^i$ where $\alpha_i \geq 0$, $\sum_{i=1}^{n+1} \alpha_i = 1$. A simple induction argument then shows that

$$\phi(\underset{\sim}{x}) \le \sum_{i=1}^{n+1} \alpha_i \phi(\underset{\sim}{x}^i)$$

and therefore,

$$\phi(\underset{\sim}{x}) \le M \sum_{i=1}^{n+1} \alpha_i = M.$$

Theorem 2: Let B be an open, convex subset of R^n and let $\phi(\underset{\sim}{x})$ be convex on B. Then $\phi(\underset{\sim}{x})$ is continuous on B.

Proof: Let $\underset{\sim}{x}^0$ be a point in B. Since B is open, there exists a simplex, $S = S(\underset{\sim}{x}^1,\ldots,\underset{\sim}{x}^{n+1})$, lying entirely in B, which contains in its interior $\underset{\sim}{x}^0$ together with a ball K, centered as $\underset{\sim}{x}^0$ and of positive radius γ. Thus, $||\underset{\sim}{y}-\underset{\sim}{x}^0|| < \gamma$ implies $\underset{\sim}{y} \in S$. We show that for arbitrary $\varepsilon > 0$, $||\underset{\sim}{y}-\underset{\sim}{x}^0|| \le \eta = \text{Min}(\gamma, \dfrac{\varepsilon\gamma}{M-\phi(\underset{\sim}{x}^0)})$ implies $|\phi(\underset{\sim}{y}) - \phi(\underset{\sim}{x}^0)| \le \varepsilon$. If $||\underset{\sim}{y}-\underset{\sim}{x}^0|| \le \eta$, then the points $\bar{\underset{\sim}{x}} = \underset{\sim}{x}^0 + \frac{\gamma}{\eta}(\underset{\sim}{y}-\underset{\sim}{x}^0)$ and $\bar{\bar{\underset{\sim}{x}}} = \underset{\sim}{x}^0 - \frac{\gamma}{\eta}(\underset{\sim}{y}-\underset{\sim}{x}^0)$ lie in the ball K (see Figure 6.3) and therefore in S, so by the Lemma, $\phi(\bar{\underset{\sim}{x}}) \le M$ and $\phi(\bar{\bar{\underset{\sim}{x}}}) \le M$. Now $\underset{\sim}{y}$ is a convex combination of $\underset{\sim}{x}^0$ and $\underset{\sim}{x}$, namely $\underset{\sim}{y} = \frac{\eta}{\gamma}\bar{\underset{\sim}{x}} + (1 - \frac{\eta}{\gamma})\underset{\sim}{x}^0$, and $\underset{\sim}{x}^0$ is a convex combination of $\underset{\sim}{y}$ and $\bar{\bar{\underset{\sim}{x}}}$, namely $\underset{\sim}{x}^0 = \frac{\eta}{\gamma+\eta}\bar{\bar{\underset{\sim}{x}}} + \frac{\gamma}{\gamma+\eta}\underset{\sim}{y}$. Therefore, $\phi(\underset{\sim}{y}) \le \frac{\eta}{\gamma} M + (1 - \frac{\eta}{\gamma})\phi(\underset{\sim}{x}^0)$ and $\phi(\underset{\sim}{x}^0) \le \frac{\eta}{\gamma+\eta} M + \frac{\gamma}{\gamma+\eta}\phi(\underset{\sim}{y})$.

From this, it follows that

$$\phi(\underset{\sim}{y}) - \phi(\underset{\sim}{x}^0) \le \frac{\eta}{\gamma}(M - \phi(\underset{\sim}{x}^0)) \le \varepsilon$$

and

$$\phi(\underset{\sim}{x}^0) - \phi(\underset{\sim}{y}) \le \frac{\eta}{\gamma}(M - \phi(\underset{\sim}{x}^0)) \le \varepsilon.$$

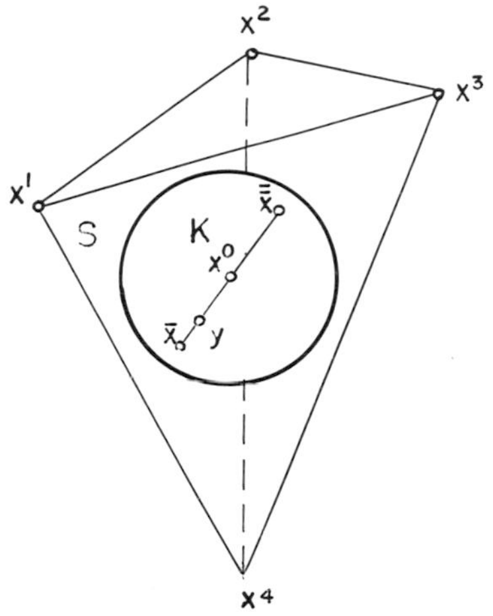

Figure 6.3

On a non-open convex set, a convex function need not be continuous. On the interval $0 \leq x \leq 1$, the function

$$\phi(x) = \begin{cases} x & \text{for} \quad 0 < x \leq 1 \\ 1 & \text{for} \quad x = 0, \end{cases}$$

is convex, but not continuous.

An example of a convex function, which will be of importance later, is that of a quadratic form of a positive, definite matrix.

<u>Definition</u>. A real, symmetric, n-by-n matrix $\underset{\sim}{A}$ is <u>positive definite</u> iff $\underset{\sim}{x}'\underset{\sim}{A}\underset{\sim}{x} > 0$ for all non-zero $\underset{\sim}{x} \in R^n$; it is <u>positive semi-definite</u> iff $\underset{\sim}{x}'\underset{\sim}{A}\underset{\sim}{x} \geq 0$ for all $\underset{\sim}{x} \in R^n$.

For positive, semi-definite matrices we may have $\underset{\sim}{x}'\underset{\sim}{A}\underset{\sim}{x} = 0$ even if $\underset{\sim}{x} \neq \underset{\sim}{0}$, but in that case, we nevertheless have $\underset{\sim}{A}\underset{\sim}{x} = \underset{\sim}{0}$. For if $\underset{\sim}{y} \in R^n$ and t real are arbitrary, then

$$(\underset{\sim}{x}+t\underset{\sim}{y})'\underset{\sim}{A}(\underset{\sim}{x}+t\underset{\sim}{y}) = \underset{\sim}{x}'\underset{\sim}{A}\underset{\sim}{x} + 2t\underset{\sim}{y}'\underset{\sim}{A}\underset{\sim}{x} + t^2\underset{\sim}{y}'\underset{\sim}{A}\underset{\sim}{y} \geq \underset{\sim}{0}.$$

If $\underset{\sim}{x}'\underset{\sim}{A}\underset{\sim}{x} = 0$, this implies $\underset{\sim}{y}'\underset{\sim}{A}\underset{\sim}{x} = 0$. Now let $\underset{\sim}{y} = \underset{\sim}{A}\underset{\sim}{x}$, so $(\underset{\sim}{A}\underset{\sim}{x})'\underset{\sim}{A}\underset{\sim}{x} = \underset{\sim}{0}$ whence, finally, $\underset{\sim}{A}\underset{\sim}{x} = \underset{\sim}{0}$.

<u>Theorem 3</u>: Let $\underset{\sim}{A}$ be a real, symmetric, positive definite, n-by-n matrix. Then $\phi(\underset{\sim}{x}) = \underset{\sim}{x}'\underset{\sim}{A}\underset{\sim}{x}$ is strongly convex on R^n. If $\underset{\sim}{A}$ is positive semi-definite, $\phi(\underset{\sim}{x})$ is convex on R^n.

<u>Proof</u>: Let $\underset{\sim}{A}$ be positive definite and let $0 < \alpha < 1$. Since $\alpha > \alpha^2$, we have, for $\underset{\sim}{x},\underset{\sim}{y} \in R^n$ with $\underset{\sim}{x} \neq \underset{\sim}{y}$, that

$$\begin{aligned}\alpha\phi(\underset{\sim}{x}) + (1-\alpha)\phi(\underset{\sim}{y}) &= \alpha\underset{\sim}{x}'\underset{\sim}{A}\underset{\sim}{x} + (1-\alpha)\underset{\sim}{y}'\underset{\sim}{A}\underset{\sim}{y} = \\ &= \alpha(\underset{\sim}{x}-\underset{\sim}{y})'\underset{\sim}{A}(\underset{\sim}{x}-\underset{\sim}{y}) + \alpha\underset{\sim}{y}'\underset{\sim}{A}(\underset{\sim}{x}-\underset{\sim}{y}) + \alpha(\underset{\sim}{x}-\underset{\sim}{y})'\underset{\sim}{A}\underset{\sim}{y} + \underset{\sim}{y}'\underset{\sim}{A}\underset{\sim}{y} > \\ &> \alpha^2(\underset{\sim}{x}-\underset{\sim}{y})'\underset{\sim}{A}(\underset{\sim}{x}-\underset{\sim}{y}) + 2\alpha\underset{\sim}{y}'\underset{\sim}{A}(\underset{\sim}{x}-\underset{\sim}{y}) + \underset{\sim}{y}'\underset{\sim}{A}\underset{\sim}{y} = \\ &= [\alpha(\underset{\sim}{x}-\underset{\sim}{y})+\underset{\sim}{y}]'\underset{\sim}{A}[\alpha(\underset{\sim}{x}-\underset{\sim}{y})+\underset{\sim}{y}] = \phi(\alpha\underset{\sim}{x} + (1-\alpha)\underset{\sim}{y}).\end{aligned}$$

If $\underset{\sim}{A}$ is positive semi-definite, our estimate can only be $\geq$, since we might have $(\underset{\sim}{x}-\underset{\sim}{y})'\underset{\sim}{A}(\underset{\sim}{x}-\underset{\sim}{y}) = 0$ for $\underset{\sim}{x} \neq \underset{\sim}{y}$.

<u>Remark</u>. Theorem 3 also follows from theorem 5.

If the function $\phi(\underset{\sim}{x})$ has first, or even second partial derivatives, we may test it for convexity with the aid of the following two theorems. The notation grad $\phi(\underset{\sim}{x})$ denotes the vector with components $\partial\phi(\underset{\sim}{x})/\partial x_i$.

A function of one variable is convex if every tangent

line lies "below" the curve. In general, we have

Theorem 4: Let $\phi(\underset{\sim}{x})$ be defined on a convex set B in R^n and have first order partial derivatives there. $\phi(\underset{\sim}{x})$ is convex iff

$$\phi(\underset{\sim}{y}) \geq \phi(\underset{\sim}{x}) + (\underset{\sim}{y}-\underset{\sim}{x})'\text{grad }\phi(\underset{\sim}{x}) \qquad (6.8)$$

for all $\underset{\sim}{x},\underset{\sim}{y} \in B$. $\phi(\underset{\sim}{x})$ is strongly convex iff we have strict inequality in (6.8) whenever $\underset{\sim}{x} \neq \underset{\sim}{y}$.

Proof: I. Suppose (6.8) holds. If $\underset{\sim}{y},\underset{\sim}{z} \in B$, and if $\underset{\sim}{x} = \alpha\underset{\sim}{y} + (1-\alpha)\underset{\sim}{z}$, where $0 < \alpha < 1$, then

$$\begin{aligned}\alpha\phi(\underset{\sim}{y}) + (1-\alpha)\phi(\underset{\sim}{z}) &\geq \phi(\underset{\sim}{x}) + [\alpha(\underset{\sim}{y}-\underset{\sim}{x})+(1-\alpha)(\underset{\sim}{z}-\underset{\sim}{x})]'\text{grad }\phi(\underset{\sim}{x}) \\ &= \phi(\underset{\sim}{x}),\end{aligned}$$

because the expression inside the square brackets vanishes. Thus, $\phi(\underset{\sim}{x})$ is convex. The assertion for strict convexity is proven similarly.

II. Let $\phi(\underset{\sim}{x})$ be convex. Define an auxiliary function

$$\Phi(\alpha) = (1-\alpha)\phi(\underset{\sim}{x}) + \alpha\phi(\underset{\sim}{y}) - \phi((1-\alpha)\underset{\sim}{x}+\alpha\underset{\sim}{y}).$$

If $\underset{\sim}{x} \neq \underset{\sim}{y}$, and $0 < \alpha < 1$, then $\Phi(\alpha) \geq 0$; in any case, $\Phi(0) = 0$. Therefore, $\Phi'(0) \geq 0$, which means that $-\phi(\underset{\sim}{x}) + \phi(\underset{\sim}{y}) - (\underset{\sim}{y}-\underset{\sim}{x})'\text{grad }\phi(\underset{\sim}{x}) \geq 0$. If $\phi(\underset{\sim}{x})$ is strongly convex, $-\Phi(\alpha)$ is strongly convex as a function of the real variable α. Since $\Phi(1/2) > 0$ and $\Phi(\alpha) > 2\cdot\Phi(1/2)\cdot\alpha$ for $0 < \alpha < 1/2$, it follows that $\Phi'(0) > 0$.

Theorem 5: Let $\phi(\underset{\sim}{x})$ be defined on the convex set

B in R^n and have continuous second order partial derivatives there. If the matrix

$$A(x) = \left(\frac{\partial^2 \phi(x)}{\partial x_i \partial x_k}\right) \quad (i,k = 1,\ldots,n)$$

is positive semi-definite (positive definite) for all $x \in B$, then $\phi(x)$ is convex (strongly convex) on B.

Proof: Let $A(x)$ be positive semi-definite (positive definite). For $x,y \in B$ with $x \neq y$, Taylor's Theorem says that there is a point $\bar{x}$ on the interval from x to y such that

$$\phi(y) = \phi(x) + (y-x)'\text{grad } \phi(x) + \tfrac{1}{2}(y-x)'A(\bar{x})(y-x), \quad (6.9)$$

which implies (6.8) (with strict inequality).

The proofs of the following theorems on the minimal points of convex functions are simple and left to the reader.

Theorem 6: Let $\phi(x)$ be defined and convex on a convex set B in R^n. Every relative minimum of $\phi(x)$ is an absolute minimum and the set of minimal points is convex.

Theorem 7: If $\phi(x)$ is strongly convex on a convex set in R^n, then there exists at most one minimal point.

6.3. Convex Optimization Problems

We continue to consider the problem formulated in (6.1):

$$f_j(\underset{\sim}{x}) \leq 0 \quad (j = 1,\ldots,m), \quad \underset{\sim}{x} \geq \underset{\sim}{0}, \quad F(\underset{\sim}{x}) = \text{Min!}$$

This is called a <u>convex optimization problem</u> if the functions $F(\underset{\sim}{x})$, $f_1(\underset{\sim}{x}),\ldots,f_m(\underset{\sim}{x})$ are defined and convex for $x \in R^n$. By theorem 2, they are also continuous. As in the case of linear optimization, we denote by M the set of <u>feasible points</u>, which is the set of all those points satisfying all of the constraints $f_j(\underset{\sim}{x}) \leq 0$, $j = 1,\ldots,m$, and $\underset{\sim}{x} \geq \underset{\sim}{0}$. M is a convex set, because the $f_j(\underset{\sim}{x})$ are convex functions, so that any point on the interval connecting two feasible points will also satisfy all constraints. A point $\underset{\sim}{x}^0 \in M$ such that $F(\underset{\sim}{x}^0) \leq F(\underset{\sim}{x})$ for all $\underset{\sim}{x} \in M$ is called a <u>minimal solution</u> of the convex optimization problem. In contrast to linear optimization, the minimal solution no longer necessarily lies in the boundary of M. The example indicated by Figure 6.4 shows that $F(\underset{\sim}{x})$ may attain its minimum in the interior of M.

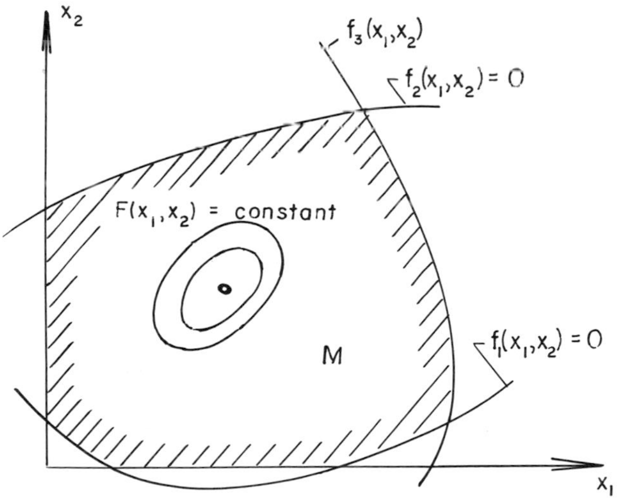

Figure 6.4.

If the set M is bounded, the continuous function $F(\tilde{x})$ attains its minimum on M, since M is also closed. As in the case of linear optimization, the set of minimal solutions is convex.

6.4. Further Types of Non-linear Optimization Problems

We will now name some variants of convex optimization problems. We consider the optimization problem (6.1), without positivity constraints $\tilde{x} \geq \tilde{0}$, and determine whether the functions F and f_j share a certain property, e.g. quasiconvexity. If they do, the optimization problem is given the same name, and we speak of a e.g. quasiconvex optimization problem. For simplicity's sake, we assume that the domain B of all the real-valued functions we consider, is a convex subset of R^n, just as in §6.2. In the definitions of the various classes of functions, which we give in the following table, $\tilde{x}$ and $\tilde{y}$ are to be arbitrary points in B.

If	for	then, on B $\phi(\tilde{x})$ is called
$\phi(\alpha\tilde{x}+(1-\alpha)\tilde{y}) \leq \alpha\phi(\tilde{x}) + (1-\alpha)\phi(\tilde{y})$	all α such that $0 \leq \alpha \leq 1$	convex
$\phi(\alpha\tilde{x}+(1-\alpha)\tilde{y}) \geq \alpha\phi(\tilde{x}) + (1-\alpha)\phi(\tilde{y})$	all α such that $0 \leq \alpha \leq 1$	concave
$\phi(\alpha\tilde{x}+(1-\alpha)\tilde{y}) < \alpha\phi(\tilde{x}) + (1-\alpha)\phi(\tilde{y})$	all α such that $0 < \alpha < 1$	strongly convex
$\phi(\alpha\tilde{x}+(1-\alpha)\tilde{y}) > \alpha\phi(\tilde{x}) + (1-\alpha)\phi(\tilde{y})$	all α such that $0 < \alpha < 1$	strongly concave

If	for		then, on B $\phi(\underset{\sim}{x})$ is called
$\phi(\alpha\underset{\sim}{x}+(1-\alpha)\underset{\sim}{y}) \leq \phi(\underset{\sim}{y})$	$\phi(\underset{\sim}{x}) \leq \phi(\underset{\sim}{y})$	all α s.t.	quasiconvex
$\phi(\alpha\underset{\sim}{x}+(1-\alpha)\underset{\sim}{y}) \geq \phi(\underset{\sim}{y})$	$\phi(\underset{\sim}{x}) \geq \phi(\underset{\sim}{y})$	$0 \leq \alpha \leq 1$	quasiconcave
$\phi(\alpha\underset{\sim}{x}+(1-\alpha)\underset{\sim}{y}) < \phi(\underset{\sim}{y})$	$\phi(\underset{\sim}{x}) < \phi(\underset{\sim}{y})$	all α s.t.	strongly quasiconvex
$\phi(\alpha\underset{\sim}{x}+(1-\alpha)\underset{\sim}{y}) > \phi(\underset{\sim}{y})$	$\phi(\underset{\sim}{x}) > \phi(\underset{\sim}{y})$	$0 < \alpha < 1$	strongly quasiconcave

If the function $\phi(\underset{\sim}{x})$ is also differentiable on B, we have two more definitions.

$\phi(\underset{\sim}{x})$ is ____ on B	if	for
pseudoconvex	$\phi(\underset{\sim}{x}) \geq \phi(\underset{\sim}{y})$	$(\underset{\sim}{x}-\underset{\sim}{y})'\text{grad}\ \phi(\underset{\sim}{y}) \geq 0$
pseudoconcave	$\phi(\underset{\sim}{x}) \leq \phi(\underset{\sim}{y})$	$(\underset{\sim}{x}-\underset{\sim}{y})'\text{grad}\ \phi(\underset{\sim}{y}) \leq 0$

A function $\phi(\underset{\sim}{x})$ which is simultaneously quasiconvex and quasiconcave, e.g. $\phi(\underset{\sim}{x}) = x^3$, is called quasilinear. A function which is both pseudoconvex and pseudoconcave is called pseudolinear. The properties of these functions are investigated in Stoer-Witzgall, 1970.

Figure 6.5 illustrates these various types of functions for the case of one function of only one real variable, f(x).

A function of the form

$$\phi(\underset{\sim}{x}) = \sum_{j=1}^{q} c_j x_1^{a_{j1}} x_2^{a_{j2}} \cdots x_m^{a_{jm}},$$

with real $a_{j\mu}$ and positive constants c_j, is called (a) posinomial (function). Sometimes we dispense with the condition $c_j > 0$.

Type	Examples of continuously differentiable functions $f(x)$	Not everywhere differentiable functions $f(x)$
strongly convex	e^x ; x^2	$|x| + x^2$
convex, not strongly convex	0 for $|x| \leq 1$; $(|x|-1)^2$ for $|x| \geq 1$	$|x|$; $x + |x|$
pseudoconvex, not convex	$x \cdot e^x$; $\dfrac{-1}{1+x^2}$	contradicts the definition
strongly quasi-convex, not pseudoconvex	x^3	$x + |x|$
quasiconvex, not strongly quasi-convex	0 for $|x| \leq 1$; $(x-1)^3$ for $x \geq 1$; $(x+1)^3$ for $x \leq -1$	$x - |x|$
not quasiconvex	$-x^2$	$-|x|$

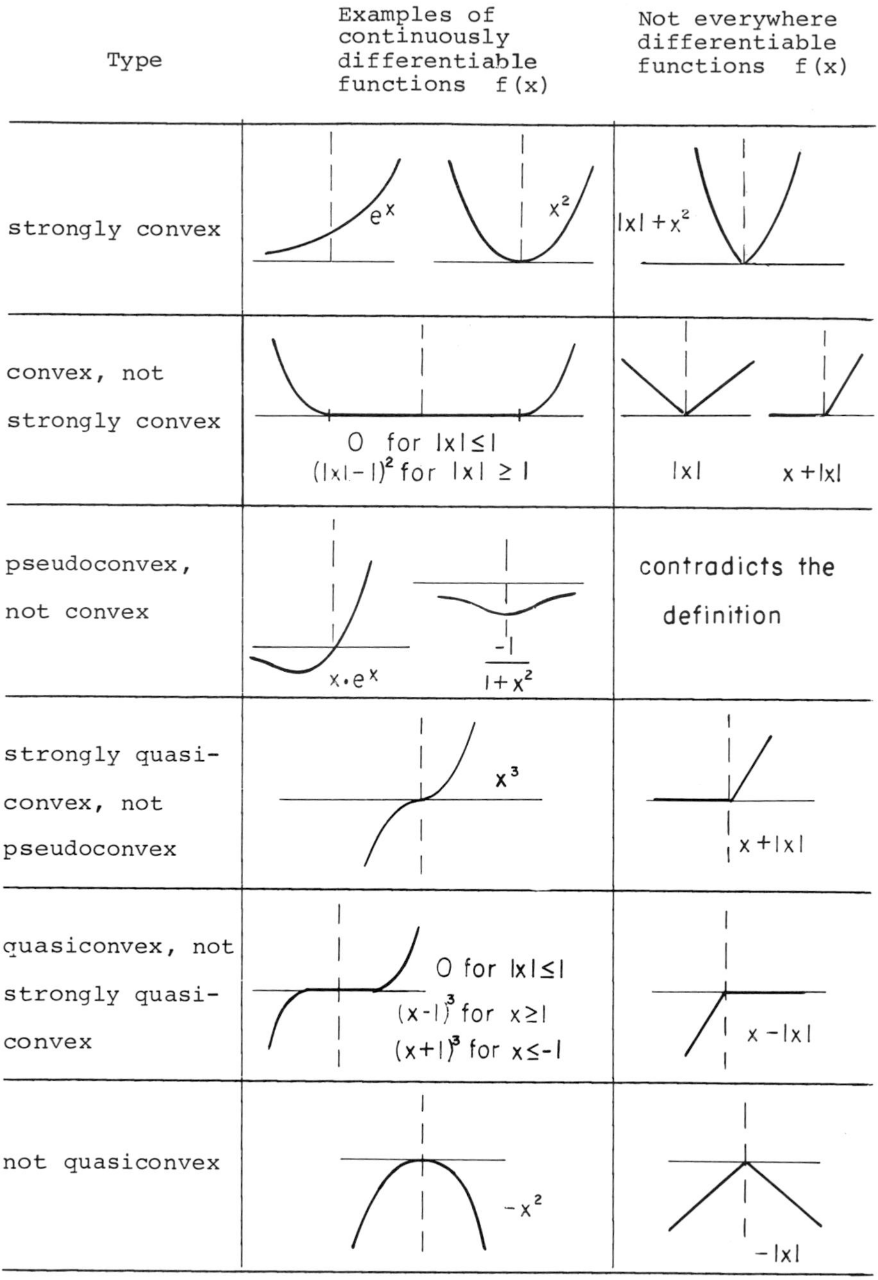

Figure 6.5. Examples of various types of functions

The optimization is called separable if the functions F and f_j in (6.1) may be expressed as a sum of functions $F_k(x_k)$ and $f_{jk}(x_k)$, each of which depends on only one variable, x_k. Then the F_k and f_{jk} either have the form

$$F = \sum_{k=1}^{n} F_k(x_k), \qquad f_j = \sum_{k=1}^{n} f_{jk}(x_k),$$

or may be brought to this form by a linear, non-singular transformation of the independent variables. See Bracken-McCormick (1968), p. 15.

If the constraints are linear, and if the objective function has the form

$$F = \underset{\sim}{x}'\underset{\sim}{A}\underset{\sim}{x} + \underset{\sim}{p}'\underset{\sim}{x} + \alpha \tag{6.10}$$

the problem is called a general quadratic optimization problem, while if the objective function has the form

$$F = \frac{\gamma + \underset{\sim}{c}'\underset{\sim}{x}}{\delta + \underset{\sim}{d}'\underset{\sim}{x}} \tag{6.11}$$

it is called a hyperbolic optimization problem.

The general quadratic optimization problem has two special cases:

a) special quadratic optimization, which is often described simply as quadratic optimization, where the matrix $\underset{\sim}{A}$ is positive semidefinite; and

b) bilinear optimization, where $x = (y_1,\ldots,y_r, z_1,\ldots,z_s)'$ consists of two sorts of variables, and the quadratic form $\underset{\sim}{x}'\underset{\sim}{A}\underset{\sim}{x}$ is a sum of only the elements

$$\sum_{j=1}^{r} \sum_{k=1}^{s} b_{jk} y_j z_k,$$

and the two kinds of variables appear separately in the constraints:

$\underset{\sim}{D}\underset{\sim}{y} \leq \underset{\sim}{d}$, $\underset{\sim}{H}\underset{\sim}{z} \leq \underset{\sim}{h}$ (with given matrices $\underset{\sim}{D}$, $\underset{\sim}{H}$, and given vectors $\underset{\sim}{d}$, $\underset{\sim}{h}$).

6.5. Simple Theorems on the Variants of Convexity

Let $\phi(\underset{\sim}{x})$ be a real-valued function defined on a neighborhood of the point $\underset{\sim}{x}^0$; then we say that the function has a "strong local minimum" or an "isolated minimum" at that point if there exists a punctured ball $K_\rho(\underset{\sim}{x}^0)$ (= the set of points $\underset{\sim}{x}$ such that $0 < ||\underset{\sim}{x} - \underset{\sim}{x}^0|| < \rho$) such that

$$\phi(\underset{\sim}{x}) > \phi(\underset{\sim}{x}^0) \quad \text{for all} \quad \underset{\sim}{x} \in K_\rho(\underset{\sim}{x}^0). \tag{6.12}$$

Theorem 8: Let $\phi(\underset{\sim}{x})$ be a quasiconvex function defined on a convex region B. Then every strong local minimum $\underset{\sim}{x}^0$ is also a global (strong) minimum.

Proof: (By contradiction) Suppose there exists an $\underset{\sim}{x}^1 \neq \underset{\sim}{x}^0$, such that $\phi(\underset{\sim}{x}^1) \leq \phi(\underset{\sim}{x}^0)$. Let ρ be the radius of a ball $K_\rho(\underset{\sim}{x}^0)$, for which (6.12) holds for all $\underset{\sim}{x} \in K_\rho(\underset{\sim}{x}^0) \cap B$. Since ϕ is quasiconvex, we have

$$\phi((1-\alpha)\underset{\sim}{x}^0 + \alpha\underset{\sim}{x}^1) \leq \phi(\underset{\sim}{x}^0) \quad \text{for} \quad 0 < \alpha < 1. \tag{6.13}$$

But if $0 < \alpha < \rho/||\underset{\sim}{x}^1-\underset{\sim}{x}^0||$, this contradicts (6.12).

We have the very similar

Theorem 9: Let $\phi(\underset{\sim}{x})$ be a strongly quasiconvex (-concave) function defined on a convex region B. Then every local minimum (maximum) $\underset{\sim}{x}^0$ is also a global extremum on the region B.

Proof: Let $\underset{\sim}{x}^0$ be some local minimum. Then there is a ball $K_\rho(\underset{\sim}{x}^0)$ with radius ρ centered at $\underset{\sim}{x}^0$ such that

$$\phi(\underset{\sim}{x}^0) \leq \phi(\underset{\sim}{x}) \quad \text{for} \quad \underset{\sim}{x} \in K_\rho(\underset{\sim}{x}^0) \cap B. \tag{6.14}$$

Suppose there exists an $\underset{\sim}{x}^1 \in B$ with $\underset{\sim}{x}^1 \notin K_\rho(\underset{\sim}{x}^0)$ such that $\phi(\underset{\sim}{x}^1) < \phi(\underset{\sim}{x}^0)$. Then the strong quasiconvexity of ϕ implies that

$$\phi((1-\alpha)\underset{\sim}{x}^0 + \alpha\underset{\sim}{x}^1) < \phi(\underset{\sim}{x}^0) \quad \text{for} \quad 0 < \alpha < 1, \tag{6.15}$$

which, for $\alpha < \rho/||\underset{\sim}{x}^1-\underset{\sim}{x}^0||$, contradicts (6.14) because $(1-\alpha)\underset{\sim}{x}^0 + \alpha\underset{\sim}{x}^1 \in K_\rho(\underset{\sim}{x}^0)$.

Theorem 10: Every pseudoconvex (-concave) function $\phi(\underset{\sim}{x})$ defined on a convex region B is strongly quasiconvex (-concave) on B.

Proof: (By contradiction) Suppose ϕ is pseudoconvex, and hence differentiable, but not strongly quasiconvex. Then there exist points $\underset{\sim}{x}^1$, $\underset{\sim}{x}^2$, and $\underset{\sim}{z}$ in B such that

$$\phi(\underset{\sim}{x}^2) < \phi(\underset{\sim}{x}^1) \quad \text{and} \quad \phi(\underset{\sim}{z}) \geq \phi(\underset{\sim}{x}^1) \tag{6.16}$$

Here, $\underset{\sim}{z}$ is in the interval $I = (\underset{\sim}{x}^1,\underset{\sim}{x}^2)$, i.e., there is an α, $0 < \alpha < 1$, such that $\underset{\sim}{z} = (1-\alpha)\underset{\sim}{x}^1 + \alpha\underset{\sim}{x}^2$. Let ϕ attain its maximum on the interval I at a point $\underset{\sim}{y} \neq \underset{\sim}{x}^1$. Now replace the $x_1,\ldots,x_n$ axes by a new orthogonal basis $s_1,\ldots,s_n$, for which s_1 points in the direction from $\underset{\sim}{x}^1$ to $\underset{\sim}{y}$. Since $\underset{\sim}{y}$ is an interior point of I,

$$\frac{d\phi(\underset{\sim}{y})}{ds_1} = 0. \tag{6.17}$$

Consequently,

$$(\underset{\sim}{x}-\underset{\sim}{y})'\text{grad } \phi(\underset{\sim}{y}) = 0 \quad \text{for} \quad \underset{\sim}{x} \in I, \tag{6.18}$$

since the first component of the gradient and the remaining components of $\underset{\sim}{x} - \underset{\sim}{y}$ vanish. Then we also have

$$(\underset{\sim}{x}-\underset{\sim}{y})'\text{grad } \phi(\underset{\sim}{y}) \geq 0 \quad \text{for} \quad \underset{\sim}{x} \in I.$$

Because ϕ is pseudoconvex, this implies that

$$\phi(\underset{\sim}{x}) \geq \phi(\underset{\sim}{y}) \quad \text{for} \quad \underset{\sim}{x} \in I$$

so, in particular, $\phi(\underset{\sim}{x}^2) \geq \phi(\underset{\sim}{y})$, contradicting

$$\phi(\underset{\sim}{y}) \geq \phi(\underset{\sim}{x}^1) > \phi(\underset{\sim}{x}^2);$$

see Figure 6.6.

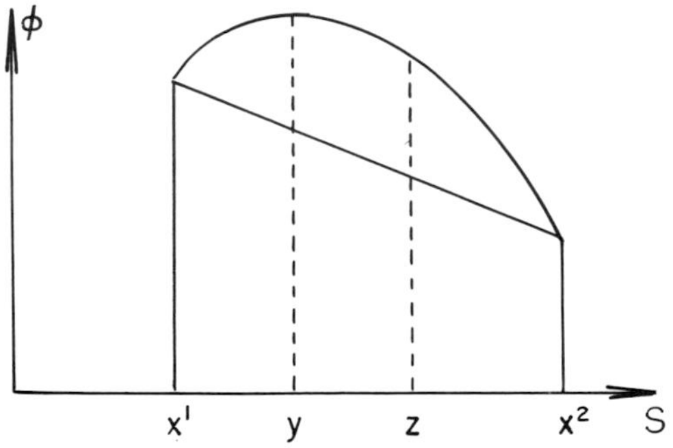

Figure 6.6. A proof of strong quasiconvexity

6.6. A Classification of Non-linear Differentiable Optimization Problems

The connections between the various classes of functions introduced in §6.4 become clearer under the assumption of differentiability. Without a differentiability hypothesis, one can, for example, display functions $\phi(\underset{\sim}{x})$ which are strongly quasiconvex, but not quasiconvex, such as the function

$$\phi(x) = \begin{cases} 1 & \text{for} \quad x = x_0 \\ 0 & \text{otherwise} \end{cases}$$

In such cases, one can still construct good theorems, with the aid of the concept of semicontinuity, cf. Mangasarian, (1969). We, however, will consider only optimization problems where the functions are differentiable, in constructing the following diagram. Here each arrow leads to a more general class.

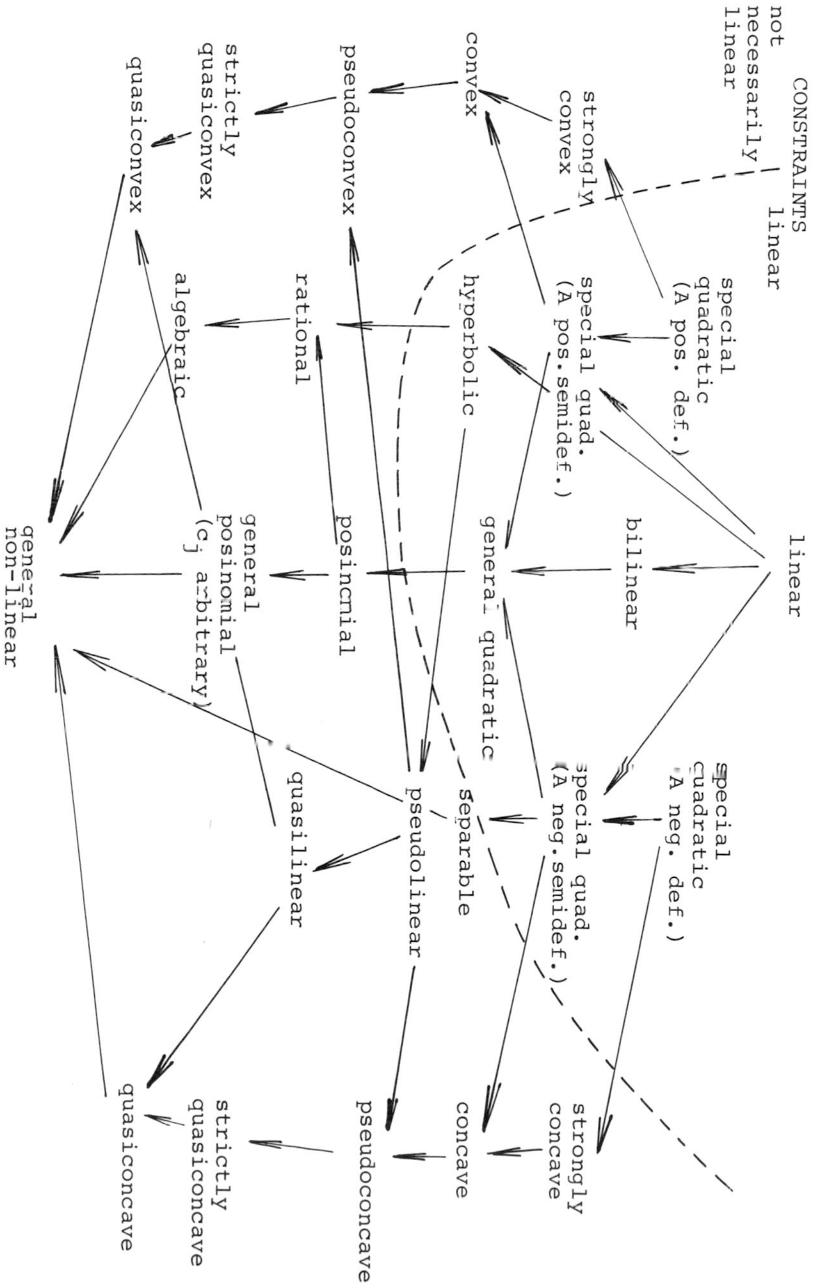

Figure 6.7. Classes of Optimization Problems

6.7. Classes of Convex and Pseudoconvex Functions

Composition with the exponential function maps convex functions to convex functions, but does not take every concave function to a concave function; see Figure 6.8. e^{x^2} is convex, but e^{-x^2} is only pseudoconcave.

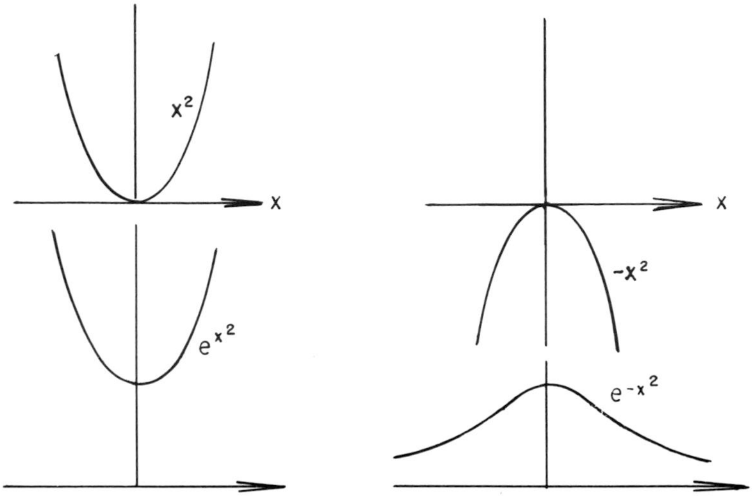

Figure 6.8. Convex and pseudoconcave functions

Further observe that the sum of convex functions is again convex. The product of convex functions is not convex, in general. See Figure 6.9.

Theorem 11: Let $\phi(\underset{\sim}{x})$ be a convex function, defined on the convex domain B, with range W of real numbers. Let h(z) be a monotone, non-decreasing and convex function on W. Then the function $\Phi(\underset{\sim}{x}) = h(\phi(\underset{\sim}{x}))$ is convex on B.

Proof: For $0 < \alpha < 1$ and arbitrary $\underset{\sim}{x},\underset{\sim}{y} \in B$, the

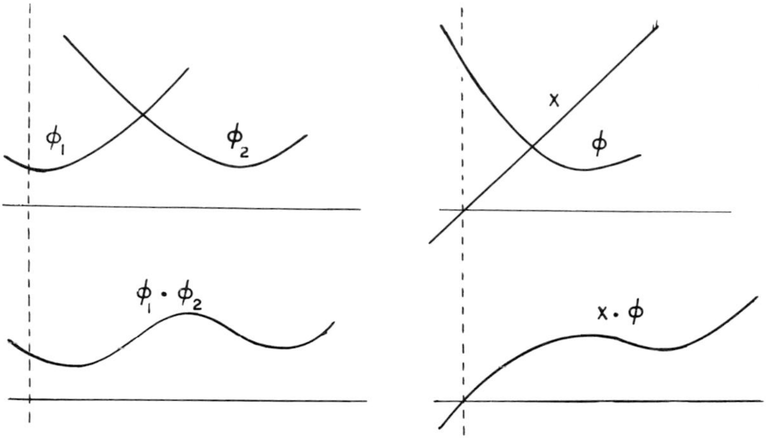

Figure 6.9. Products of convex functions.

convexity of the function h implies that

$$h(\alpha\phi(\underset{\sim}{x})+(1-\alpha)\phi(\underset{\sim}{y})) \le \alpha h(\phi(\underset{\sim}{x}))+(1-\alpha)h(\phi(\underset{\sim}{y})) =$$

$$= \alpha\Phi(\underset{\sim}{x}) + (1-\alpha)\Phi(\underset{\sim}{y}).$$

Applying the convexity of ϕ and the monotonicity of h yields the estimate

$$\begin{aligned} \Phi(\alpha\underset{\sim}{x}+(1-\alpha)\underset{\sim}{y}) &= h(\phi(\alpha x + (1-\alpha)y)) \\ &\le h(\alpha\phi(\underset{\sim}{x}) + (1-\alpha)\phi(\underset{\sim}{y})) \\ &\le \alpha\Phi(\underset{\sim}{x}) + (1-\alpha)\Phi(\underset{\sim}{y}). \end{aligned}$$

For the analogous theorem on concave functions, use the characterization of concave functions corresponding to (6.8). Let $\phi(\underset{\sim}{x})$ be differentiable on the convex set B; $\phi(\underset{\sim}{x})$ is concave iff, for all $\underset{\sim}{x},\underset{\sim}{y} \in B$,

$$\phi(\underset{\sim}{y}) \leq \phi(\underset{\sim}{x}) + (\underset{\sim}{y}-\underset{\sim}{x})'\text{grad } \phi(\underset{\sim}{x}). \tag{6.19}$$

The proof is totally analogous to the one of formula (6.8).

<u>Theorem 12</u>: Let $\phi(\underset{\sim}{x})$ be concave and differentiable on the convex set B, and real-valued with range W. Let $h(z)$ be defined on W with positive derivative: $h'(z) > 0$ for all $z \in W$. Then the function $\Phi(\underset{\sim}{x}) = h(\phi(\underset{\sim}{x}))$ is pseudoconcave on B.

<u>Proof</u>: Let $\underset{\sim}{x}$ and $\underset{\sim}{y}$ be arbitrary points of B such that

$$(\underset{\sim}{y}-\underset{\sim}{x})'\text{grad } \Phi(\underset{\sim}{x}) \leq 0. \tag{6.20}$$

Then the assertion is that $\Phi(\underset{\sim}{y}) \leq \Phi(\underset{\sim}{x})$. Now by the chain rule,

$$\text{grad } \Phi(\underset{\sim}{x}) = a\cdot\text{grad } \phi(\underset{\sim}{x}),$$

with the abbreviation $a = h'(\phi(\underset{\sim}{x}))$. Since the derivative is positive, $a > 0$. Applying (6.19) and (6.20), it now follows that

$$\phi(\underset{\sim}{y}) - \phi(\underset{\sim}{x}) \leq (\underset{\sim}{y}-\underset{\sim}{x})'\text{grad } \phi(\underset{\sim}{x}) = \frac{1}{a}(\underset{\sim}{y}-\underset{\sim}{x})'\text{grad } \Phi(\underset{\sim}{x}) \leq 0.$$

But now $\phi(\underset{\sim}{y}) \leq \phi(\underset{\sim}{x})$ implies that $\Phi(\underset{\sim}{y}) \leq \Phi(\underset{\sim}{x})$, because h is monotone.

<u>Examples for theorems 11 and 12</u>: $h(z) = e^z$ explains the initial examples, of Figure 6.8. If W contains only positive numbers, one can set $h(z) = z^k$, where $k \geq 1$ when applying theorem 11, and $k > 0$ when theorem 12 is applied.

Theorem 13: For every quasiconvex function $\phi(\tilde{x})$ defined on a convex domain B, the set M of minimal points is convex.

Proof: Let c be an arbitrary real constant, and let M_c be the set of points $\tilde{x} \varepsilon B$ such that $\phi(\tilde{x}) \leq c$. We will show, more generally, that M_c is empty or convex for every c. Let $\tilde{y}$ and $\tilde{z}$ be two distinct points in M_c and suppose $\phi(\tilde{z}) \leq \phi(\tilde{y})$. For $0 < \alpha < 1$, the quasiconvexity of ϕ implies that

$$\phi(\alpha\tilde{z} + (1-\alpha)\tilde{y}) \leq \phi(\tilde{y}) \leq c,$$

but this says that

$$\alpha\tilde{z} + (1-\alpha)\tilde{y} \subset M_c.$$

Theorem 14: Let $Z(\tilde{x})$ and $N(\tilde{x})$ be two functions defined on the convex region B such that $N(\tilde{x}) > 0$ on B, and $Z(\tilde{x})$ is convex on B. In addition, let at least one of the following two hypotheses be satisfied.

a) N is affine linear

b) N is convex and $Z(\tilde{x}) \leq 0$ on B.

If Z and N are also differentiable, then the function $\phi(\tilde{x}) = Z(\tilde{x})/N(\tilde{x})$ is pseudoconvex on B.

Proof: Let $\tilde{x}$ and $\tilde{y}$ be two arbitrary points in B such that

$$(\tilde{y}-\tilde{x})'\text{grad } \phi(\tilde{x}) \geq 0. \qquad (6.21)$$

Then we must show that $\phi(\tilde{y}) \geq \phi(\tilde{x})$, i.e. that

$$Z(\underset{\sim}{y})N(\underset{\sim}{x}) - Z(\underset{\sim}{x})N(\underset{\sim}{y}) \geq 0. \tag{6.22}$$

Computing grad $\phi(\underset{\sim}{x})$, we obtain

$$N^2(\underset{\sim}{x})\text{grad } \phi(\underset{\sim}{x}) = N(\underset{\sim}{x})\text{grad } Z(\underset{\sim}{x}) - Z(\underset{\sim}{x})\text{grad } N(\underset{\sim}{x}).$$

From this, from (6.21), and from (6.8) applied to the convex function $Z(\underset{\sim}{x})$, we obtain

$$\begin{aligned} 0 &\leq N^2(\underset{\sim}{x})(\underset{\sim}{y}-\underset{\sim}{x})'\text{grad } \phi(\underset{\sim}{x}) \\ &= N(\underset{\sim}{x})(\underset{\sim}{y}-\underset{\sim}{x})'\text{grad } Z(\underset{\sim}{x}) - Z(\underset{\sim}{x})(\underset{\sim}{y}-\underset{\sim}{x})'\text{grad } N(\underset{\sim}{x}) \qquad (6.23) \\ &\leq N(\underset{\sim}{x})[Z(\underset{\sim}{y}) - Z(\underset{\sim}{x})] - Z(\underset{\sim}{x})(\underset{\sim}{y}-\underset{\sim}{x})'\text{grad } N(\underset{\sim}{x}). \end{aligned}$$

Now if N is affine linear, $N(\underset{\sim}{x}) = \underset{\sim}{x}'\underset{\sim}{b} + \beta$, for some constant vector $\underset{\sim}{b}$ and some constant β, so that grad $N(\underset{\sim}{x}) = \underset{\sim}{b}$, and

$$N(\underset{\sim}{y}) - N(\underset{\sim}{x}) = (\underset{\sim}{y}-\underset{\sim}{x})'\underset{\sim}{b} = (\underset{\sim}{y}-\underset{\sim}{x})'\text{grad } N(\underset{\sim}{x}).$$

Substitute this in (6.23), and (6.22) follows immediately.

If instead (hypothesis b) N is convex and $Z \leq 0$, (6.8) implies that

$$-Z(\underset{\sim}{x})(\underset{\sim}{y}-\underset{\sim}{x})'\text{grad } N(\underset{\sim}{x}) \leq -Z(\underset{\sim}{x})[N(\underset{\sim}{y})-N(\underset{\sim}{x})]$$

and (6.22) again follows from (6.23).

If $Z(\underset{\sim}{x})$ and $N(\underset{\sim}{x})$ are both affine linear, all the hypotheses of theorem 14 are fulfilled, and $\phi(\underset{\sim}{x})$ has the form of a hyperbolic function, as given by (6.11). Theorem 14 also contains the important conclusion that every hyperbolic optimization problem is a special case of a pseudoconvex problem.

One can show in the same way that every hyperbolic

optimization problem is also pseudoconcave and quasilinear.

6.8. Further Examples of Continuous Optimization Problems

1. The profitability problem as hyperbolic optimization. From an economic standpoint, profitability, which is defined as the quotient of net return by invested capital, is often of interest. We will explain this with the idealized example 2 of §1.1. Suppose, for the sake of specific formulation of the problem, that the invested capital is \$200 per cow and \$20 per sheep, and that fixed costs come to \$4000, so that the total invested capital is $\$(4000 + 200x_1 + 20x_2)$. Our new objective function is then

$$Q = \frac{250x_1 + 45x_2}{4000 + 200x_1 + 20x_2} = \text{Max!}$$

The inequality constraints of (1.4) would remain, and all in all, we would have a problem of hyperbolic optimization.

2. Quadratic optimization in cost computations. One often arrives at a non-linear optimization problem in the most natural of ways. A businessman, for example, would like to sell an item at a unit price p which maximizes the total return, $Q = Np$, where N is the number of units sold. The situation is often such, that more units can be sold at a lower price. Even with the simplest of assumptions, of a linear relationship between N and p, say $N = c_1 - c_2 p$ for some constants c_1 and c_2, Q will be non-linear in p. Naturally one can easily think of broader and more complicated examples. In economics, one is thus highly interested in solutions of non-linear optimization problems, yet large scale problems have been computed primarily with linear

optimization (e.g., in the petroleum industry, with about 10,000 variables). The existing methods for handling very large scale non-linear optimization problems are not yet economically feasible.

3. Isoperimetry of triangles (convex optimization). Consider the set of all plane triangles of a given perimeter, 2s, and find a triangle with the greatest surface area, F. In the classic formulation, this becomes the following problem. Let x_1, x_2, x_3 be the lengths of the sides of a triangle. Then find $F = [s(s-x_1)(s-x_2)(s-x_3)]^{1/2} =$ Max, subject to the constraint $x_1 + x_2 + x_3 = 2s$. But this formulation is not correct, for the problem is a true optimization problem, with inequalities as constraints. These must be $0 \leq x_j \leq s$, $j = 1,2,3$,; only then is the solution -- an equilateral triangle with $x_1 = x_2 = x_3 = 2s/3$ and $F^2 = s^4/27$ -- determined, for without these constraints, no absolute maximum exists. For example, $x_1 = x_2 = -x_3 = 2s$ already yields a larger value for F^2, namely $F^2 = 3s^4$.

4. Convex and non-convex optimization in siting problems. Simple geometric minimum problems with a uniquely determined solution frequently lead to convex optimization. Consider first the following. Four towns are located at the corners, P_1, P_2, P_3, and P_4, of a square. Which point S in the interior of the square should be chosen as the site for a factory, if we want to minimize the sum of the distances, $\sum_{j=1}^{4} \overline{P_j S}$. The solution to this convex optimization problem is clearly the center point, M, of the square. Now suppose there lies a circular lake, centered at M, inside

the square, and that the connecting routes between the factory and the towns can go only around the lake; see Figure 6.10. For reasons of symmetry, there are now four minimal solutions, separated one from another. The optimization problem thus can no longer be convex, but is now algebraic.

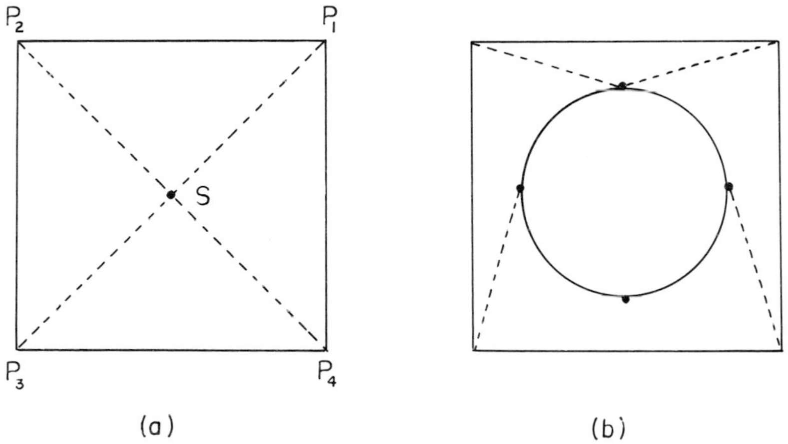

Figure 6.10. Factory sites in a region with a lake.

5. Convex and non-convex optimization problems in physics.

a) The principle of least time for light beams. In the (x,y)-plane, let a light ray originate at the point $x = 0$, $y = a_1 > 0$, and terminate at the point $x = b$, $y = -a_2 < 0$, as in Figure 6.11. In each of the half planes $y > 0$ and $y < 0$, there is a constant medium, in which the speed of light is v_1 and v_2, respectively. The light ray follows a piecewise linear path, at an angle β_1 from

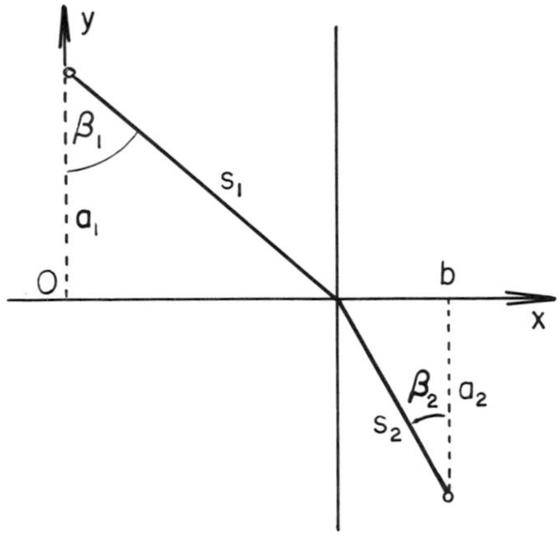

Figure 6.11. The path of least time.

the vertical (direction, which is parallel to the y-axis) in the upper half plane, and an angle β_2 from the vertical in the lower half plane. If the lengths of the light paths in the two half planes are s_1 and s_2, respectively (cf. Figure 6.11), then the travel time of the light ray is

$$Q = \frac{s_1}{v_1} + \frac{s_2}{v_2} = \sum_{j=1}^{2} \frac{a_j}{v_j \cos \beta_j} .$$

Here the variables β_j, better, $x_j = \tan \beta_j$, satisfy the constraint

$$a_1 x_1 + a_2 x_2 = b,$$

and the objective function, in x_j-variables, assumes the form

$$Q = \sum_{j=1}^{2} \frac{a_j}{v_j} (1 + x_j^2)^{1/2} = \text{Min.}$$

This algebraic convex optimization problem has a well-known solution, namely the law of refraction,

$$\frac{\sin \beta_1}{\sin \beta_2} = \frac{v_1}{v_2} .$$

b) Quadratic optimization in elasticity theory. Consider the following two-dimensional problem. A point mass m, of weight G, is suspended by n rods, of negligible weight, constant diameter F_j, length ℓ_j, and coefficient of elasticity E_j, $j = 1,\ldots,n$. The rods form an angle α_j with the horizontal; see Figure 6.12. A position

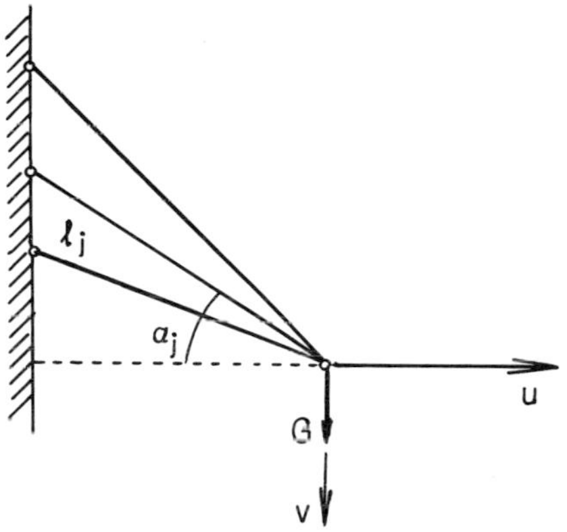

Figure 6.12. A problem in elasticity theory.

of equilibrium is reached, with a deflection of the mass m by u in the horizontal, and v in the vertical, direction. This stretches the rods by a distance δ_j, where

$$\delta_j = u \cos \alpha_j + v \sin \alpha_j \quad (j = 1,\ldots,n). \tag{6.24}$$

For the jth rod, the work of deformation,

$$A_j = \frac{1}{2} E_j F_j \frac{\delta_j^2}{\ell_j},$$

becomes the intrinsic potential energy. Remembering the potential energy of $-Gv$ of the weight in the field of gravity, we apply the principle that potential energy is minimized at a position of equilibrium, and obtain the optimization problem with objective function

$$Q = \frac{1}{2} \sum_{j=1}^{n} E_j F_j \frac{\delta_j^2}{\ell_j} = Gv$$

and constraints (6.24). If everything is arranged as in Figure 6.12, we can, if we wish, add positivity constraints for the $n + 2$ variables, namely

$$\delta_j \geq 0, \quad u \geq 0, \quad v \geq 0.$$

c) Equilibrium positions of mechanical systems. A mechanical system is to be described by generalized coordinates, $q_1,\ldots,q_n$, with an equilibrium position determined by $q_j = 0$, $j = 1,\ldots,n$. Let us consider only "small" deflections from the position of equilibrium $(q_j << 1)$. The potential energy, which may be regarded as the objective function, then becomes a quadratic form in the q_j; indeed, one which is positive semi-definite if the equilibrium position is stable. If the description of the system is to be in other coordinates, one can easily give examples where the objective function Q also has linear terms which do not alter the convexity of the objective function, and where there are additional linear constraints.

d) A non-convex optimization problem for a mechanical

system with several positions of equilibrium. A point mass is movable, without friction, on a vertical disk, and is under the influence of a constant force K directed to the highest point P of the disk (realizable by a weight suspended on a string); see Figure 6.13. By introducing

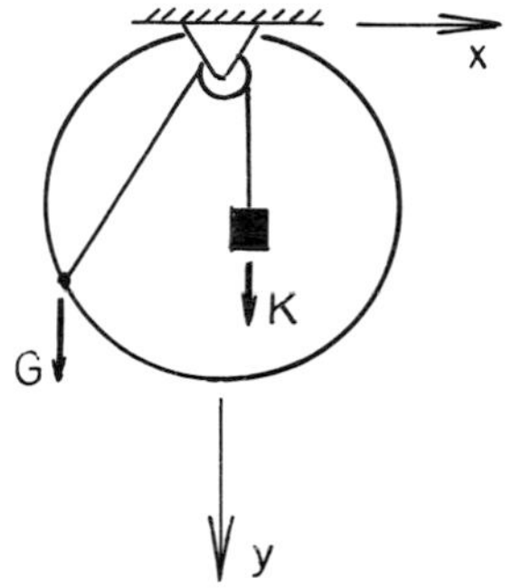

Figure 6.13. A mechanical system with 4 positions of equilibrium.

coordinates x, y and r, the radius of the disk, as in Figure 6.13, we obtain, as constraint, $x^2 + y^2 = 2ry$, and as objective function, the total potential energy,

$$Q = -Gy + K(x^2 + y^2)^{1/2}.$$

If we instead introduce new coordinates, $\xi = x^2$, and $\eta = (y-r)^2$, the constraint becomes linear, but the objective function is no longer convex. Here there are four equilibrium positions, in general, which naturally are not all stable, and which may be reduced in number by one, by adding the inequality constraint $x \geq 0$.

6. *A quasiconcave problem of computer time.* This example will demonstrate how the introduction of different coordinates can change the type of the optimization problem.

Let a computer have a memory with an average retrieval time t and a bit cost of p for that memory. (After a lecture by Dr. Jessen (Konstanz) given at Hamburg in December, 1970.) In simplified form, the cost-throughput relationship (cost per arithmetic operation) is given by

$$Q = a_0 + a_1 t + a_2 p + a_3 tp,$$

where the a_i are given positive constants. For the usual commercial computer installations, the values of t and p lie, in the (t,p)-plane, in a small, roughly elliptical region M, which represents the set of feasible points. Letting $t \geq 0$, $p \geq 0$, $(t,p) \in M$, and $Q = \text{Min}$, we have a quasiconcave optimization problem. The transformation $t = e^u$, $p = e^v$ maps M into another convex region, in the (u,v)-plane, and Q into a convex function,

$$Q = a_0 + a_1 e^u + a_2 e^v + a_3 e^{u+v},$$

so that we now have a convex optimization problem by requiring $Q = \text{Min}$.

7. Convex optimization and the quotient inclusion theorem for matrices (following Elsner, 1971). Let $\underset{\sim}{A} = (a_{jk})$ be an irreducible square matrix of non-negative elements, $a_{jk} \geq 0$, $j,k = 1,\ldots,n$. For the maximal eigenvalue λ of $\underset{\sim}{A}$, there is a corresponding eigenvector $\underset{\sim}{z}$ with positive components $z_1,\ldots,z_n$, such that the equation $\underset{\sim}{A}\underset{\sim}{z} = \lambda\underset{\sim}{z}$ is satisfied. Now let $\underset{\sim}{x}$ be an arbitrary vector with positive components x_j. If we compute the quotients

$$q_j = \frac{\sum_{k=1}^{n} a_{jk} x_k}{x_j} \quad (j = 1,\ldots,n),$$

then, by the inclusion theorem, we have

$$\operatorname*{Min}_j q_j \le \lambda \le \operatorname*{Max}_j q_j = M(\underset{\sim}{x}).$$

To obtain a good upper bound, one would like to minimize $M(\underset{\sim}{x})$. Thus we need to find $\operatorname*{Min}_{\underset{\sim}{x}>\underset{\sim}{0}} M(\underset{\sim}{x})$. In this form, this is a non-convex optimization problem. But if we replace the coordinates x_j by new coordinates r_j, where $x_j = e^{r_j}$, the q_j, as functions of $\underset{\sim}{r}$, will have the form

$$q_j(\underset{\sim}{r}) = \sum_{k=1}^{n} a_{jk} e^{(r_k - r_j)}.$$

These functions are now convex, since, for arbitrary vectors $\underset{\sim}{r}$ and $\underset{\sim}{s}$ and for $0 \le \alpha \le 1$, we have

$$q_j(\alpha\underset{\sim}{r}+(1-\alpha)\underset{\sim}{s}) = \sum_k a_{jk} e^{\alpha(r_k-r_j)+(1-\alpha)(s_k-s_j)}$$

$$\le \alpha q_j(\underset{\sim}{r}) + (1-\alpha) q_j(\underset{\sim}{s}),$$

for the convexity of the exponential function implies that

$$e^{\alpha\rho+(1-\alpha)\sigma} < \alpha e^{\rho} + (1-\alpha)e^{\sigma}.$$

Therefore, $\operatorname*{Max}_j q_j(\underset{\sim}{r})$ is also convex, in $\underset{\sim}{r}$.

Such a conclusion cannot be obtained for the minimum of the q_j.

8. <u>Optimal</u> <u>control</u>. The problem of optimal control represents a generalization of the classical variational problems. For the functions

$$\underset{\sim}{x}(t) = \{x_1(t),\ldots,x_n(t)\}' \in R^n \quad \text{(the state)},$$
$$\underset{\sim}{u}(t) = \{u_1(t),\ldots,u_m(t)\}' \in R^m \quad \text{(the control)},$$

we consider a system of ordinary differential equations,

$$\dot{\underset{\sim}{x}} = G(t,\underset{\sim}{x}(t),\underset{\sim}{u}(t)) \;, \tag{6.25}$$

an initial vector, $\underset{\sim}{x}(t_1) = \underset{\sim}{a}$, and perhaps a terminal vector, $\underset{\sim}{x}(t_2) = \underset{\sim}{b}$, and we are to minimize an integral (the cost integral),

$$F = \int_{t_1}^{t_2} \phi(t,\underset{\sim}{x}(t),\underset{\sim}{u}(t))dt, \tag{6.26}$$

by a suitable choice of the control $\underset{\sim}{u}(t)$, which is to lie between the bounds $\underset{\sim}{u}$ and $\overline{\underset{\sim}{u}}$,

$$\underset{\sim}{u} \le \underset{\sim}{u}(t) \le \overline{\underset{\sim}{u}}. \tag{6.27}$$

If the time, t_2, and the terminal vector, $\underset{\sim}{x}(t_2)$, are given, we speak of a "fixed-end problem" and otherwise, of a "free-end problem". ϕ is a given function, of its arguments. In the simplest case, $\phi = 1$, we have the problem of minimizing time. The given terminal state is to be reached in the least time possible.

An exposition of the theory of optimal control would burst the seams of this little treatise. We will content ourselves in giving a simple, typical example and in making a few remarks about the numerical treatment.

<u>The problem of least travel time</u>. A train is to travel from a place P_0 (place coordinate $x = 0$) to a place P_1 (place coordinate $x = p$) in the shortest time possible.

We consider a highly idealized version of the problem,

in that friction, air resistance, etc., are all ignored. As a control, $u(t)$, we choose the acceleration, $\ddot{x}(t)$, allowable for the train, which cannot exceed some positive value a, nor some negative value $-b$. The train begins traveling at time $t = t_1 = 0$, and reaches its goal at a time $t = t_2$, which is still unknown, but is to be as small as possible. Thus the problem reads

$$\ddot{x}(t) = u(t), \quad x(0) = \dot{x}(0) = 0, \quad x(t_2) = p, \ \dot{x}(t_2) = 0$$

$$F = \int_0^{t_2} dt = t_2 = \text{Min}, \quad -b \le u(t) \le a. \tag{6.28}$$

By the so-called Pontrjagin Maximum Principle (which is not proven here; see Converse, 1970, or Melsa-Schultz, 1970), the train will go fastest, if it is under maximum acceleration a from time $t = 0$ to time $t = t_z$, and then under maximum deceleration $-b$ from time $t = t_z$ to time $t = t_2$. The intermediate time t_z can be expressed at once as a function of t_2; we have $t_z = \frac{b}{a+b} t_2$. With the aid of the equation

$$x(t_2) = \frac{a}{2} t_z^2 + a t_z (t_2 - t_z) - \frac{b}{2}(t_2 - t_z)^2 = p$$

we obtain

$$p = \frac{ab}{2(a+b)} t_2^2,$$

from which the minimal time t_2 is easily found.

The switch of the control variable u from one bound, a, to the other bound, $-b$, which we observe here (see Figure 6.14), is in fact typical for optimal control. Also observe that the problem is not sensible without the in-

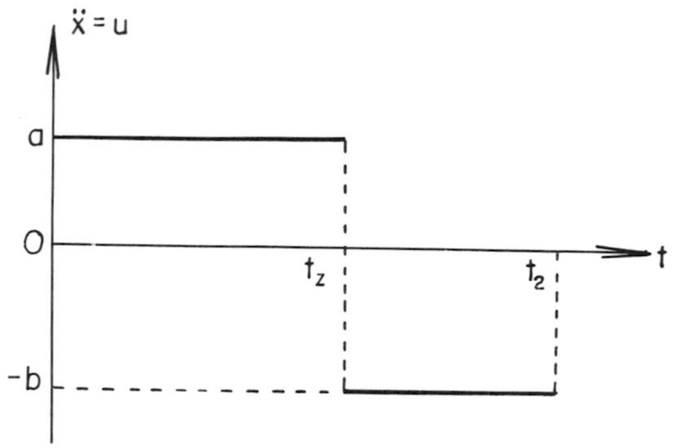

Figure 6.14. Optimal control.

(The travel of a train)

equalities (the constraints $-b \leq u(t) \leq a$), for without these constraints, there would be no solution.

As to numerical computation, because of the great importance of optimal control problems, many different methods of approximating the solutions of such problems have been developed, although we will not be able to delve into this subject here. Let us only mention the obvious method of discretization, in which the time interval under consideration is partitioned into a finite number of smaller intervals, the differential quotients are replaced by difference quotients or other, better approximating expressions, and the integral to be minimized is replaced by a finite sum. In this way, one obtains a finite optimization problem, with finitely many variables, and finitely many constraints. This optimization problem is linear if G and ϕ are affine-linearly dependent on $\underset{\sim}{x}$ and $\underset{\sim}{u}$, and nonlinear otherwise. In any case, the methods of this book, for approximating the solutions of such optimization problems, may be applied.

9. Algebraic Optimization. The design of a street network. In the plane, the points P_j, $j = 1,\ldots,n$, represent towns with coordinates (x_j, y_j). Let f_{jk} be the "traffic flow" between P_j and P_k, that is, the number of vehicles traveling between P_j and P_k in a year. The costs per kilometer of street (building, maintenance, etc.) are assessed in the form $k(f) = a + bf$, where a and b are constants and f is the appropriate traffic flow. The problem is to design a network of roads for which the total cost is as small as possible.

The treatment of this problem for larger values of n runs into curious topological difficulties. For values of n which are not too great (see Figure 6.15), it is

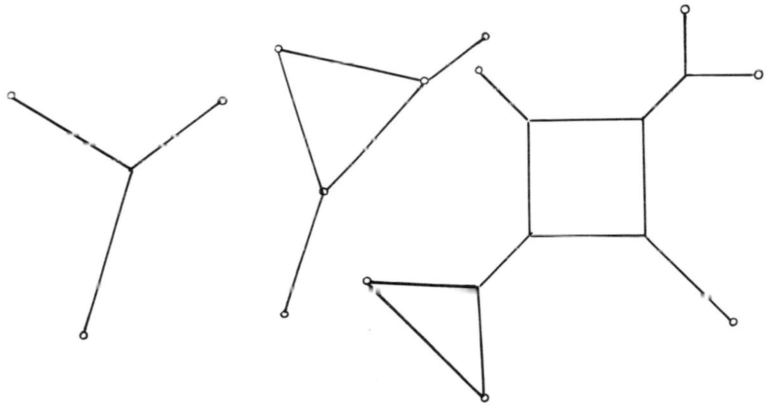

Figure 6.15. Design of a street network.

still possible to discuss the various arrays, but with a larger number of towns, there is no way of knowing, a priori, which arrangements of the road net to consider, and therefore, no way of explicitly formulating the optimization problem.

6.9. Examples of Integer Optimization

For these, it may be that

a) some or all of the independent variables can assume integer values only, or

b) the objective function can assume integer values only, or

c) both a) and b) occur simultaneously.

We again give a few simple examples.

1) Integer quadratic optimization. (Proximate office problem, quadratic assignment problem, wiring a computer installation, etc.)

Let an office have n rooms and n persons. We want to assign people who have much interaction to offices which are highly proximate (see Figure 6.16), while longer

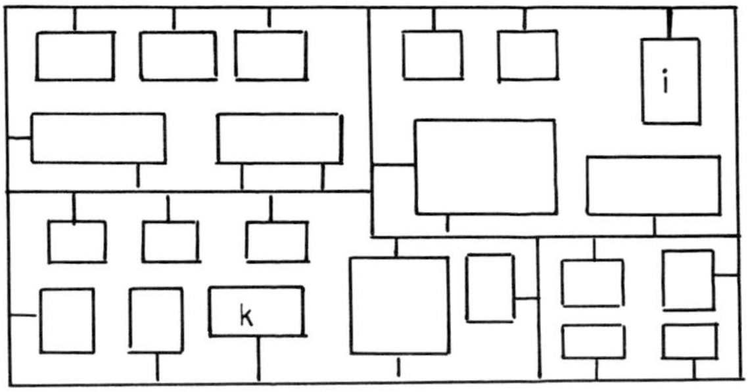

Figure 6.16. A proximate office problem.

paths are acceptable between those with little interaction. Let the (not necessarily straight) path between room number i and room number k be of length a_{ik}, and let $b_{j\ell}$ be the frequency of contact between person number j and person number ℓ. We want to find the numbers x_{jk}, each of which is 0 or 1 and which form a permutation, or doubly stochastic, matrix, and therefore satisfy

$$\sum_j x_{jk} = \sum_k x_{jk} = 1.$$

Under these constraints, we want

$$\phi = \sum_{i,j,k,\ell} a_{ik} b_{j\ell} x_{ij} x_{k\ell} = \text{Min.}$$

2) Cutting waste problem.

A circular metal disk of some given radius, say $R = 50$ cm, is to be cut up into (mutually non-overlapping) disks, each of radius $r_1 = 1$ cm, $r_2 = 2$ cm, or $r_3 = 3$ cm, so as to minimize the waste (= the remaining irregular piece), Q; see Figure 6.17. Here $\frac{1}{\pi} Q$ is an integer.

3) Stamp problem.

For given integers p and q, with $p > 0$ and $q > 0$, define $s(p,q)$ to be the smallest number of natural numbers, $n_1, \ldots, n_s$, with the property that, for every given natural number $i \leq p$, there exist non-negative integers x_{ij} such that

$$\sum_{j=1}^{s} x_{ij} \leq q \quad \text{and} \quad i = \sum_{j=1}^{s} x_{ij} n_j.$$

In other words, how many denominations of stamps are required so that every postage $\leq p$ can be made with no more

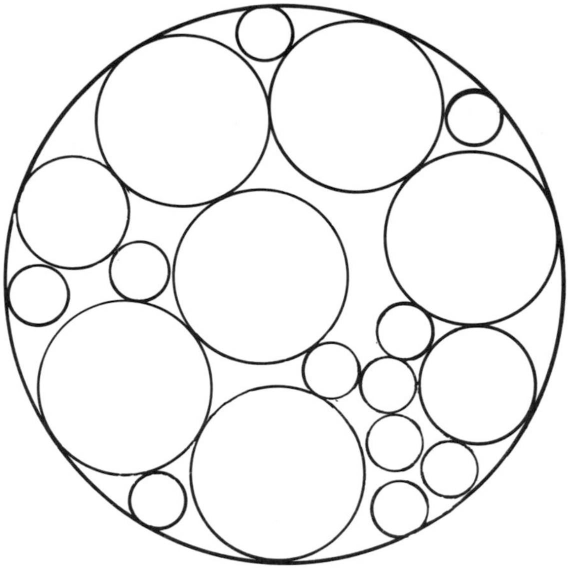

Figure 6.17. A cutting waste problem for circular disks.

than q stamps.

Numerical example: $s(20, 3) = 4$, e.g. $\{n_i\} = \{1, 4, 6, 7\}$,

but $s(20, 4) = 3$, e.g. $\{n_i\} = \{1, 4, 6\}$.

4) Convex integer optimization. (A statistical example).

Let two populations, I and II, be given, where these are assumed to be normally distributed with unknown means a and b respectively, and known deviations, σ_0 and σ_1 respectively. A sample of size m and n, respectively, is to be taken. Let the cost of each observation from I, respectively II, be A, respectively B.

We want to find the sample sizes m and n, for

which the confidence level is within given bounds, while the resulting costs are minimized. As a constraint for m and n, one obtains

$$Am + Bn = \text{Min}, \quad \frac{\sigma_0^2}{m} + \frac{\tau_0^2}{n} \le K,$$

where K is a given constant. For more details, see Pfanzagl, 1966.

§7. A Characterization of Minimal Solutions for Convex Optimization

7.1. The Kuhn-Tucker Saddle-point Theorem

In the case of linear optimization, theorem 4 of §5 provided a characterization of the minimal solution for the original problem in terms of the maximal solution to the dual problem. The generalization of this idea to convex optimization constitutes the content of the Kuhn-Tucker Theorem.

As in §6.3, the problem under consideration is

$$\left.\begin{array}{l} f_j(\underset{\sim}{x}) \le 0 \quad (j = 1,\dots,m), \quad \underset{\sim}{x} \ge \underset{\sim}{0}, \\ F(\underset{\sim}{x}) = \text{Min!} \end{array}\right\} \tag{7.1}$$

The functions $F(\underset{\sim}{x})$ and $f_j(\underset{\sim}{x})$ are defined and convex for $\underset{\sim}{x} \in R^n$. As with the Lagrange multiplier method for determining extremal values subject to constraints, we introduce the function

$$\Phi(\underset{\sim}{x},\underset{\sim}{u}) = F(\underset{\sim}{x}) + \sum_{j=1}^{m} u_j f_j(\underset{\sim}{x}). \tag{7.2}$$

Here, $\underset{\sim}{u} = (u_1,\ldots,u_m)'$ is a vector in R^m. The components u_j are also called <u>multipliers</u>, and the function $\Phi(\underset{\sim}{x},\underset{\sim}{u})$ of $n + m$ variables, the <u>Lagrange function</u>, for problem (7.1).

If the functions $f_j(\underset{\sim}{x})$ are also collected, and formed into a vector, $\underset{\sim}{f}(\underset{\sim}{x}) = (f_1(\underset{\sim}{x}),\ldots,f_m(\underset{\sim}{x}))'$, (7.2) can be rewritten as

$$\phi(\underset{\sim}{x},\underset{\sim}{u}) = F(\underset{\sim}{x}) + \underset{\sim}{u}'\underset{\sim}{f}(\underset{\sim}{x}).$$

<u>Definition</u>. A point $\begin{pmatrix}\underset{\sim}{x}^0\\ \underset{\sim}{u}^0\end{pmatrix}$ in R^{n+m}, with $\underset{\sim}{x}^0 \geq \underset{\sim}{0}$ and $\underset{\sim}{u}^0 \geq \underset{\sim}{0}$, is called a <u>saddle point</u> of $\Phi(\underset{\sim}{x},\underset{\sim}{u})$ iff

$$\Phi(\underset{\sim}{x}^0,\underset{\sim}{u}) \leq \Phi(\underset{\sim}{x}^0,\underset{\sim}{u}^0) \leq \Phi(\underset{\sim}{x},\underset{\sim}{u}^0) \tag{7.3}$$

for all $\underset{\sim}{x} \geq \underset{\sim}{0}$ and all $\underset{\sim}{u} \geq \underset{\sim}{0}$.

<u>Remark</u>. This is a saddle point with respect to the subset of R^{n+m} given by $\underset{\sim}{x} \geq \underset{\sim}{0}$ and $\underset{\sim}{u} \geq \underset{\sim}{0}$; whenever we use the concept of saddle point, it will be with this meaning.

The following theorem on the functions $F(\underset{\sim}{x})$ and $f_j(\underset{\sim}{x})$ holds without further qualification (not even convexity).

<u>Theorem 1</u>: If $\begin{pmatrix}\underset{\sim}{x}^0\\ \underset{\sim}{u}^0\end{pmatrix}$ is a saddle point of $\Phi(\underset{\sim}{x},\underset{\sim}{u})$, then $\underset{\sim}{x}^0$ is a minimal solution of problem (7.1).

<u>Proof</u>: (7.3) implies that, for $\underset{\sim}{x} \geq \underset{\sim}{0}$ and $\underset{\sim}{u} \geq \underset{\sim}{0}$,

$$F(\underset{\sim}{x}^0) + \underset{\sim}{u}'\underset{\sim}{f}(\underset{\sim}{x}^0) \leq F(\underset{\sim}{x}^0) + \underset{\sim}{u}^{0'}\underset{\sim}{f}(\underset{\sim}{x}^0) \leq F(\underset{\sim}{x}) + \underset{\sim}{u}^{0'}\underset{\sim}{f}(\underset{\sim}{x}).$$

It follows that $\underset{\sim}{u}'\underset{\sim}{f}(\underset{\sim}{x}^0) \leq \underset{\sim}{u}^{0'}\underset{\sim}{f}(\underset{\sim}{x}^0)$ for all $\underset{\sim}{u} \in R^m$ with $\underset{\sim}{u} \geq \underset{\sim}{0}$. This is only possible if $\underset{\sim}{f}(\underset{\sim}{x}^0) \leq \underset{\sim}{0}$. This makes $\underset{\sim}{u}^{0'}\underset{\sim}{f}(\underset{\sim}{x}^0) \leq 0$. Setting $\underset{\sim}{u} = \underset{\sim}{0}$ implies $\underset{\sim}{u}^{0'}\underset{\sim}{f}(\underset{\sim}{x}^0) \geq 0$, and therefore, $\underset{\sim}{u}^{0'}\underset{\sim}{f}(\underset{\sim}{x}^0) = 0$. $\underset{\sim}{x}^0$ thus satisfies all of the constraints of problem (7.1), and $F(\underset{\sim}{x}^0) \leq F(\underset{\sim}{x}) + \underset{\sim}{u}^{0'}\underset{\sim}{f}(\underset{\sim}{x})$ for all $\underset{\sim}{x} \geq \underset{\sim}{0}$. If $\underset{\sim}{x}$ is a feasible point, $\underset{\sim}{f}(\underset{\sim}{x}) \leq 0$, and so then $F(\underset{\sim}{x}^0) \leq F(\underset{\sim}{x})$; $\underset{\sim}{x}^0$ is a minimal solution of problem (7.1).

We will show next that with suitable constraint qualifications, a converse of theorem 1 holds, so that a saddle point of $\Phi(\underset{\sim}{x},\underset{\sim}{u})$ can be found for every minimal solution of problem (7.1). That we need additional qualifications is shown by the following example.

Let $n = m = 1$, $F(x) = -x$, and $f_1(x) = x^2$. The constraints $x^2 \leq 0$ and $x \geq 0$ are satisfied only by $x = 0$. Therefore $x = 0$ is also the minimal solution. The Lagrange function $\Phi(x,u) = -x + ux^2$. If it had a saddle point for $x = 0$ and corresponding $u \geq 0$, it would follow that $0 \leq -x + ux^2$ for $x \geq 0$, which is clearly impossible.

One qualification which excludes such cases and assures the converse of theorem 1 is the following.

(V): There exists a feasible point $\underset{\sim}{\tilde{x}}$ such that $f_j(\underset{\sim}{\tilde{x}}) < 0$ for $j = 1,\ldots,m$.

In the proof of the Kuhn-Tucker theorem we use the following important theorem, which is proven in the appendix.

<u>The Separation Theorem for Convex Sets</u>. Let B_1 and B_2 be two proper convex subsets of R^n which have no points

in common. Let B_2 be open. Then there exists a hyperplane, $a'x = \beta$, which separates B_1 and B_2; i.e., there is a vector $a \neq 0$ and a real number β such that $a'x \leq \beta < a'y$ for all $x \in B_1$ and $y \in B_2$.

Theorem 2: (The Kuhn-Tucker Theorem) Let condition (V) be satisfied for problem (7.1). Then $x^0 \geq 0$ is a minimal solution of problem (7.1) iff there exists a $u^0 \geq 0$ such that $\begin{pmatrix} x^0 \\ u^0 \end{pmatrix}$ is a saddle point of $\Phi(x,u)$.

Proof: Theorem 1 already shows that a saddle point leads to a minimal solution. Conversely, let x^0 be a minimal solution of problem (7.1). Letting $y = (y_0, y_1, \ldots, y_m)'$ be vectors in R^{m+1}, define two sets, B_1 and B_2, by

$$B_1 = \{y \mid y_0 \geq F(x), y_j \geq f_j(x) \quad (j = 1, \ldots, m) \text{ for at least one } x \geq 0\},$$

$$B_2 = \{y \mid y_0 < F(x^0), y_j < 0 \quad (j = 1, \ldots, m)\}$$

B_1 and B_2 are convex. B_2 is open. Since x^0 is a minimal solution, there is no y which lies in both B_1 and B_2. B_2 is a proper subset of R^{m+1} and is not empty. Therefore B_1 is also a proper subset of R^{m+1}. The separation theorem for convex sets is thus applicable, and says that there is a vector $v = (v_0, v_1, \ldots, v_m)'$ $(v \neq 0)$ such that

$$v'y > v'z \quad \text{for } y \in B_1, \ z \in B_2. \tag{7.4}$$

Since the components of $z \in B_2$ may be negative of arbitrarily large size, it follows that $v \geq 0$. If in (7.4) we allow $\geq$, the inequality is still valid if $\underset{\sim}{y}$ is in B_1 and $\underset{\sim}{z}$ is in the boundary of B_2; in particular, if $\underset{\sim}{z} = (F(\underset{\sim}{x}^0), 0,\ldots,0)'$ and $\underset{\sim}{y} = (F(\underset{\sim}{x}), f_1(\underset{\sim}{x}),\ldots,f_m(\underset{\sim}{x}))'$, in which case it becomes

$$v_0 F(\underset{\sim}{x}) + \sum_{j=1}^{m} v_j f_j(\underset{\sim}{x}) \geq v_0 F(\underset{\sim}{x}^0) \quad \text{for all} \quad \underset{\sim}{x} \geq \underset{\sim}{0}. \tag{7.5}$$

From this, one concludes that $v_0 > 0$. For if $v_0 = 0$, then $\sum_{j=1}^{m} v_j f_j(\underset{\sim}{x}) \geq 0$, for all $\underset{\sim}{x} \geq \underset{\sim}{0}$ and at least one $v_j > 0$, $j = 1,\ldots,m$, contradicting condition (V), $\sum_{j=1}^{m} v_j f_j(\tilde{\underset{\sim}{x}}) < 0$. Now set $\underset{\sim}{u}^0 = \frac{1}{v_0}(v_1,\ldots,v_m)'$, so $\underset{\sim}{u}^0 \geq \underset{\sim}{0}$ and

$$F(\underset{\sim}{x}) + \underset{\sim}{u}^{0'} \underset{\sim}{f}(\underset{\sim}{x}) \geq F(\underset{\sim}{x}^0) \quad \text{for all} \quad \underset{\sim}{x} \geq \underset{\sim}{0}. \tag{7.6}$$

Letting $\underset{\sim}{x} = \underset{\sim}{x}^0$ here yields $\underset{\sim}{u}^{0'} \underset{\sim}{f}(\underset{\sim}{x}^0) \geq 0$. Since $\underset{\sim}{x}^0$ is a feasible vector for problem (7.1), also $\underset{\sim}{f}(\underset{\sim}{x}^0) \leq 0$. Since $\underset{\sim}{u}^0 \geq \underset{\sim}{0}$, it follows that

$$\underset{\sim}{u}^{0'} \underset{\sim}{f}(\underset{\sim}{x}^0) = 0 \tag{7.7}$$

and also that

$$\underset{\sim}{u}' \underset{\sim}{f}(\underset{\sim}{x}^0) \leq 0 \quad \text{for} \quad \underset{\sim}{u} \geq \underset{\sim}{0}. \tag{7.8}$$

(7.6), (7.7), and (7.8) imply that

$$F(\underset{\sim}{x}^0) + \underset{\sim}{u}' \underset{\sim}{f}(\underset{\sim}{x}^0) \leq F(\underset{\sim}{x}^0) + \underset{\sim}{u}^{0'} \underset{\sim}{f}(\underset{\sim}{x}^0) \leq F(\underset{\sim}{x}) + \underset{\sim}{u}^{0'} \underset{\sim}{f}(\underset{\sim}{x})$$
$$(\underset{\sim}{x} \geq \underset{\sim}{0}, \quad \underset{\sim}{u} \geq \underset{\sim}{0}),$$

so $\begin{pmatrix} \underset{\sim}{x}^0 \\ \underset{\sim}{u}^0 \end{pmatrix}$ is a saddle point of $\Phi(\underset{\sim}{x},\underset{\sim}{u}) = F(\underset{\sim}{x}) + \underset{\sim}{u}' \underset{\sim}{f}(\underset{\sim}{x})$.

It follows from the proof of the Kuhn-Tucker theorem that condition (V) may be replaced by the following qualification (V'), which however only appears to be less restrictive.

(V'): For each index $j = 1,\ldots,m$, there exists some feasible point $\tilde{x}^j$ such that $f_j(\tilde{x}^j) < 0$.

If (V') is satisfied, set $\tilde{x} = \frac{1}{m}\sum_{i=1}^{m}\tilde{x}^i$. Then $\tilde{x}$ is feasible because it is a convex combination of feasible points, $\tilde{x}^i$, and

$$f_j(\tilde{x}) \leq \frac{1}{m}\sum_{i=1}^{m} f_j(\tilde{x}^i) \leq \frac{1}{m} f_j(\tilde{x}^j) < 0 \quad (j = 1,\ldots,m).$$

(V') therefore implies (V). Conversely, (V') follows from (V); merely let $\tilde{x}^j = \tilde{x}$, $j = 1,\ldots,m$. (V') and (V) are thus equivalent.

Condition (V) excludes the possibility that there occur inequality constraints, $f_j(x) \leq 0$, in (7.1), which are actually $= 0$ for all $x \in M$. In particular, constraints $g(x) = 0$ are excluded, where $g(x)$ is an affine linear function which might appear in (7.1) in the form $g(x) \leq 0$, $-g(x) \leq 0$. In §9 we will show that condition (V) may be dropped for convex optimization problems which contain only constraints of this type.

7.2. An Inclusion Theorem

As in the case of linear problems (cf. §5.1) we can find upper and lower bounds for the minimal value of the objective function. If x^0 is a minimal solution of problem (7.1) and x^1 is an arbitrary feasible point,

$$F(x^0) \leq F(x^1).$$

Thus we have an upper bound for $F(x^0)$. A lower bound is

found as follows. Let $\underset{\sim}{u}^* \in R^m$ be ≥ 0. Then if the problem

$$F(\underset{\sim}{x}) + \underset{\sim}{u}^{*\prime}\underset{\sim}{f}(\underset{\sim}{x}) = \text{Min! for } \underset{\sim}{x} \geq \underset{\sim}{0} \qquad (7.9)$$

(with no further constraints) is solvable, and if $\underset{\sim}{x}^2$ is a solution,

$$F(\underset{\sim}{x}^2) + \underset{\sim}{u}^{*\prime}\underset{\sim}{f}(\underset{\sim}{x}^2) \leq F(\underset{\sim}{x}^0) + \underset{\sim}{u}^{*\prime}\underset{\sim}{f}(\underset{\sim}{x}^0) \leq F(\underset{\sim}{x}^0).$$

A lower bound for $F(\underset{\sim}{x}^0)$ is then found by solving this simpler problem (7.9). If $\underset{\sim}{u}^* = \underset{\sim}{u}^0$ (= the second component of the saddle point vector in theorem 1), $\underset{\sim}{x}^0$ solves problem (7.9). If $\underset{\sim}{u}^*$ is a good approximation to $\underset{\sim}{u}^0$, we can expect the solution of (7.9), therefore, to be a good lower bound; similarly, if $\underset{\sim}{x}^1$ is a close approximation to $\underset{\sim}{x}^0$, we expect a close upper bound.

Example (n = 2, m = 1):

$$F(\underset{\sim}{x}) = x_1^2 + x_2^2 = \text{Min!} \quad f_1(\underset{\sim}{x}) - e^{-x_1} - x_2 \leq 0$$
$$x_1, x_2 \geq 0.$$

The minimal point will be the point on the curve $x_2 = e^{-x_1}$ which is closest to the origin. This leads to the equation $x_1 e^{2x_1} = 1$ for x_1. The solution, and the minimal value of F, can be computed up to the desired number of decimals on a machine. But even with paper and pencil only, and a table of the exponential function in intervals of 0.001 (e.g., Abramowitz & Stegun, Handbook of Mathematical Functions), we can achieve the following result, without great effort or even bothering to interpolate. An approximate solution of the above equation is given by $x_1 = 0.430$.

Letting $x_2 = 0.651$ makes $(x_1,x_2)'$ a feasible point, and therefore $F(\underset{\sim}{x}^0) \leq (0.430)^2 + (0.651)^2 \leq 0.609$. Next we have to find $\underset{\sim}{u}^*$. §8.1 would lead us to expect that $\Phi_x(\underset{\sim}{x}^0,\underset{\sim}{u}^0) = 0$. Thus $\underset{\sim}{u}^*$ should satisfy the equations $2x_1 - u^*e^{-x_1} = 0$ and $2x_2 - u^* = 0$ approximately, so let $\underset{\sim}{u}^* = 1.3$.

Problem (7.9) now is to find the minimum of

$$\xi_1^2 + \xi_2^2 + 1.3(e^{-\xi_1} - \xi_2) \quad (\xi_1,\xi_2 \geq 0).$$

The equations $2\xi_1 - 1.3e^{-\xi_1} = 0$, and $2\xi_2 - 1.3 = 0$ yield $\xi_2 = 0.65$ and $0.424 \leq \xi_1 \leq 0.425$, so that

$$0.607 \leq F(\underset{\sim}{x}^0) \leq 0.609.$$

§8. Convex Optimization for Differentiable Functions

8.1. Local Kuhn-Tucker Conditions

The characterization of the solution of a convex optimization problem, given in §7 by the Kuhn-Tucker theorem, contains the saddle point condition. This is a global condition for the Lagrange function. $\Phi(\underset{\sim}{x}^0,\underset{\sim}{u}^0)$ must be compared to $\Phi(\underset{\sim}{x},\underset{\sim}{u}^0)$ and to $\Phi(\underset{\sim}{x}^0,\underset{\sim}{u})$ for all $\underset{\sim}{x} \geq \underset{\sim}{0}$ and all $\underset{\sim}{u} \geq \underset{\sim}{0}$. But if the objective function $F(\underset{\sim}{x})$ and the constraints $f_j(\underset{\sim}{x})$ are differentiable, the saddle point condition may be replaced by equivalent local conditions. The optimization problem under consideration still is

$$F(\underset{\sim}{x}) = \text{Min!}, \quad f_j(\underset{\sim}{x}) \leq 0, \ j = 1,\ldots,m, \quad \underset{\sim}{x} \geq \underset{\sim}{0}. \tag{8.1}$$

Let the functions $F(\underset{\sim}{x})$, $f_1(\underset{\sim}{x}),\ldots,f_m(\underset{\sim}{x})$ be convex for $\underset{\sim}{x} \in R^n$, and let them have first partial derivatives. As before, the Lagrange function Φ is defined by

$$\Phi(\underset{\sim}{x},\underset{\sim}{u}) = F(\underset{\sim}{x}) + \underset{\sim}{u}'\underset{\sim}{f}(\underset{\sim}{x}) \quad (\underset{\sim}{x} \in R^n,\ \underset{\sim}{u} \in R^m), \tag{8.2}$$

where $\underset{\sim}{f}(\underset{\sim}{x})$ is the vector with components $f_1(\underset{\sim}{x}),\ldots,f_m(\underset{\sim}{x})$. Let Φ_x and Φ_u denote the gradients of Φ with respect to $\underset{\sim}{x}$ and $\underset{\sim}{u}$:

$$\Phi_x = \left(\frac{\partial\Phi}{\partial x_1},\ldots,\frac{\partial\Phi}{\partial x_n}\right)', \quad \Phi_u = \left(\frac{\partial\Phi}{\partial u_1},\ldots,\frac{\partial\Phi}{\partial u_m}\right)'.$$

From (8.2) we see that $\Phi_u(\underset{\sim}{x},\underset{\sim}{u}) = \underset{\sim}{f}(\underset{\sim}{x})$.

<u>Theorem 1</u>: Suppose condition (V) holds, so there is a feasible point $\tilde{\underset{\sim}{x}}$ such that $f_j(\tilde{\underset{\sim}{x}}) < 0$, $j = 1,\ldots,m$. Then $\underset{\sim}{x}^0 \geq \underset{\sim}{0}$ is a minimal solution of (8.1) iff there exists a $\underset{\sim}{u}^0 \geq \underset{\sim}{0}$ such that

$$\Phi_x(\underset{\sim}{x}^0,\underset{\sim}{u}^0) \geq \underset{\sim}{0}, \quad \underset{\sim}{x}^{0'}\Phi_x(\underset{\sim}{x}^0,\underset{\sim}{u}^0) = 0, \tag{8.3}$$

$$\Phi_u(\underset{\sim}{x}^0,\underset{\sim}{u}^0) \leq \underset{\sim}{0}, \quad \underset{\sim}{u}^{0'}\Phi_u(\underset{\sim}{x}^0,\underset{\sim}{u}^0) = 0. \tag{8.4}$$

<u>Proof</u>: We will show that conditions (8.3) and (8.4) are equivalent to the saddle point condition

$$\Phi(\underset{\sim}{x}^0,\underset{\sim}{u}) \leq \Phi(\underset{\sim}{x}^0,\underset{\sim}{u}^0) \leq \Phi(\underset{\sim}{x},\underset{\sim}{u}^0) \quad (\underset{\sim}{x} \geq \underset{\sim}{0},\ \underset{\sim}{u} \geq \underset{\sim}{0}). \tag{8.5}$$

I. ((8.5) implies (8.3) and (8.4)). Suppose there is a negative component of $\Phi_x(\underset{\sim}{x}^0,\underset{\sim}{u}^0)$, say $\partial\Phi/\partial x_k < 0$, so there exists a vector $\underset{\sim}{x} \geq \underset{\sim}{0}$ with components $x_\ell = x_\ell^0$ for $\ell \neq k$ and $x_k > x_k^0$ such that $\Phi(\underset{\sim}{x},\underset{\sim}{u}^0) < \Phi(\underset{\sim}{x}^0,\underset{\sim}{u}^0)$, contradicting (8.5). (8.5) thus implies that $\Phi_x(\underset{\sim}{x}^0,\underset{\sim}{u}^0) \geq \underset{\sim}{0}$. Because $\underset{\sim}{x}^0 \geq \underset{\sim}{0}$, all of the summands $x_k^0 \cdot \partial\Phi(\underset{\sim}{x}^0,\underset{\sim}{u}^0)/\partial x_k$ in the inner product $\underset{\sim}{x}^{0'}\Phi_x(\underset{\sim}{x}^0,\underset{\sim}{u}^0)$ also are non-negative. Now if there were an index k such that $\partial\Phi(\underset{\sim}{x}^0,\underset{\sim}{u}^0)/\partial x_k > 0$ and $x_k^0 > 0$, there would also be a vector $\underset{\sim}{x}$ with components $x_\ell = x_\ell^0$, $\ell \neq k$, and $0 \leq x_k < x_k^0$, such that $\Phi(\underset{\sim}{x},\underset{\sim}{u}^0) < \Phi(\underset{\sim}{x}^0,\underset{\sim}{u}^0)$, again in contradiction to (8.5). The assumption that (8.4) is false leads similarly to a contradiction to (8.5).

II. ((8.3) and (8.4) imply (8.5)). Since $\underset{\sim}{u}^0 \geq \underset{\sim}{0}$, $\Phi(\underset{\sim}{x},\underset{\sim}{u}^0)$ is a convex function of $\underset{\sim}{x} \in R^n$. By theorem 4, §6.2, this implies

$$\Phi(\underset{\sim}{x},\underset{\sim}{u}^0) \geq \Phi(\underset{\sim}{x}^0,\underset{\sim}{u}^0) + (\underset{\sim}{x}-\underset{\sim}{x}^0)'\Phi_x(\underset{\sim}{x}^0,\underset{\sim}{u}^0) \text{ for } \underset{\sim}{x} \geq \underset{\sim}{0}. \quad (8.6)$$

Since $\Phi(\underset{\sim}{x}^0,\underset{\sim}{u})$ is affine linear in $\underset{\sim}{u}$,

$$\Phi(\underset{\sim}{x}^0,\underset{\sim}{u}) = \Phi(\underset{\sim}{x}^0,\underset{\sim}{u}^0) + (\underset{\sim}{u}-\underset{\sim}{u}^0)'\Phi_u(\underset{\sim}{x}^0,\underset{\sim}{u}^0) \text{ for } \underset{\sim}{u} \geq \underset{\sim}{0}. \quad (8.7)$$

(8.3) and (8.4), together with (8.6) and (8.7), then imply (8.5).

In order to formulate the following theorem in more uniform notation, we introduce the functions $g_k(\underset{\sim}{x}) = -x_k$, $k = 1,\ldots,n$, thereby allowing us to rewrite the positivity constraints $\underset{\sim}{x} \geq \underset{\sim}{0}$ in the form $g_k(\underset{\sim}{x}) \leq \underset{\sim}{0}$, $k = 1,\ldots,n$. The gradient of $g_k(\underset{\sim}{x})$ is $-\underset{\sim}{e}^k$, where $\underset{\sim}{e}^k$ is the kth unit vector of R^n.

If $\underset{\sim}{x}^0 \geq \underset{\sim}{0}$ is a feasible point, we let Q^0 be that subset of the indices $j = 1,\ldots,m$, for whose elements $\underset{\sim}{x}^0$ satisfies the corresponding inequality constraints, $f_j(\underset{\sim}{x}) \leq 0$, with an equality sign, and similarly, we let P^0 be that subset of the indices $k = 1,\ldots,n$, for whose elements $\tilde{x}^0$ satisfies the corresponding sign constraint, $g_k(\tilde{x}) \leq 0$, with an equality sign. Thus we have

$$f_j(\tilde{x}^0) \begin{cases} = 0 & \text{for } j \in Q^0, \\ < 0 & \text{for } j \notin Q^0, \end{cases}$$

$$g_k(\tilde{x}^0) = -x_k^0 \begin{cases} = 0 & \text{for } k \in P^0, \\ < 0 & \text{for } k \notin P^0. \end{cases}$$

Theorem 2: Suppose condition (V) holds, so there is a feasible point $\tilde{\tilde{x}}$ such that $f_j(\tilde{\tilde{x}}) < 0$, $j = 1,\ldots,m$. Then a feasible point $\tilde{x}^0$ is a minimal solution of problem (8.1) iff there exist vectors $\tilde{u}^0 \geq \tilde{0}$ $(\tilde{u}^0 \in R^m)$ and $\tilde{v}^0 \geq \tilde{0}$ $(\tilde{v}^0 \in R^n)$ such that

$$-\text{grad } F(\tilde{x}^0) = \sum_{j\in Q^0} u_j^0 \text{grad } f_j(\tilde{x}^0) + \sum_{k\in P^0} v_k^0 \text{grad } g_k(\tilde{x}^0); \tag{8.8}$$

$$u_j^0 \begin{cases} \geq 0 & \text{for } j \in Q^0, \\ = 0 & \text{for } j \notin Q^0; \end{cases} \tag{8.9}$$

$$v_k^0 \begin{cases} \geq 0 & \text{for } k \in P^0, \\ = 0 & \text{for } k \notin P^0. \end{cases} \tag{8.10}$$

Proof: I. (The saddle point condition (8.5) implies (8.8), (8.9), and (8.10).) Let $\tilde{u}^0 \geq \tilde{0}$ be a vector for which (8.5) holds. By theorem 1, (8.3) and (8.4) also hold. Since $\Phi_u(\tilde{x}^0,\tilde{u}^0) = \tilde{f}(\tilde{x}^0)$, we cannot, by (8.4), simultaneously have $f_j(\tilde{x}^0) < 0$ and $u_j^0 > 0$. This implies (8.9). Setting

$$\underset{\sim}{v}^0 = \Phi_x(\underset{\sim}{u}^0,\underset{\sim}{x}^0) = \text{grad } F(\underset{\sim}{x}^0) + \sum_{j=1}^{m} u_j^0 \text{grad } f_j(\underset{\sim}{x}^0), \quad (8.11)$$

so that $\underset{\sim}{v}^0 \geq \underset{\sim}{0}$, by (8.3), we see that we cannot simultaneously have $x_k^0 > 0$ and $v_k^0 > 0$. This implies (8.10). If we substitute $\underset{\sim}{v}^0 = -\sum_{k=1}^{n} v_k^0 \text{grad } g_k(\underset{\sim}{x}^0)$ in (8.11), and omit the (vanishing) summands with $j \not\in Q^0$ and $k \not\in P^0$, we obtain (8.8).

II. ((8.8), (8.9), and (8.10) imply (8.5).) Because of $\Phi_u(\underset{\sim}{x}^0,\underset{\sim}{u}^0) = \underset{\sim}{f}(\underset{\sim}{x}^0)$ and (8.9), (8.4) holds. By the definition of $g_k(\underset{\sim}{x})$ and (8.8),

$$\underset{\sim}{v}_0 = \text{grad } F(\underset{\sim}{x}^0) + \sum_{j=1}^{m} u_j^0 \text{grad } f_j(\underset{\sim}{x}^0) = \Phi_x(\underset{\sim}{x}^0,\underset{\sim}{u}^0).$$

This, and (8.10), implies (8.3). But (8.3) and (8.4) together imply (8.5), by theorem 1.

Theorem 2 admits a geometric interpretation. A minimal solution, $\underset{\sim}{x}^0$, is characterized by the property that the vector $-\text{grad } F(\underset{\sim}{x}^0)$ is a non-negative linear combination (i.e., some multiple of $-\text{grad } F(\underset{\sim}{x}^0)$ is a convex combination) of those gradients, which belong to the hypersurfaces, $f_j(\underset{\sim}{x}) = 0$ and $g_k(\underset{\sim}{x}) = 0$, in which $\underset{\sim}{x}^0$ lies. If a minimal solution, $\underset{\sim}{x}^0$, occurs in the interior of the set M, $\text{grad } F(\underset{\sim}{x}^0) = 0$ (as for a minimum without constraints).

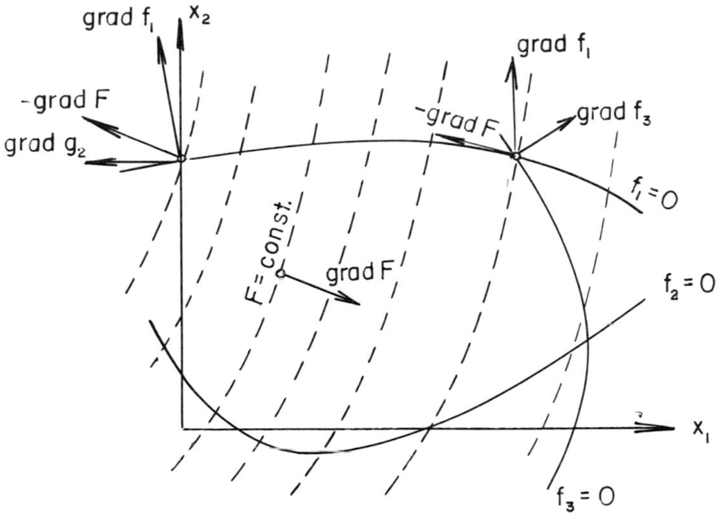

Figure 8.1

8.2. A Characterization of the Set of Minimal Solutions

The set of minimal solutions for a convex optimization problem is easily seen to be convex. In case the optimization is with differentiable functions, a more precise description of the set of minimal solutions is possible.

Lemma. Let the function $F(\underset{\sim}{x})$ be convex and differentiable for $\underset{\sim}{x} \in R^n$. Then

(a) $\underset{\sim}{y}'\text{grad } F(\underset{\sim}{x}) > 0$ implies $F(\underset{\sim}{x}+\lambda\underset{\sim}{y}) > F(\underset{\sim}{x})$ for all $\lambda > 0$;

(b) $\underset{\sim}{y}'\text{grad } F(\underset{\sim}{x}) < 0$ implies there exists a $\lambda_0 > 0$ such that $F(\underset{\sim}{x}+\lambda\underset{\sim}{y}) < F(\underset{\sim}{x})$ for $0 < \lambda \leq \lambda_0$.

Proof: (a) By theorem 4, §6.2, $F(\underset{\sim}{x}+\lambda\underset{\sim}{y}) \geq F(\underset{\sim}{x})$ +

$\lambda y'\text{grad } F(x)$ for all $\lambda \geq 0$.

(b) Set $\psi(\lambda) = F(x+\lambda y)$; then $\frac{d\psi}{d\lambda}_{(\lambda=0)} = y'\text{grad } F(x)$. This implies (b).

Theorem 3: Let x^0 be a minimal solution of problem (8.1), where the function $F(x)$ is convex and differentiable. Then the set of all minimal solutions is the set of those feasible points z for which

$$\text{grad } F(z) = \text{grad } F(x^0), \tag{8.12}$$

$$(z-x^0)'\text{grad } F(x^0) = 0. \tag{8.13}$$

Proof: I. Suppose z is feasible and satisfies (8.12) and (8.13). By theorem 4, §6.2, $F(x^0) \geq F(z) + (x^0-z)'\text{grad } F(z) = F(z)$. Since x^0 is a minimal solution, this forces $F(x^0) = F(z)$, and z is also a minimal solution.

II. Suppose z is a minimal solution. Then $F(z) = F(x^0) = F(x^0 + \lambda(z-x^0))$ for $0 \leq \lambda \leq 1$, because $F(x)$ is convex and cannot attain smaller values than $F(x^0)$ on the convex set of feasible points.

By the previous lemma, this means that $(z-x^0)'\text{grad } F(x^0) = 0$; see Figure 8.2. The function defined by

$$G(y) = F(y) - (y-x^0)'\text{grad } F(x^0)$$

is convex in y. We have $G(z) = F(z) = G(x^0)$ and $\text{grad } G(z) = \text{grad } F(z) - \text{grad } F(x^0)$. If $\text{grad } F(z) \neq \text{grad } F(x^0)$, then $\text{grad } G(z) \neq 0$, and there exists a w such that $w'\text{grad } G(z) < 0$, so by the lemma again, $G(z+\lambda w) < G(z) = G(x^0)$ for sufficiently small $\lambda > 0$. But by theorem 4, §6.2,

$$G(\underset{\sim}{z}+\lambda\underset{\sim}{w}) = F(\underset{\sim}{z}+\lambda\underset{\sim}{w}) = (\underset{\sim}{z}+\lambda\underset{\sim}{w}-\underset{\sim}{x}^0)'\text{grad } F(\underset{\sim}{x}^0)$$
$$\geq F(\underset{\sim}{x}^0) = G(\underset{\sim}{x}^0).$$

Thus we obtain a contradiction from the assumption that (8.12) does not hold.

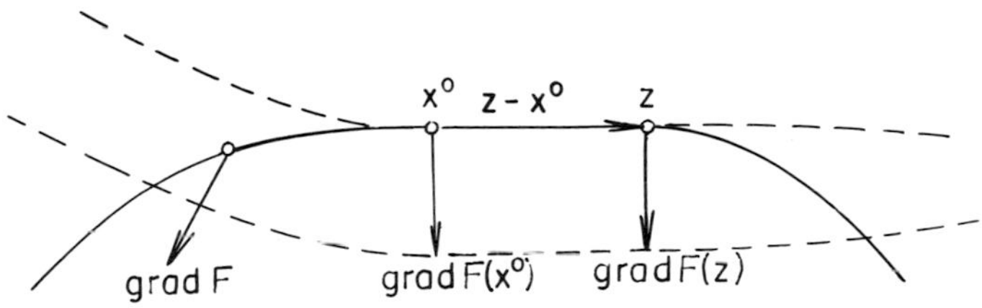

Figure 8.2.

8.3. Convex Optimization with Differentiable Functions

We now let problem D^0 denote the convex optimization problem 7.1, less the constraint $\underset{\sim}{x} \geq \underset{\sim}{0}$. For problem D^0 the set of feasible points is then

$$M^0: \quad \underset{\sim}{x} \in R^n, \quad f_j(\underset{\sim}{x}) \leq 0 \quad (j = 1,\ldots,m) \tag{8.14}$$

and the objective function is

$$F(\underset{\sim}{x}) = \text{Min!} \tag{8.15}$$

where the f_j and F are convex and differentiable.

For the dual problem, we have the variables

$x_1,\dots,x_n$, $u_1,\dots,u_m$, and for this problem, D^1, the set of feasible points, M^1, satisfies the conditions

$$M^1:\ \underset{\sim}{x} \in R^n,\ \underset{\sim}{u} \in R^m,\ \underset{\sim}{u} \geq 0,\ \text{grad } F + \underset{\sim}{u}'\text{grad } \underset{\sim}{f} = 0 \tag{8.16}$$

while the objective function is

$$\Phi(\underset{\sim}{x},\underset{\sim}{u}) = F(\underset{\sim}{x}) + \underset{\sim}{u}'\underset{\sim}{f}(\underset{\sim}{x}) = \text{Max}. \tag{8.17}$$

For clarity's sake, let us write out the vector equation appearing in (8.16) in complete detail:

$$\frac{\partial F}{\partial x_k} + \underset{\sim}{u}'\frac{\partial \underset{\sim}{f}}{\partial x_k} = \frac{\partial F}{\partial x_k} + \sum_{j=1}^{m} u_j \frac{\partial f_j}{\partial x_k} = 0$$

$$(k = 1,\dots,n).$$

We then have the

<u>Weak Duality Theorem</u> (Wolfe, 1961): For arbitrary $\underset{\sim}{x}^1 \in M^0$ and an arbitrary pair $\underset{\sim}{x}^2,\underset{\sim}{u}^2 \in M^1$,

$$F(\underset{\sim}{x}^1) \geq \Phi(\underset{\sim}{x}^2,\underset{\sim}{u}^2). \tag{8.18}$$

If we now use (5.37) and (5.38) of §5.7, we see that the existence of feasible points for both problems implies the existence of an infimum and a supremum for the values in (8.18), so that

$$\operatorname*{Inf}_{\underset{\sim}{x}\in M^0} F(\underset{\sim}{x}) \geq \operatorname*{Sup}_{(\underset{\sim}{x},\underset{\sim}{u})\in M^1} \Phi(\underset{\sim}{x},\underset{\sim}{u}). \tag{8.19}$$

In addition, $F(\underset{\sim}{x}^1)$ and $\Phi(\underset{\sim}{x}^2,\underset{\sim}{u}^2)$ are bounds for these two numbers.

<u>Proof</u>: By the convexity property (6.8),

$$F(\underset{\sim}{x}^1) \geq F(\underset{\sim}{x}^2) + (\underset{\sim}{x}^1-\underset{\sim}{x}^2)'\text{grad } F(\underset{\sim}{x}^2)$$
$$= F(\underset{\sim}{x}^2) - (\underset{\sim}{x}^1-\underset{\sim}{x}^2)'[\underset{\sim}{u}^{2'}\text{grad } \underset{\sim}{f}(\underset{\sim}{x}^2)];$$

which, if written out in detail, reads

$$F(\underset{\sim}{x}^1) \geq F(\underset{\sim}{x}^2) + \sum_k (x_k^1-x_k^2)\frac{\partial F(x^2)}{\partial x_k} =$$
$$= F(\underset{\sim}{x}^2) - \sum_j \sum_k (x_k^1-x_k^2)u_j^2 \frac{\partial f_j(\underset{\sim}{x}^2)}{\partial x_k} .$$

Again by the convexity property (6.8),

$$(\underset{\sim}{x}^1 - \underset{\sim}{x}^2)'\text{grad } \underset{\sim}{f}(\underset{\sim}{x}^2) \leq \underset{\sim}{f}(\underset{\sim}{x}^1) - \underset{\sim}{f}(\underset{\sim}{x}^2)$$

which, again detailed by components, reads

$$\sum_k (x_k^1-x_k^2)\frac{\partial f_j(\underset{\sim}{x}^2)}{\partial x_k} \leq f_j(\underset{\sim}{x}^1) - f_j(\underset{\sim}{x}^2).$$

In toto, we have

$$F(\underset{\sim}{x}^1) \geq F(\underset{\sim}{x}^2) + \underset{\sim}{u}^{2'}(\underset{\sim}{f}(\underset{\sim}{x}^2)-\underset{\sim}{f}(\underset{\sim}{x}^1)).$$

Since $\underset{\sim}{u}^{2'} \geq \underset{\sim}{0}$, $\underset{\sim}{f}(\underset{\sim}{x}^1) \leq 0$, and $-\underset{\sim}{u}^{2'}\underset{\sim}{f}(\underset{\sim}{x}^1) \geq 0$, it follows that

$$F(\underset{\sim}{x}^1) \geq F(\underset{\sim}{x}^2) + \underset{\sim}{u}^{2'}\underset{\sim}{f}(\underset{\sim}{x}^2) = \Phi(\underset{\sim}{x}^2,\underset{\sim}{u}^2).$$

But this is assertion (8.18).

Numerical Example.

In the (x,y)-plane, consider the curve, C, defined by $f_1(x,y) = (1/2)x^4 + x + 2 - y = 0$; see Figure 8.3. We are to compute which point on this curve lies closest to the origin. Thus we have the convex optimization problem

$$D^0: \quad F = x^2 + y^2 = \text{Min}, \quad f_1 = \frac{1}{2}x^4+x+2-y \leq 0. \tag{8.20}$$

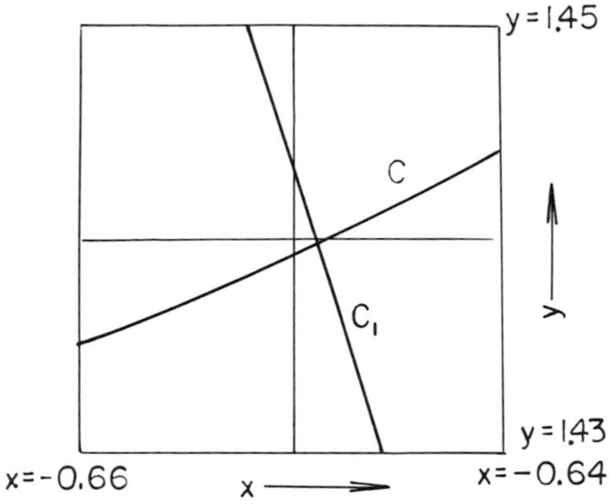

Figure 8.3. Towards Duality

For the dual problem, D^1, we have to satisfy the constraint in (8.16),

$$\text{grad } F + \underset{\sim}{u}'\text{grad } \underset{\sim}{f} = (2x + u(2x^3+1), 2y - u)' = 0.$$

In parametric form, this yields the equations

$$u = \frac{-x}{x^3 + \frac{1}{2}}, \quad y = \frac{u}{2},$$

of a curve, C_1, which intersects the curve C almost orthogonally, and thus numerically determines the minimal point most conveniently as the intersection of the curves C and C_1. Taking an approximation of $x = -0.65$ from the illustration, we obtain the table

x	u	y	$z = \frac{1}{2}x^4 + x + 2$	$f_1(x,y)$	$F(x,y)$	$\Phi = F + uf_1$	x^2+z^2
-0.65	2.8841	1.442	1.43925	-0.028	2.5020	2.4939	2.4967

Thus the minimal value of F is included in bounds, $2.4939 \leq \text{Min } F \leq 2.4967$.

One might consider it most fortunate that in this example the constraint in (8.16) was so easily fulfilled. But in any case where computational difficulties arise one can add artificial constraints -- in the instant case, e.g., $f_2(x,y) = -y \leq 0$ -- and still have one free parameter, u_2, available for satisfying the constraints.

The inclusion principle described in §7.2, when applied to a convex optimization problem with differentiable functions F and f_j, agrees with the one given here if we disregard the positivity constraint $\underset{\sim}{x} \geq \underset{\sim}{0}$ considered there. Here, as there, the equations

$$\text{grad } F + \underset{\sim}{u}'\text{grad } \underset{\sim}{f} = \underset{\sim}{0}$$

must be satisfied.

$F(\underset{\sim}{x}) + \underset{\sim}{u}'\underset{\sim}{f}(\underset{\sim}{x})$ then provides a lower bound for the minimal value of the problem. The derivation and the assumptions, however, are different. Here we obtain the bounds in terms of a duality theory, while in §7.2 they were derived directly from the saddle point theorem, and therefore remain valid without any differentiability assumptions.

8.4. Positivity Conditions for Non-linear Optimization Problems

Consider a general non-linear optimization problem

$$F(\underset{\sim}{x}) = \text{Min!} \quad f_j(\underset{\sim}{x}) \leq 0 \quad (j = 1,\ldots,m) \qquad (8.21)$$

without positivity constraints on $\underset{\sim}{x}$. No convexity conditions are demanded of F and f_j. Even then, we can still find conditions for deciding whether a point $\underset{\sim}{x}^0$ is a local minimum. These criteria contain positivity conditions for the matrices of the second partial derivatives (whose existence is assumed) of F and f_j at the point $\underset{\sim}{x}^0$, and thus may be regarded as convexity conditions in the small.

A local minimum occurs at $\underset{\sim}{x}^0 \in M$ (= set of feasible points) if (see §6.5) there exists a ball $K_\rho(\underset{\sim}{x}^0)$, centered at $\underset{\sim}{x}^0$ with positive radius ρ, so that

$$F(\underset{\sim}{x}) \geq F(\underset{\sim}{x}^0) \quad \text{for} \quad \underset{\sim}{x} \in K_\rho(\underset{\sim}{x}^0) \cap M \tag{8.22}$$

A strong local minimum occurs at $\underset{\sim}{x}^0$ if such a ball exists with

$$F(\underset{\sim}{x}) > F(\underset{\sim}{x}^0) \quad \text{for} \quad \underset{\sim}{x} \in K_\rho(\underset{\sim}{x}^0) \cap M, \quad \underset{\sim}{x} \neq \underset{\sim}{x}^0. \tag{8.23}$$

<u>Theorem 4</u>: (sufficient conditions for a strong local minimum). Let $\underset{\sim}{x}^0 \in M$ be a point at which the second partial derivatives of F and f_j exist. Let $J \subset \{1,2,...,m\}$ be a subset of the indices j, for which $f_j(\underset{\sim}{x}^0) = 0$. For $j \in J$, let there be numbers $u_j > 0$ such that

$$\text{grad}(F(\underset{\sim}{x}^0) + \sum_{j\in J} u_j f_j(\underset{\sim}{x}^0)) = 0 \tag{8.24}$$

(local Kuhn-Tucker conditions). Let the quadratic form

$$q(\underset{\sim}{y}) = \sum_{i,k=1}^{n} \frac{\partial^2}{\partial x_i \partial x_k} (F(\underset{\sim}{x}^0) + \sum_{j\in J} u_j f_j(\underset{\sim}{x}^0)) y_i y_k$$

be positive definite (i.e., $q(\underset{\sim}{y}) > 0$ for $\underset{\sim}{y} \neq \underset{\sim}{0}$ and $\underset{\sim}{y} \in H$) on the linear subspace $H \subset R^n$ of all vectors $\underset{\sim}{y}$ with $\underset{\sim}{y}'\text{grad } f_j(\underset{\sim}{x}^0) = 0$, $j \in J$. Then there is a strong (= isolated) local minimum at $\underset{\sim}{x}^0$.

Proof: (indirect). If there is no strong local minimum at $\underset{\sim}{x}^0$, there exists a sequence of points $\underset{\sim}{x}^\nu \in M$ with $\lim \underset{\sim}{x}^\nu = \underset{\sim}{x}^0$, $\underset{\sim}{x}^\nu \neq \underset{\sim}{x}^0$, and $F(\underset{\sim}{x}^\nu) \leq F(\underset{\sim}{x}^0)$. $\underset{\sim}{x}^\nu$ can then be written as $\underset{\sim}{x}^\nu = \underset{\sim}{x}^0 + \delta_\nu \underset{\sim}{y}^\nu$, where the $\underset{\sim}{y}^\nu$ have euclidean length $||\underset{\sim}{y}^\nu|| = 1$ and the δ_ν are positive numbers with $\lim \delta_\nu = 0$. The sequence of vectors $\underset{\sim}{y}^\nu$ contains a convergent subsequence, by Bolzano-Weierstrass. We may as well assume that this is the original sequence, so that $\lim \underset{\sim}{y}^\nu = \underset{\sim}{y}$, where $||\underset{\sim}{y}|| = 1$. Since $F(\underset{\sim}{x}^\nu) \leq F(\underset{\sim}{x}^0)$ and $f_j(\underset{\sim}{x}^\nu) \leq f_j(\underset{\sim}{x}^0) = 0$, $j \in J$, we have

$$(F(\underset{\sim}{x}^0 + \delta_\nu \underset{\sim}{y}^\nu) - F(\underset{\sim}{x}^0))/\delta_\nu \leq 0,$$
$$(f_j(\underset{\sim}{x}^0 + \delta_\nu \underset{\sim}{y}^\nu) - f_j(\underset{\sim}{x}^0))/\delta_\nu \leq 0 \qquad (j \in J).$$

The limiting values of these quotients, namely the directional derivatives in the direction $\underset{\sim}{y}$, are also ≤ 0:

$$\underset{\sim}{y}'\text{grad } F(\underset{\sim}{x}^0) \leq 0, \quad \underset{\sim}{y}'\text{grad } f_j(\underset{\sim}{x}^0) \leq 0 \quad (j \in J).$$

From (8.24) and $u_j > 0$, $j \in J$, it follows that these directional derivatives are actually $= 0$, so that, in particular, $\underset{\sim}{y} \in H$. Now by Taylor's theorem,

$$\underbrace{F(\underset{\sim}{x}^\nu) + \sum_{j\in J} u_j f_j(\underset{\sim}{x}^\nu)}_{\leq \underset{\sim}{0}} = F(\underset{\sim}{x}^0) + \underbrace{\sum_{j\in J} u_j f_j(\underset{\sim}{x}^0)}_{= \underset{\sim}{0}} +$$

$$\underbrace{+\delta_\nu \underset{\sim}{y}^{\nu\prime}\text{grad}(F(\underset{\sim}{x}^0) + \sum_{j\in J} u_j f_j(\underset{\sim}{x}^0))}_{= \underset{\sim}{0}} + \tfrac{1}{2}\delta_\nu^2 q(\underset{\sim}{y}^\nu) + o(\delta_\nu^2),$$

where $o(\delta_\nu^2)/\delta_\nu^2 \to 0$ as $\delta_\nu \to 0$. This implies that

$$0 \geq \frac{1}{\delta_\nu^2}(F(\underset{\sim}{x}^\nu) - F(\underset{\sim}{x}^0)) \geq \frac{1}{2} q(\underset{\sim}{y}^\nu) + o(\delta_\nu^2)/\delta_\nu^2.$$

As $\nu \to \infty$ the right side has the positive limit $(1/2)q(\underset{\sim}{y})$; the assumption that $\underset{\sim}{x}^0$ is not a strong local minimum thus leads to a contradiction.

Remarks. 1. This theorem, as well as the following theorem 5, were published by McCormick (1967). His paper also handles the case where some of the constraints are equalities. A direct proof of theorem 4 may be found in Wetterling (1970a); there the radius ρ of the ball in (8.23) is expressed in terms of bounds for the third derivatives of the functions F and f_j.

2. When the vectors grad $f_j(\underset{\sim}{x}^0)$, $j \in J$, span R^n, H collapses to the zero vector. That $q(\underset{\sim}{y})$ is positive definite is then trivial, and the local Kuhn-Tucker conditions (8.24) already are sufficient for a strong local minimum.

The following theorem 5 gives a necessary condition for a local minimum. As before, in §7.1, additional assumptions are needed. Letting $\underset{\sim}{x}^0$ be a point in M, and J be exactly the set of indices for which $f_j(\underset{\sim}{x}^0) = 0$, we formulate two constraint qualifications, (V_1) and (V_2).

(V_1) If $\underset{\sim}{y} \in R^n$ is a vector with $\underset{\sim}{y}'\text{grad } f_j(\underset{\sim}{x}^0) \leq 0$, $j \in J$, then there exists a $t_0 > 0$ and a vector valued function $\underset{\sim}{x}(t)$ defined for $0 \leq t \leq t_0$, such that

(a) $\underset{\sim}{x}(0) = \underset{\sim}{x}^0$,

(b) $d\underset{\sim}{x}(0)/dt$ exists and $= \underset{\sim}{y}$, and

(c) $\underset{\sim}{x}(t) \in M$ for $0 \leq t \leq t_0$.

(V_2) If $\underset{\sim}{y} \in R^n$ is a vector with $\underset{\sim}{y}'\text{grad } f_j(\underset{\sim}{x}^0) = 0$, $j \in J$, then there exists a t_0 and a vector valued function $\underset{\sim}{x}(t)$ for which (a), (b), and (c) hold, and additionally,

(d) $f_j(\underset{\sim}{x}(t)) = 0$, $j \in J$, and

(e) $\underset{\sim}{z} = d^2\underset{\sim}{x}(0)/dt^2$ exists.

<u>Theorem 5</u>: (necessary conditions for a local minimum). Let there be a local minimum of problem (8.21) at $\underset{\sim}{x}^0$, and let J be that set of indices for which $f_j(\underset{\sim}{x}^0) = 0$. Let the qualifications (V_1) and (V_2) be met at $\underset{\sim}{x}^0$. Let F and f_j be twice differentiable at $\underset{\sim}{x}^0$. Then there exist numbers $u_j \geq 0$, $j \in J$, such that the local Kuhn-Tucker conditions,

$$\text{grad}(F(\underset{\sim}{x}^0) + \sum_{j\in J} u_j f_j(\underset{\sim}{x}^0)) = 0, \tag{8.25}$$

are satisfied, and the quadratic form

$$q(\underset{\sim}{y}) = \sum_{i,k=1}^{n} \frac{\partial^2}{\partial x_i \partial x_k}(F(\underset{\sim}{x}^0) + \sum_{j\in J} u_j f_j(\underset{\sim}{x}^0))y_i y_k$$

is positive semi-definite (i.e., $q(\underset{\sim}{y}) \geq 0$ for $\underset{\sim}{y} \in H$) on the linear subspace $H \subset R^n$ of vectors $\underset{\sim}{y}$ such that $\underset{\sim}{y}'\text{grad } f_j(\underset{\sim}{x}^0) = 0$, $j \in J$.

<u>Remark</u>. Note that J is defined differently here, and that therefore $q(\underset{\sim}{y})$ agrees only formally with the corresponding quadratic form in theorem 4.

<u>Proof</u>: Let $\underset{\sim}{y}$ be a vector with $\underset{\sim}{y}'\text{grad } f_j(\underset{\sim}{x}^0) \leq 0$,

$j \in J$, and let $\underset{\sim}{x}(t)$ be the corresponding function provided by (V_1). Since there is a local minimum at $\underset{\sim}{x}^0$,

$$\frac{d}{dt} F(\underset{\sim}{x}(t))_{(t=0)} = \underset{\sim}{y}'\text{grad } F(\underset{\sim}{x}^0) \geq 0.$$

Thus there exists no vector $\underset{\sim}{y}$ for which $\underset{\sim}{y}'\text{grad } f_j(\underset{\sim}{x}^0) \leq 0$, $j \in J$, and $\underset{\sim}{y}'\text{grad } F(\underset{\sim}{x}^0) < 0$. The theorem of the alternative (10 in §5.5) is applicable, with $\underset{\sim}{b} = \text{grad } F(\underset{\sim}{x}^0)$ and $A = (-\text{grad } f_j(\underset{\sim}{x}^0))_{(j \in J)}$. Then there exist $u_j \geq 0$, $j \in J$, such that (8.25) holds.

Now let $\underset{\sim}{y} \in H$ and $\underset{\sim}{x}(t)$ be the corresponding function provided by (V_2). Then

$$\frac{d}{dt} F(\underset{\sim}{x}(t))_{(t=0)} = \underset{\sim}{y}'\text{grad } F(\underset{\sim}{x}^0) =$$

$$= - \sum_{j \in J} u_j \underset{\sim}{y}'\text{grad } f_j(\underset{\sim}{x}^0) = 0$$

and therefore, since there is a minimum at $\underset{\sim}{x}^0$,

$$\frac{d^2}{dt^2} F(\underset{\sim}{x}(t))_{(t=0)} = \underset{\sim}{z}'\text{grad } F(\underset{\sim}{x}^0) + \sum_{i,k=1}^{n} \frac{\partial^2 F(\underset{\sim}{x}^0)}{\partial x_i \partial x_k} y_i y_k \geq 0.$$

Furthermore, since $f_j(\underset{\sim}{x}(t)) \equiv 0$ for $j \in J$,

$$\frac{d^2}{dt^2} f_j(\underset{\sim}{x}(t))_{(t=0)} = \underset{\sim}{z}'\text{grad } f_j(\underset{\sim}{x}^0) + \sum_{i,k=1}^{n} \frac{\partial^2 f_j(\underset{\sim}{x}^0)}{\partial x_i \partial x_k} y_i y_k = 0$$

$$(j \in J).$$

Together with (8.25), this implies the asserted positive definiteness of q.

<u>Remarks</u>. 1. Conditions (V_1) and (V_2) are not as restrictive as they might appear. McCormick (1967) proved

that they are satisfied when the vectors $\text{grad } f_j(\underset{\sim}{x}^0)$, $j \in J$, are linearly independent. This means that $\underset{\sim}{x}^0$ is not a boundary point of M of the type of the degenerate vertices in linear optimization.

    2. Theorems 4 and 5 are correspondingly valid for non-linear optimization problems with infinitely many constraints of the type

$$F(\underset{\sim}{x}) = \text{Min!} \quad f(\underset{\sim}{x},\underset{\sim}{y}) \leq 0 \quad \text{for} \quad \underset{\sim}{y} \in Y$$

where Y is defined by finitely many inequalities $g_\nu(\underset{\sim}{y}) \leq 0$, $\nu = 1,\ldots,m$ (Wetterling, 1970).

    This is exactly the problem type which appears in continuous Tchebychev approximation and in finding bounds for boundary value problems (§15). Y then is the region on which the approximation, or boundary value, problem is formulated.

    Example (n = 2, m = 1):

$$F(\underset{\sim}{x}) = -(x_1 + cx_2^3) = \text{Min!} \quad f_1(\underset{\sim}{x}) - x_1^2 + x_2^2 - 1 < 0.$$

The local Kuhn-Tucker conditions may be satisfied at the point $\underset{\sim}{x}^0 = (1,0)'$: $\text{grad } F(\underset{\sim}{x}^0) = (-1,0)'$, $\text{grad } f_1(\underset{\sim}{x}^0) = (2,0)'$, and therefore, $u_1 = 1/2$. The subspace H consists of those vectors $\underset{\sim}{y}$ with $y_1 = 0$. On H, therefore, $q(\underset{\sim}{y}) = (-2c + 1)y_2^2$. This is positive definite for $c < 1/2$, and positive semi-definite for $c \leq 1/2$. In fact there is a strong local minimum at $\underset{\sim}{x}^0$ iff $c \leq 1/2$; see Figure 8.4. The minimal value of F then is -1.

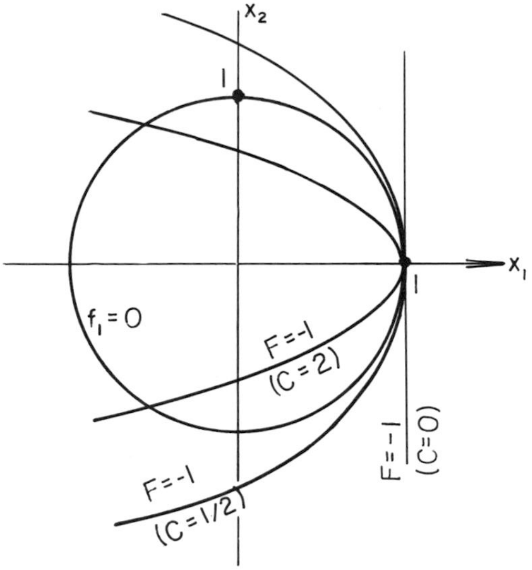

Figure 8.4. A local minimum.

§9. Convex Optimization with Affine Linear Constraints

We remarked in §7, while discussing the Kuhn-Tucker theorem, that when the constraints consist of affine linear functions $f_j(\underset{\sim}{x})$, the restrictive qualification (V) can be dropped. In the following, we will expand on this, and consider optimization problems with convex objective functions $F(\underset{\sim}{x})$ and <u>all</u> functions $f_j(\underset{\sim}{x})$ affine linear. The results obtained can later be applied to the treatment of quadratic optimization. We will not consider the case where some of the functions $f_j(\underset{\sim}{x})$ are affine linear, and the rest not, so that a condition similar to (V) need be required only of the latter. Instead, we refer the interested reader to H. Uzawa, 1958.

9.1. A Theorem on Convex Functions

In the derivation of the Kuhn-Tucker theorem in §7, we used the separation theorem for convex sets at a crucial step. Here we will also use it, and in the form of the following theorem, which is derived from the separation theorem.

$F(x)$ is to be a convex function defined for $x \in R^n$, and N, a convex subset of R^n containing the origin.

Theorem 1: Let $x^0 \in R^n$. Suppose $F(x^0+t) \geq F(x^0)$ for all $t \in N$. Then there exists a vector $p \in R^n$ such that

$$F(x) \geq F(x^0) + p'(x-x^0) \quad \text{for } x \in R^n, \tag{9.1}$$

$$p't \geq 0 \qquad \text{for } t \in N. \tag{9.2}$$

Remark. The vector p introduced here takes the place of the (generally non-existent) gradient of $F(x)$ at $x = x^0$. The theorem will be used in the following in this sense.

If $p \neq 0$, condition (9.2) implies that the origin cannot be an interior point of the convex set N, but must be a boundary point.

Proof: In order to apply the separation theorem for convex sets, we define the following subsets of R^{n+1}:

$$B_1 = \left\{ u = \begin{pmatrix} x^0 + t \\ F(x^0) \end{pmatrix} \middle| \; t \in N \right\},$$

$$B_2 = \left\{ v = \begin{pmatrix} x \\ \rho \end{pmatrix} \middle| \; \tilde{x} \in R^n, \quad \rho > F(x) \right\}.$$

B_1 is convex because N is; B_2 is convex because $F(\underset{\sim}{x})$ is a convex function. Since $F(\underset{\sim}{x})$ is a continuous function, by Theorem 2, §6.2, B_2 is an open set. Furthermore, B_1 and B_2 are proper subsets of R^{n+1}. Now let $\underset{\sim}{v} = \begin{pmatrix} \underset{\sim}{x} \\ \rho \end{pmatrix} \varepsilon B_2$. If $\underset{\sim}{x} - \underset{\sim}{x}^0 \notin N$, then $\underset{\sim}{v} \notin B_1$. If $\underset{\sim}{x} - \underset{\sim}{x}^0 = \underset{\sim}{t} \varepsilon N$, then by assumption $\rho > F(\underset{\sim}{x}) = F(\underset{\sim}{x}^0 + \underset{\sim}{t}) \geq F(\underset{\sim}{x}^0)$, and again, $\underset{\sim}{v} \notin B_1$. Therefore, B_1 and B_2 have no points in common. By the separation theorem, there exists a vector $\underset{\sim}{a} \neq \underset{\sim}{0}$ such that $\underset{\sim}{a}'\underset{\sim}{u} < \underset{\sim}{a}'\underset{\sim}{v}$ for $\underset{\sim}{u} \varepsilon B_1$ and $\underset{\sim}{v} \varepsilon B_2$. Let $\underset{\sim}{a} = \begin{pmatrix} \underset{\sim}{z} \\ \zeta \end{pmatrix}$, where $\underset{\sim}{z} \varepsilon R^n$ and ζ is real. Then, for $\underset{\sim}{t} \varepsilon N$, $\underset{\sim}{x} \varepsilon R^n$, and $\rho > F(\underset{\sim}{x})$,

$$\underset{\sim}{z}'(\underset{\sim}{x}^0 + \underset{\sim}{t}) + \zeta F(\underset{\sim}{x}^0) < \underset{\sim}{z}'\underset{\sim}{x} + \zeta\rho. \tag{9.3}$$

Specializing to $\underset{\sim}{x} = \underset{\sim}{x}^0$ and $\underset{\sim}{t} = \underset{\sim}{0} \varepsilon N$, we have $\zeta F(\underset{\sim}{x}^0) < \zeta\rho$ for all $\rho > F(\underset{\sim}{x}^0)$. It follows that $\zeta > 0$. Set $\underset{\sim}{p} = -\frac{1}{\zeta}\underset{\sim}{z}$ and note that (9.3) holds with $\leq$ instead of $<$ if $\rho = F(\underset{\sim}{x})$. This leads to $F(\underset{\sim}{x}) \geq F(\underset{\sim}{x}^0) + \underset{\sim}{p}'(\underset{\sim}{x} - \underset{\sim}{x}^0 - \underset{\sim}{t})$, for $\underset{\sim}{x} \varepsilon R^n$ and $\underset{\sim}{t} \varepsilon N$. Specializing this with $\underset{\sim}{t} = \underset{\sim}{0}$ yields (9.1), and (9.2) follows with $\underset{\sim}{x} = \underset{\sim}{x}^0$.

9.2. The Kuhn-Tucker Theorem for Optimization Problems with Affine Linear Constraints and Convex Objective Function

The optimization problem to be considered is

$$F(\underset{\sim}{x}) = \text{Min!}, \quad f_j(\underset{\sim}{x}) \leq 0, \quad j = 1,\dots,m, \quad \underset{\sim}{x} \geq \underset{\sim}{0}. \tag{9.4}$$

Here $F(\underset{\sim}{x})$ is defined and convex for $\underset{\sim}{x} \in R^n$, and therefore also continuous, by theorem 2, §6. The $f_j(\underset{\sim}{x})$ are affine linear functions:

$$f_j(\underset{\sim}{x}) = \underset{\sim}{a}^{j'}\underset{\sim}{x} - b_j, \quad j = 1,\ldots,m.$$

Let $\underset{\sim}{A}$ denote the m-by-n matrix whose row vectors are the $\underset{\sim}{a}^{j'}$, and $\underset{\sim}{b}$, the vector with components b_j. Then the constraints $f_j(\underset{\sim}{x}) \leq 0$ read

$$\underset{\sim}{A}\underset{\sim}{x} - \underset{\sim}{b} \leq \underset{\sim}{0}.$$

We then have $\operatorname{grad} f_j(\underset{\sim}{x}) = \underset{\sim}{a}^j$, and also

$$f_j(\underset{\sim}{x}) = f_j(\underset{\sim}{x}^0) + (\underset{\sim}{x} - \underset{\sim}{x}^0)'\operatorname{grad} f_j(\underset{\sim}{x}^0). \tag{9.5}$$

The Lagrange function Φ becomes

$$\Phi(\underset{\sim}{x},\underset{\sim}{u}) = F(\underset{\sim}{x}) + \underset{\sim}{u}'(\underset{\sim}{A}\underset{\sim}{x} - \underset{\sim}{b})$$

with $\underset{\sim}{u} \in R^m$.

<u>Theorem 2</u>: Let $\underset{\sim}{x}^0 \in R^n$ be a vector ≥ 0. $\underset{\sim}{x}^0$ is a minimal solution of problem (9.4) iff there exists a $\underset{\sim}{u}^0 \geq \underset{\sim}{0}$, $\underset{\sim}{u}^0 \in R^m$, such that

$$\Phi(\underset{\sim}{x}^0,\underset{\sim}{u}) \leq \Phi(\underset{\sim}{x}^0,\underset{\sim}{u}^0) \leq \Phi(\underset{\sim}{x},\underset{\sim}{u}^0) \text{ for } \underset{\sim}{x} \geq \underset{\sim}{0},\ \underset{\sim}{u} \geq \underset{\sim}{0}, \tag{9.6}$$

i.e., iff $(\underset{\sim}{x}^0,\underset{\sim}{u}^0)$ is a saddle point of the function $\Phi(\underset{\sim}{x},\underset{\sim}{u})$.

<u>Proof</u>: I. Suppose there exists a $\underset{\sim}{u}^0 \geq \underset{\sim}{0}$ satisfying (9.6). Then $\underset{\sim}{x}^0$ is a minimal solution by theorem 1, §7, since the condition (V) was not used in that proof.

II. Suppose $\underset{\sim}{x}^0$ is a minimal solution of (9.4). As in §8.1, define functions $g_k(\underset{\sim}{x}) = -x_k$, let $\underset{\sim}{e}^k$ be the kth unit vector of R^n, and define index sets Q^0 and P^0 by

$$f_j(\underset{\sim}{x}^0) = \underset{\sim}{a}^{j'}\underset{\sim}{x}^0 - b_j \begin{cases} = 0 & \text{for } j \in Q^0 \\ < 0 & \text{for } j \notin Q^0 \end{cases}$$

$$g_k(\underset{\sim}{x}^0) = -x_k^0 \begin{cases} = 0 & \text{for } k \in P^0 \\ < 0 & \text{for } k \notin P^0. \end{cases}$$

And let the set N be defined as the set of all $\underset{\sim}{t} \in R^n$ such that

$$\underset{\sim}{a}^{j'}\underset{\sim}{t} \leq 0 \quad \text{for } j \in Q^0, \quad -t_k \leq 0 \quad \text{for } k \in P^0.$$

N is convex and contains the origin, $\underset{\sim}{0}$, of R^n. To apply theorem 1, we must first show that $F(\underset{\sim}{x}^0 + \underset{\sim}{t}) \geq F(\underset{\sim}{x}^0)$ for $\underset{\sim}{t} \in N$. The assumption that there exists a $\underset{\sim}{t} \in N$ for which $F(\underset{\sim}{x}^0 + \underset{\sim}{t}) < F(\underset{\sim}{x}^0)$ implies that $F(\underset{\sim}{x}^0 + \lambda\underset{\sim}{t}) \leq (1-\lambda)F(\underset{\sim}{x}^0) + \lambda F(\underset{\sim}{x}^0 + \underset{\sim}{t}) < F(\underset{\sim}{x}^0)$ for $0 < \lambda \leq 1$. Furthermore, $x_k^0 + \lambda t_k > x_k^0 = 0$ for $k \in P^0$ and $\lambda \geq 0$. Additionally, by (9.5),

$$f_j(\underset{\sim}{x}^0 + \lambda\underset{\sim}{t}) = f_j(\underset{\sim}{x}^0) + \lambda\underset{\sim}{a}^{j'}\underset{\sim}{t} \leq f_j(\underset{\sim}{x}^0) = 0 \quad \text{for } j \in Q^0$$
$$\text{and } \lambda \geq 0.$$

Since $f_j(\underset{\sim}{x}^0) < 0$ for $j \notin Q^0$ and $x_k^0 > 0$ for $k \notin P^0$, $\underset{\sim}{x}^0 + \lambda\underset{\sim}{t}$ is a feasible vector for problem (9.4) for sufficiently small $\lambda > 0$. Then $F(\underset{\sim}{x}^0 + \lambda\underset{\sim}{t}) < F(\underset{\sim}{x}^0)$ contradicts the hypothesis that $\underset{\sim}{x}^0$ is a minimal solution.

Theorem 1 may thus be applied. There is a vector $\underset{\sim}{p} \in R^n$ such that

$$F(\underset{\sim}{x}) \geq F(\underset{\sim}{x}^0) + \underset{\sim}{p}'(\underset{\sim}{x}-\underset{\sim}{x}^0) \quad \text{for} \quad \underset{\sim}{x} \in R^n, \tag{9.7}$$

$$\underset{\sim}{p}'\underset{\sim}{t} \geq 0 \quad \text{for} \quad \underset{\sim}{t} \in N. \tag{9.8}$$

We now consider the following "linearized" problem:

$$\begin{aligned} \tilde{F}(\underset{\sim}{x}) &= \underset{\sim}{p}'(\underset{\sim}{x}-\underset{\sim}{x}^0) + F(\underset{\sim}{x}^0) = \text{Min!} \\ \underset{\sim}{f}(\underset{\sim}{x}) &= \underset{\sim}{A}\underset{\sim}{x} - \underset{\sim}{b} \leq \underset{\sim}{0}, \quad \underset{\sim}{x} \geq \underset{\sim}{0}. \end{aligned} \tag{9.9}$$

The set M of feasible points is one and the same for problems (9.4) and (9.9). The Lagrange function for the linearized problem is

$$\tilde{\Phi}(\underset{\sim}{x},\underset{\sim}{u}) = \tilde{F}(\underset{\sim}{x}) + \underset{\sim}{u}'\underset{\sim}{f}(\underset{\sim}{x}) = \Phi(\underset{\sim}{x},\underset{\sim}{u}) + (\tilde{F}(\underset{\sim}{x}) - F(\underset{\sim}{x})).$$

Now let B be the matrix whose row vectors are the $\underset{\sim}{a}^{j'}$ with $j \in Q^0$ and the $-\underset{\sim}{e}^{k'}$ with $k \in P^0$. The column number of B is then n, and the row number is the total number of indices appearing in Q^0 or P^0. The set N defined above is the set of all $\underset{\sim}{t} \in R^n$ for which $\underset{\sim}{B}\underset{\sim}{t} \leq \underset{\sim}{0}$. By (9.8) there is no $\underset{\sim}{t} \in N$ such that $\underset{\sim}{p}'\underset{\sim}{t} < 0$. The system of inequalities, $-\underset{\sim}{B}\underset{\sim}{t} \geq \underset{\sim}{0}$, $\underset{\sim}{p}'\underset{\sim}{t} < 0$, has no solution $\underset{\sim}{t} \in R^n$. Therefore, the system $-\underset{\sim}{B}'\underset{\sim}{w} = \underset{\sim}{p}$, $\underset{\sim}{w} \geq \underset{\sim}{0}$, has a solution, $\underset{\sim}{w}$, by theorem 10, §5, so that there exist numbers $w_j \geq 0$, $j \in Q^0$, and $\hat{w}_k \geq 0$, $k \in P^0$, such that

$$-\underset{\sim}{p} = \sum_{j \in Q^0} \underset{\sim}{a}^j w_j + \sum_{k \in P^0} (-\underset{\sim}{e}^k)\hat{w}_k. \tag{9.10}$$

We now define vectors $\underset{\sim}{u}^0 \in R^m$ and $\underset{\sim}{v}^0 \in R^n$ by

$$u_j^0 \begin{cases} = w_j & \text{for } j \in Q^0, \\ = 0 & \text{for } j \notin Q^0, \end{cases} \qquad v_k^0 \begin{cases} = \hat{w}_k & \text{for } k \in P^0, \\ = 0 & \text{for } k \notin P^0. \end{cases}$$

(9.10) then implies

$$-\text{grad } \tilde{F}(\underset{\sim}{x}^0) = \sum_{j\varepsilon Q^0} u_j^0 \text{grad } f_j(\underset{\sim}{x}^0) + \sum_{k\varepsilon P^0} v_k^0 \text{grad } g_k(\underset{\sim}{x}^0).$$

As one can see from the proof of theorem 2, §8.1, this statement is equivalent to the saddle point condition for $\Phi(\underset{\sim}{x},\underset{\sim}{u})$:

$$\tilde{\Phi}(\underset{\sim}{x}^0,\underset{\sim}{u}) \le \tilde{\Phi}(\underset{\sim}{x}^0,\underset{\sim}{u}^0) \le \tilde{\Phi}(\underset{\sim}{x},\underset{\sim}{u}^0) \quad \text{for} \quad \underset{\sim}{x} \ge \underset{\sim}{0}, \; \underset{\sim}{u} \ge \underset{\sim}{0}. \tag{9.11}$$

Since $F(\underset{\sim}{x}^0) = \tilde{F}(\underset{\sim}{x}^0)$, $\Phi(\underset{\sim}{x}^0,\underset{\sim}{u}) = \tilde{\Phi}(\underset{\sim}{x}^0,\underset{\sim}{u})$ for $\underset{\sim}{u} \varepsilon R^m$, and by (9.7), $F(\underset{\sim}{x}) \ge F(\underset{\sim}{x}^0) + \underset{\sim}{p}'(\underset{\sim}{x}-\underset{\sim}{x}^0) = \tilde{F}(\underset{\sim}{x})$, so that $\Phi(\underset{\sim}{x},\underset{\sim}{u}^0) \ge \tilde{\Phi}(\underset{\sim}{x},\underset{\sim}{u}^0)$ for $\underset{\sim}{x} \varepsilon R^n$. Thus (9.11) implies the saddle point condition for $\Phi(\underset{\sim}{x},\underset{\sim}{u})$:

$$\Phi(\underset{\sim}{x}^0,\underset{\sim}{u}) \le \Phi(\underset{\sim}{x}^0,\underset{\sim}{u}^0) \le \Phi(\underset{\sim}{x},\underset{\sim}{u}^0) \quad \text{for} \quad \underset{\sim}{x} \ge \underset{\sim}{0},\; \underset{\sim}{u} \ge \underset{\sim}{0}.$$

Remark. If the constraints $f_j(\underset{\sim}{x})$ are affine linear, then theorems 1 and 2 of §8.1 are valid without the additional condition (V) on the existence of an $\underset{\sim}{\tilde{x}}$ with $f_j(\underset{\sim}{\tilde{x}}) < 0$, $j = 1,\ldots,m$, which was required there.

§10. The Numerical Treatment of Convex Optimization Problems

It is not the goal of this section to describe individually the numerous, well-known numerical methods for handling convex optimization problems. A survey of various such methods, along with commentary on their utility, and also on the original papers, can be found in P. Wolfe, 1963. One of the methods described there, the cutting plane method of J. E. Kelley, Jr., 1960, will be reproduced here. It distinguishes itself by being easily described, and graphically illustrated, and also applied with numerical convenience.

10.1. The Cutting Plane Method: Derivation and Proof of Convergence

The cutting plane method presupposes the following problem type.

Find an $\underset{\sim}{x} \in R^n$ such that

$$F(\underset{\sim}{x}) = \underset{\sim}{p}'\underset{\sim}{x} = \text{Min!}, \quad f_j(\underset{\sim}{x}) \leq 0 \quad (j = 1,\ldots,m). \qquad (10.1)$$

In contrast to the problem types previously handled, here the objective function is linear, and there are no positivity constraints, $\underset{\sim}{x} \geq \underset{\sim}{0}$. Nevertheless, this does not imply any loss of generality. For problems with positivity constraints, $\underset{\sim}{x} \geq \underset{\sim}{0}$, these can be added to the inequality constraints, $f_j(\underset{\sim}{x}) \leq 0$. For a problem with a given, non-linear convex objective function, $\tilde{F}(\underset{\sim}{x})$, add the inequality constraint $\tilde{F}(\underset{\sim}{x}) - x_{n+1} \leq 0$ to the constraints $f_j(\underset{\sim}{x}) \leq 0$, $j = 1,\ldots,m$, and in R^{n+1}, consider the problem with these $m+1$ constraints and the linear objective function, x_{n+1} = Min!. If $\underset{\sim}{\tilde{x}} \in R^n$ is a minimal solution of the original problem, $\begin{pmatrix} \underset{\sim}{\tilde{x}} \\ \tilde{x}_{n+1} \end{pmatrix}$, where $\tilde{x}_{n+1} = \tilde{F}(\underset{\sim}{\tilde{x}})$, is a minimal solution of the extended problem, and vice versa.

We will assume that the following conditions are satisfied; we shall need them in our investigation of the cutting plane method.

(A) The set of feasible points for problem (10.1),

$$M = \{\underset{\sim}{x} | f_j(\underset{\sim}{x}) \le 0 \quad (j = 1,\ldots,m)\},$$

is contained in a polyhedron $S_0 = \{\underset{\sim}{x} | \underset{\sim}{A}\underset{\sim}{x} \le \underset{\sim}{b}\}$. (By the definition in §2, a polyhedron is a bounded point set; M therefore is also bounded.) Here $\underset{\sim}{A}$ is a q-by-n matrix, and $\underset{\sim}{b} \in R^q$ (and $q > n$).

(B) The functions $f_j(\underset{\sim}{x})$ are convex and differentiable on the set S_0. The partial derivatives of the functions $f_j(\underset{\sim}{x})$ are bounded on S_0:

$$||\text{grad } f_j(\underset{\sim}{x})|| \le K \quad (\underset{\sim}{x} \in S_0;\ j = 1,\ldots,m). \tag{10.2}$$

(A knowledge of the bound K is not required for a numerical application of the method.)

The following rules generate a sequence of sets $S_t \subset R^n$ and a sequence of points $\underset{\sim}{x}^t \in S_t$ for $t = 0, 1, 2,\ldots$.

(I) For $t = 0, 1, 2,\ldots,$ let $\underset{\sim}{x}^t$ be a minimal solution of the linear optimization problem

$$F(\underset{\sim}{x}) = \underset{\sim}{p}'\underset{\sim}{x} = \text{Min!}, \quad \underset{\sim}{x} \in S_t. \tag{10.3}$$

(II) If k is an index for which $f_k(\underset{\sim}{x}^t) = \max\limits_{j=1,\ldots,m} f_j(\underset{\sim}{x}^t)$, let S_{t+1} be the intersection of the set S_t with the set of points, $\underset{\sim}{x} \in R^n$, which satisfy the linear inequality

$$f_k(\underset{\sim}{x}^t) + (\underset{\sim}{x}-\underset{\sim}{x}^t)'\text{grad } f_k(\underset{\sim}{x}^t) \le 0. \tag{10.4}$$

Rule (II) says that S_{t+1} is created from S_t by cutting off a piece of S_t with some hyperplane; hence the name "cutting plane method". This hyperplane in R^{n+1} is obtained as the intersection of the tangent hyperplane to the surface $z = f_k(\tilde{x})$ at the point $\tilde{x}^t$, with the hyperplane $z = 0$. In this sense, there is a similarity with Newton's method of iteration. If there are several indices j for which a maximum in rule (II) is attained, choose an arbitrary one of these as k. The set S_t will be described by the q linear inequalities, $\tilde{A}\tilde{x} \leq \tilde{b}$, and the t linear inequalities, (10.4); thus (10.3) truly is a linear optimization problem. It follows from rule (II) that S_{t+1} is contained in S_t, so that the set of feasible points is bounded for all these linear optimization problems, and is, in fact, contained in the bounded set S_0. If S_t is not empty, there exists at least one minimal solution, $\tilde{x}^t$. Should several minimal solutions exist, choose one arbitrarily as $\tilde{x}^t$. If one of the sets, S_t, is empty, it will follow that the set M also is empty, and thus, no solutions for problem (10.1) exist. If none of the sets S_t is empty, then by rules (I) and (II) one obtains sets S_t and points $\tilde{x}^t$, for every integer $t \geq 0$, for which the following assertions hold.

(a) $M \subset S_t$.

(b) Either $\tilde{x}^t$ is a minimal solution of (10.1), or $\tilde{x}^t \notin M$ and S_{t+1} is a proper subset of S_t.

Assertions (a) and (b) will be proven by induction on t.

<u>$t = 0$.</u> (a) is satisfied because of the choice of S_0. $\tilde{x}^0$ is a minimal solution of (10.3) for $t = 0$. If

$\underset{\sim}{x}^0 \in M$, $\underset{\sim}{x}^0$ is also a minimal solution of (10.1) because $M \subset S_0$. If $\underset{\sim}{x}^0 \notin M$, $f_k(\underset{\sim}{x}^0) > 0$ in (II), and $\underset{\sim}{x}^0$ does not satisfy inequality (10.4) for $t = 0$, and hence is not in S_1. Thus (b) is satisfied for $t = 0$. Figure 10.1 illustrates several steps of the process for the case $n = 2$; a square was chosen for S_0.

<u>Induction Step</u>. Let assertions (a) and (b) hold for some integer $t \geq 0$. If $\underset{\sim}{x}^t$ is a minimal solution of (10.1), $f_k(\underset{\sim}{x}^t) = 0$ in (II) (the minimum of a linear objective function is attained on the boundary of M); if $\underset{\sim}{x}^t \notin M$, $f_k(\underset{\sim}{x}^t) > 0$. In any case, $f_k(\underset{\sim}{x}^t) \geq 0$. If $\underset{\sim}{x} \in M$, $\underset{\sim}{x} \in S_t$ by the induction hypothesis; in addition, because $f_k(\underset{\sim}{x})$ is convex and theorem 4, §6.2 applies, $f_k(\underset{\sim}{x}^t) + (\underset{\sim}{x}-\underset{\sim}{x}^t)'\text{grad } f_k(\underset{\sim}{x}^t) \leq f_k(\underset{\sim}{x}) \leq 0$. Thus $\underset{\sim}{x} \in S_{t+1}$ also, and (a) holds for $t + 1$. That (b) also holds for $t+1$ is shown as for $t = 0$ above.

The following three cases are possible.

(1) Some S_t is empty. Assertion (a) clearly is still valid, and the set M must also be empty. The process should be terminated, since (10.1) has no solution.

(2) Some $\underset{\sim}{x}^t$ lies in M, and thus is a minimal solution of (10.1), by (b). The process may also be terminated.

(3) One obtains an infinite sequence of points $\underset{\sim}{x}^t$ and also a sequence of sets S_t, each of which is contained in the preceding.

In case (3), all the points $\underset{\sim}{x}^t$ of the sequence lie in the bounded set S_0. Thus there exists a convergent subsequence, whose limit we may denote by $\bar{\underset{\sim}{x}}$. $\bar{\underset{\sim}{x}}$ lies in

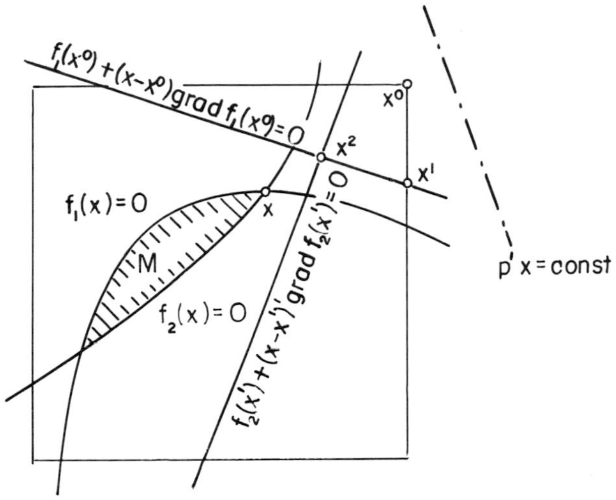

Figure 10.1. The cutting plane method

the intersection, S, of all the sets S_t. S is closed because it is the intersection of closed sets. It will become clear that $\bar{x}$ is a minimal solution of (10.1). First we see that $\bar{x} \in M$. For suppose that $\bar{x} \notin M$; then $\eta = \max_{j=1,\ldots,m} f_j(\bar{x}) > 0$. Let $\bar{k}$ be an index with $f_{\bar{k}}(\bar{x}) = \eta$. Since $f_{\bar{k}}(x)$ is continuous, there exists a point x^t in the sequence we have constructed for which

$$||x^t-\bar{x}|| < \frac{\eta}{2K} \quad \text{and} \quad f_k(x^t) > \frac{\eta}{2} .$$

(where $||x^t-x||$ is the euclidean length of the vector $x^t - x$, and K is the bound given in (10.2)). Let k be an index associated with x^t by rule (II), so that $f_k(x^t) > \eta/2$ also, and therefore

$$f_k(\underset{\sim}{x}^t) + (\bar{\underset{\sim}{x}}-\underset{\sim}{x}^t)'\text{grad } f_k(\underset{\sim}{x}^t) > \frac{\eta}{2} - \frac{\eta}{2K}\cdot K = 0.$$

$\bar{\underset{\sim}{x}}$ then does not belong to the set S_{t+1}, contradicting $\bar{\underset{\sim}{x}} \in S$. So we have proven that $\bar{\underset{\sim}{x}} \in M$.

We still must show that the objective function $\underset{\sim}{p}'\underset{\sim}{x}$ attains its minimum with respect to M at $\bar{\underset{\sim}{x}}$. This is clear, because it attains its minimum with respect to S at $\bar{\underset{\sim}{x}}$, and $M \subset S$.

Also note the following. Every accumulation point of the sequence of the $\underset{\sim}{x}^t$ is a minimal solution of (10.1). If the minimal solution $\bar{\underset{\sim}{x}}$ of (10.1) is uniquely determined, there is then only one accumulation point, and the sequence of the $\underset{\sim}{x}^t$ converges to $\bar{\underset{\sim}{x}}$. In the one dimensional case, the cutting plane method becomes Newton's method for the iterative determination of zeros.

10.2. On the Numerical Application of the Cutting Plane Method

At every step of the cutting plane process, we have to solve a linear optimization problem. The number of constraints increases by one at each of these steps. However, at various stages of the process, certain constraints may become dispensable. The apparently substantial computational effort may be contained within reasonable bounds by a transition to the dual linear optimization problem. Then problem (10.3) becomes

$$\left.\begin{aligned} F(\underset{\sim}{x}) &= \sum_{\ell=1}^{n} p_\ell x_\ell = \text{Min!}, \\ \sum_{\ell=1}^{n} a_{i\ell} x_\ell &\le b_i \quad (i = 1,\ldots,q), \\ \sum_{\ell=1}^{n} g_{\tau\ell} x_\ell &\le d_\tau \quad (\tau = 0,1,\ldots,t-1). \end{aligned}\right\} \tag{10.5}$$

Here the $a_{i\ell}$, $i = 1,\ldots,q$, $\ell = 1,\ldots,n$, are the elements of matrix $\underset{\sim}{A}$, the $g_{\tau\ell}$, $\tau = 0,1,\ldots,t-1$, $\ell = 1,\ldots,n$, are the components of the vector grad $f_k(\underset{\sim}{x}^\tau)$, and we have defined

$$d_\tau = \sum_{\ell=1}^{n} g_{\tau\ell} x_\ell^\tau - f_k(\underset{\sim}{x}^\tau) \qquad (\tau = 0,1,\ldots,t-1).$$

Note that the index k also depends on τ. We thus have a minimum problem with inequalities as constraints, and without positivity constraints. If we make this a maximum problem with objective function $\sum(-p_\ell)x_\ell$, we have exactly the problem type D^1 of §5.1. The dual problem D^0 then reads

$$\left.\begin{aligned} &\sum_{i=1}^{q} b_i u_i + \sum_{\tau=0}^{t-1} d_\tau v_\tau = \text{Min!} \\ &\sum_{i=1}^{q} a_{i\ell} u_i + \sum_{\tau=0}^{t-1} g_{\tau\ell} v_\tau = -p_\ell \quad (\ell = 1,\ldots,n) \\ &u_i \ge 0 \ (i = 1,\ldots,q), \quad v_\tau \ge 0 \ (\tau = 0,1,\ldots,t-1). \end{aligned}\right\} \tag{10.6}$$

This is a problem of the type for which we developed the simplex process in §3 and §4. This problem is to be solved for $t = 0,1,2,\ldots$. When $t = 0$, no v_τ appear. The number of constraints now remains the same, but a new variable v_τ appears at each step.

We then can use the process described in §4.5, taking the terminal tableau of the simplex method, expanding it by one column, and using this tableau as the initial tableau for the following step. If we choose the constraints, $Ax = b$, which define the polyhedron S_0, so as to include the inequalities $x \le c$ (e.g., if we choose S_0 to be the box defined by $\bar{c} \le x \le c$), then A will contain an identity matrix as a submatrix, and we can apply the simplification given in §4.5 when filling in the new column of the simplex tableau. Similarly, we can apply the method of §5.1, and easily determine the vector x^t, which is the solution of problem (10.5), the dual of (10.6).

This method, which is well suited to machine computations, is best for the general case of a convex optimization problem. For the special case of quadratic optimization, one would use the methods described in §14.

Example. We want to optimize the design and manufacture of a vehicle. It is to have a top speed of x_1, a performance of x_2, and development and production costs of x_3 (all measured in the appropriate units). For x_1, x_2, and x_3 we have the constraints

$$x_2 \ge \phi(x_1)$$
$$x_3 \ge \psi(x_2)$$

where $\phi(x)$ and $\psi(x)$ are convex functions of one variable. In addition, the costs should not exceed an amount a:

$$x_3 \le a.$$

We want to minimize a linear function,

$$bx_3 - cx_2 - dx_1,$$

with (non-negative) constant coefficients b, c, and d, which represents a trade-off between the expense and the gain, which is determined by the performance of the vehicle. This convex optimization problem, i.e.

$$\begin{aligned} F(\underset{\sim}{x}) &= bx_3 - cx_2 - dx_1 = \text{Min!} \\ \phi(x_1) - x_2 &\le 0 \\ \psi(x_2) - x_3 &\le 0 \\ x_3 - a &\le 0 \\ x_i &\ge 0 \quad (i = 1,2,3), \end{aligned}$$

was attacked with the cutting plane method for particular ϕ and ψ, namely $\phi(x) = \psi(x) = e^x$, and $a = 10$, and various values of b, c, and d. For S_0, the box determined by $0 \le x_1 \le 2$, $0 \le x_2 \le 3$, and $0 \le x_3 \le 10$, was chosen. The process was programmed, and the machine provided the following results, among others.

$b = c = 0, \quad d = 1$

t	x_1^t	x_2^t	x_3^t	$F(\underset{\sim}{x}^t)$
0	2	3	0	-2
1	2	2.498	10	-2
2	1.338	2.498	10	-1.338
3	1.314	2.320	10	-1.314
4	0.938	2.320	10	-0.938
5	0.846	2.320	10	-0.846
6	0.839	2.303	10	-0.839
7	0.834	2.303	10	-0.834
8	0.834	2.303	10	-0.834

Solution: $x_1 = \log\log 10 = 0.83403, \quad x_2 = \log 10 = 2.30259,$

$x_3 = 10.$

$b = 0.2, \quad c = 0, \quad d = 0.8$

t	x_1^t	x_2^t	x_3^t	$F(\underset{\sim}{x}^t)$
0	2	3	0	-1.6
1	2	2	0	-1.6
2	2	1	0	-1.6
3	1.135	1	0	-0.908
4	1	0	0	-0.8
5	0	0	0	0
6	0.214	0.582	1.582	0.145
7	0.491	1.582	4.300	0.467
8	0.279	1.319	3.586	0.494
9	0.157	1.168	3.175	0.509
10	0.219	1.245	3.465	0.517
11	0.189	1.207	3.412	0.517
12	0.173	1.188	3.279	0.517
.	.	.	.	.
.	.	.	.	.
.	.	.	.	.

Solution: $x_1 = 0.18413, \quad x_2 = 1.20217, \quad x_3 = 3.32733$

$(x_1 + e^{x_1} = \log 4).$

III. QUADRATIC OPTIMIZATION

Optimization problems with affine linear constraints and an objective function which is the sum of a linear function and a quadratic form with a positive semi-definite matrix, are in an intermediate position between linear and convex optimization problems. On the one hand, they are a special case of convex optimization, and all of the theorems of Chapter II naturally apply. On the other hand, they have certain properties which we recall from linear optimization, and which are no longer found in general convex optimization.

There also are a number of examples of applications which lead directly to such quadratic optimization problems, e.g., the example of milk utilization in the Netherlands, discussed in §6.

§11. Introduction

11.1. Definitions

We are given:

a real m-by-n matrix, $\underset{\sim}{A}$;

a vector, $\underset{\sim}{b} \in R^m$;

a vector, $\underset{\sim}{p} \in R^n$; and

a real, symmetric positive semi-definite n-by-n matrix, $\underset{\sim}{C}$.

This (finite amount of) data determines a quadratic optimization problem:

Find a vector $\underset{\sim}{x} \in R^n$ such that

$$Q(\underset{\sim}{x}) = \underset{\sim}{p}'\underset{\sim}{x} + \underset{\sim}{x}'\underset{\sim}{C}\underset{\sim}{x} = \text{Min!}, \quad \underset{\sim}{A}\underset{\sim}{x} \le \underset{\sim}{b}, \quad \underset{\sim}{x} \ge \underset{\sim}{0}. \qquad (11.1)$$

The objective function, which we denote here by $Q(\underset{\sim}{x})$ (instead of $F(\underset{\sim}{x})$), is convex by theorem 3, §6. It is also differentiable (arbitrarily often, in fact).

We will also consider quadratic optimization problems for which the constraints are in the form of equalities, $\underset{\sim}{A}\underset{\sim}{x} = \underset{\sim}{b}$, and also those for which some of the positivity constraints, $\underset{\sim}{x} \ge \underset{\sim}{0}$, are lacking. Problems for which the constraints no longer are affine linear, but also contain quadratic forms, do not fall under the concept of quadratic optimization.

11.2. Assignment Problems and Quadratic Optimization

There also exist quadratic optimization problems which do not satisfy the last-named hypothesis (the matrix $\underset{\sim}{C}$ is positive semi-definite) of §11.1. Examples include certain assignment problems. Such problems arise in creating class schedules for elementary and secondary schools, or in scheduling exams, as the following example will demonstrate.

Examinees a, b, c, ... are to be tested by examiners A, B, C, ...; we can show who will be examined by whom by simply marking X's in a table, e.g.,

	a	b	c	d	e	...
A	×	×	×			
B	×		×	×		
C	×	×			×	
⋮						

(11.2)

Thus B, for example, must examine persons a, c, and d.

In this way, every exam is described by a couple, e.g., Bc, and these exams can be numbered consecutively as "events": $x_1, x_2, \ldots, x_n$. Let there be q exam periods available, e.g., Monday from 9:00 to 10:00 a.m., from 10:00 to 11:00 a.m., etcetera. The exams now have to be distributed among the q periods so as to have the smallest possible number of "collisions", where a collision occurs when a given plan schedules two exams for one examinee at the same period, or when an examiner has two examinees assigned to her in one period. In either case, a new exam in a new period is required. Should two events, x_j and x_k, collide, we assign them a positive "resistance" r_{jk}; if they do not collide, let $r_{jk} = 0$. Events x_j which occur at the same exam period define a class, K_s, and have a class resistance of

$$\sum_{x_j, x_k \varepsilon K_s} r_{jk}.$$

Thus the events, $x_1, x_2, \ldots, x_n$, are to be distributed among q classes, $K_1, K_2, \ldots, K_q$, so as to minimize the total resistance,

$$r = \sum_{s=1}^{q} \sum_{x_j, x_k \varepsilon K_s} r_{jk}. \tag{11.3}$$

We now introduce an n-by-q matrix, $\underset{\sim}{X} = (x_{js})$, containing only 0's and 1's, which are determined by

$$x_{js} = \begin{cases} 1, & \text{if } x_j \text{ belongs to class } K_s \\ 0, & \text{if } x_j \text{ does not belong to } K_s. \end{cases}$$

Naturally every x_j must belong to exactly one class, i.e.,

$$\sum_{s=1}^{q} x_{js} = 1 \quad (j = 1, \ldots, n); \tag{11.4}$$

then the total resistance can be expressed by the comprehensive formula,

$$r = \sum_{j,k,s} r_{jk} x_{js} x_{ks}, \tag{11.5}$$

and we have a quadratic optimization problem. Given numbers $r_{jk} \geq 0$, find numbers $x_{js} = 0$ or 1, such that the quadratic function (11.5) is minimized, subject to the constraints (11.4). This problem is not one considered in §11.1; generally (11.5) is not a positive definite quadratic form. One method for handling such problems, which was used successfully for exams with many hundreds of examinees, and executed on a large computer, was constructed by Kirchgässner,

1965. He uses results from graph theory, of which we will mention only the connection with the coloring problem for graphs. Draw a graph in which the nodes are the events, $x_1,\ldots,x_n$. Connect x_j and x_k with an edge iff they could be in collision, i.e., iff they are in the same row or in the same column of table (11.2). For example, the part of table (11.2) which is complete would correspond to the part of a graph shown in Figure 11.1. Now, to every finite

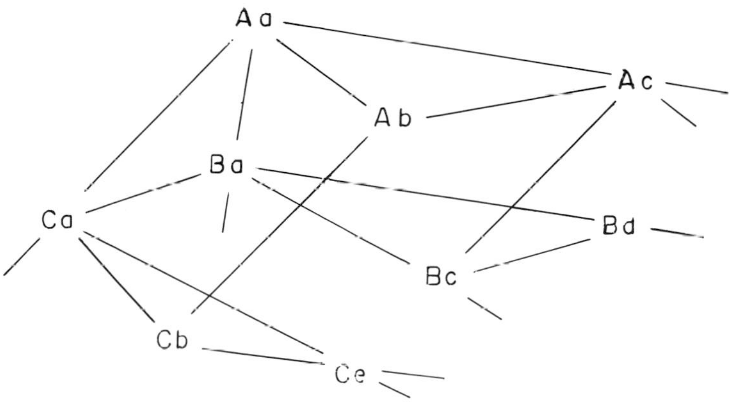

Figure 11.1

graph one can assign a "chromatic number" γ: γ is the smallest number with the property that every node can be colored with one of a total of γ colors in such a way that every edge has differently colored endpoints. If the chromatic number of the graph corresponding to the assignment problem is $\gamma \le q$, the problem has an ideal solution. One achieves $r = 0$, and every collision is avoided, by assigning every color to one of the classes, $K_1,\ldots,K_q$. But if $\gamma > q$, collisions are inevitable and the graph must have sufficient edges removed to reduce its chromatic number to q.

Which edges must be removed to minimize r depends on the numbers r_{jk}. Kirchgässner, 1965, provides a theory for an effective process with the help of the "critical" $(q+1)$-chromatic subgraphs of the complete graph.

§12. The Kuhn-Tucker Theorem and Applications

12.1. The Specialization of the Kuhn-Tucker Theorem to Quadratic Optimization Problems

Since the objective function and the constraints in problem (11.1) are differentiable, theorem 1, §8.1, is applicable. Since the constraints are affine linear, condition (V) of that theorem is dispensable, by §9.

For the case of problem (11.1), the Lagrange function, defined in §7 for convex optimization problems, becomes

$$\Phi(\underset{\sim}{x},\underset{\sim}{u}) = \underset{\sim}{p}'\underset{\sim}{x} + \underset{\sim}{x}'\underset{\sim}{C}\underset{\sim}{x} + \underset{\sim}{u}'(\underset{\sim}{A}\underset{\sim}{x} - \underset{\sim}{b}), \qquad (12.1)$$

where $\underset{\sim}{u}$ is a vector in R^m. As gradients Φ_x and Φ_u, one obtains

$$\underset{\sim}{v} = \Phi_x = \underset{\sim}{p} + 2\underset{\sim}{C}\underset{\sim}{x} + \underset{\sim}{A}'\underset{\sim}{u},$$
$$\underset{\sim}{y} = -\Phi_u = -\underset{\sim}{A}\underset{\sim}{x} + \underset{\sim}{b}.$$

Conditions (8.3) and (8.4) then read

$$\underset{\sim}{v}^0 = \Phi_x(\underset{\sim}{x}^0,\underset{\sim}{u}^0) \geq \underset{\sim}{0}, \quad \underset{\sim}{x}^{0'}\underset{\sim}{v}^0 = 0,$$
$$\underset{\sim}{y}^0 = -\Phi_u(\underset{\sim}{x}^0,\underset{\sim}{u}^0) \geq \underset{\sim}{0}, \quad \underset{\sim}{u}^{0'}\underset{\sim}{y}^0 = 0.$$

The two conditions $\underset{\sim}{x}^{0'}\underset{\sim}{v}^0 = \underset{\sim}{u}^{0'}\underset{\sim}{y}^0 = 0$ may be summarized as $\underset{\sim}{x}^{0'}\underset{\sim}{v}^0 + \underset{\sim}{u}^{0'}\underset{\sim}{y}^0 = 0$, since all the summands in the inner products are non-negative.

Thus we obtain

<u>Theorem 1</u>: A vector $x^0 \geq 0$ $(x^0 \in R^n)$ is a minimal solution of the quadratic optimization problem (11.1) iff there exist vectors, $u^0 \in R^m$, $v^0 \in R^n$, and $y^0 \in R^m$, such that

$$\left.\begin{array}{c} Ax^0 + y^0 = b, \quad v^0 - 2Cx^0 - A'u^0 = p, \\ u^0 \geq 0, \quad v^0 \geq 0, \quad y^0 \geq 0, \end{array}\right\} \tag{12.2}$$

$$x^{0\prime}v^0 + u^{0\prime}y^0 = 0. \tag{12.3}$$

<u>Remark</u>. (12.2) contains only affine linear conditions and (12.3) is the only non-linear condition.

12.2. Existence of a Solution and an Inclusion Theorem

For completeness, we include the following theorem on the existence of a solution for the quadratic optimization problem (11.1).

<u>Theorem 2</u>: The quadratic optimization problem (11.1) has a minimal solution iff there is a solution for (12.2) with vectors $x^0, v^0 \in R^n$ and $u^0, y^0 \in R^m$ where $x^0 \geq 0$.

<u>Proof</u>. I. Let x^0 be a minimal solution of (11.1). By theorem 1, (12.2) is solvable.

II. Let $x^0 \geq 0$, v^0, u^0, y^0 be a solution of (12.2). $y^0 \geq 0$ implies $Ax^0 \leq b$, so that x^0 is a feasible vector, and the set of feasible vectors is not empty. Since $Q(x)$ is convex, we have for feasible x that

$$Q(x) - Q(x^0) \geq (x-x^0)'\text{grad } Q(x^0) = (x-x^0)'(p+2Cx^0)$$

$$= (\tilde{x}-\tilde{x}^0)'(\tilde{v}^0 - \tilde{A}'\tilde{u}^0)$$
$$= \tilde{x}'\tilde{v}^0 - \tilde{x}^{0'}\tilde{v}^0 - (\tilde{A}\tilde{x}-\tilde{b})'\tilde{u}^0 + (\tilde{A}\tilde{x}^0-\tilde{b})'\tilde{u}^0$$
$$\geq -\tilde{x}^0\tilde{v}^0 - \tilde{y}^{0'}\tilde{u}^0.$$

$Q(\tilde{x})$ is thus bounded below on the set M of feasible vectors. This implies that $Q(\tilde{x})$ attains its minimum on M, by a theorem of Barankin and Dorfman, 1958, which we prove in the appendix.

<u>Corollary</u>. Let $\tilde{x}^0 \geq \tilde{0}$, $\tilde{v}^0$, $\tilde{u}^0$, $\tilde{y}^0$ be a solution of (12.2). Let $\tilde{x}^1$ be a minimal solution of problem (11.1). Then

$$Q(\tilde{x}^0) - (\tilde{x}^{0'}\tilde{v}^0 + \tilde{y}^{0'}\tilde{u}^0) \leq Q(\tilde{x}^1) \leq Q(\tilde{x}^0). \qquad (12.4)$$

This corollary is an inclusion theorem for the minimal value of the quadratic optimization problem (11.1). The expression $\tilde{x}^{0'}\tilde{v}^0 + \tilde{y}^{0'}\tilde{u}^0$, which determines the precision of the inclusion, is exactly the one which vanishes, by theorem 1, in the presence of a minimal solution $\tilde{x}^0$ and appropriately chosen $\tilde{v}^0$, $\tilde{y}^0$, and $\tilde{u}^0$. As the following example shows, this inclusion theorem is even of numerical use.

<u>Example</u>. Find an $\tilde{x} = \begin{pmatrix} x_1 \\ x_2 \end{pmatrix} \varepsilon R^2$ such that

$$\begin{aligned} x_1 + x_2 &\leq 8, \qquad x_1 \geq 0,\ x_2 \geq 0, \\ x_1 &\leq 6, \\ x_1 + 3x_2 &\leq 18, \\ Q(\tilde{x}) = 2x_1^2 + x_2^2 - 48x_1 - 40x_2 &= \text{Min!} \end{aligned}$$

Using the notation introduced in §11, we have

$$\underset{\sim}{A} = \begin{pmatrix} 1 & 1 \\ 1 & 0 \\ 1 & 3 \end{pmatrix}, \quad \underset{\sim}{C} = \begin{pmatrix} 2 & 0 \\ 0 & 1 \end{pmatrix}, \quad \underset{\sim}{b} = \begin{pmatrix} 8 \\ 6 \\ 18 \end{pmatrix}, \quad \underset{\sim}{p} = \begin{pmatrix} -48 \\ -40 \end{pmatrix}.$$

If we choose $\underset{\sim}{x}^0 = \begin{pmatrix} 3 \\ 5 \end{pmatrix}$ (a vertex of M, Figure 12.1),

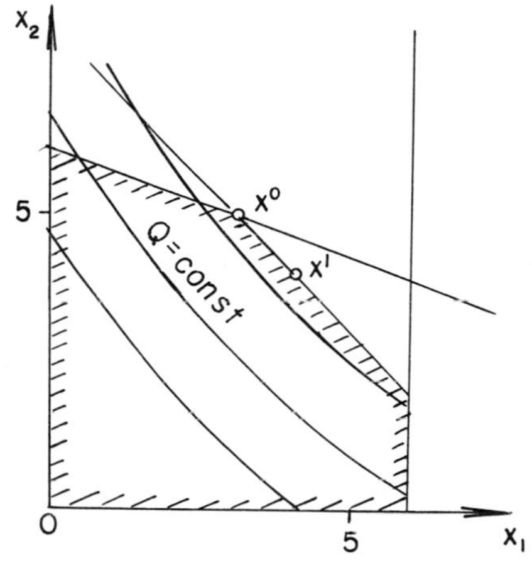

Figure 12.1

then $\underset{\sim}{y}^0 = \underset{\sim}{b} - \underset{\sim}{A}\underset{\sim}{x}^0 = (0, 3, 0)'$. Letting $\underset{\sim}{v}^0 = \underset{\sim}{0}$, one nonnegative solution of $\underset{\sim}{v}^0 - 2\underset{\sim}{C}\underset{\sim}{x}^0 - \underset{\sim}{A}'\underset{\sim}{u}^0 = \underset{\sim}{p}$ turns out to be the vector $\underset{\sim}{u}^0 = (30, 6, 0)'$. Since $Q(x^0) = -301$, we obtain the following bounds for the minimal value $Q(\underset{\sim}{x}^1)$.

$$-301 - 3 \cdot 6 = -319 \leq Q(\underset{\sim}{x}^1) \leq -301.$$

The minimal solution, as is easily seen from theorem 1, is $\underset{\sim}{x}^1 = \begin{pmatrix} 4 \\ 4 \end{pmatrix}$ with $Q(\underset{\sim}{x}^1) = -304$.

With the help of theorem 3, §8.2, we can obtain a good

overview of the totality of solutions for a quadratic optimization problem. Let x^0 be a minimal solution. Since the gradient of the objective function is

$$\operatorname{grad} Q(x) = p + 2Cx,$$

a feasible point $x^0 + y$ will be a minimal solution iff $\operatorname{grad} Q(x^0) = \operatorname{grad} Q(x^0+y)$, so that $Cy = 0$, and also $y'\operatorname{grad} Q(x^0) = 0$, so that $y'p + 2y'Cx^0 = 0$. Since the matrix C is symmetric, $y'C = 0$. Thus we obtain

Theorem 3: If x^0 is a minimal solution, then a feasible point $x^0 + y$ is also a minimal solution iff $Cy = 0$ and $p'y = 0$.

The set of minimal solutions is thus the intersection of a linear manifold with the set of feasible points.

If the matrix C is positive definite, $Cy = 0$ only if $y = 0$. In that case there is at most one minimal solution. This also follows, incidentally, from the fact that the objective function is then strongly convex.

12.3. The Kuhn-Tucker Theorem for Quadratic Optimization Problems with Various Types of Constraints

A. Constraints in the Form of Equalities. With the notation of §11, we have the following theorem for the problem

$$Q(x) = p'x + x'Cx = \text{Min!}, \quad Ax = b, \; x \geq 0. \tag{12.5}$$

Theorem 4: A vector $x^0 \geq 0$ $(x^0 \in R^n)$ is a minimal solution of (12.5) iff $Ax^0 = b$ and there exist vectors,

$\underset{\sim}{u}^0 \in R^m$ and $\underset{\sim}{v}^0 \in R^n$, such that

$$\underset{\sim}{v}^0 - 2\underset{\sim}{C}\underset{\sim}{x}^0 - \underset{\sim}{A}'\underset{\sim}{u}^0 = \underset{\sim}{p}, \quad \underset{\sim}{v}^0 \geq \underset{\sim}{0}, \tag{12.6}$$

$$\underset{\sim}{x}^{0'}\underset{\sim}{v}^0 = 0 \tag{12.7}$$

(no positivity constraints on $\underset{\sim}{u}^0$!).

<u>Proof</u>: Replace $\underset{\sim}{A}\underset{\sim}{x} = \underset{\sim}{b}$ by the inequalities $\underset{\sim}{A}\underset{\sim}{x} \leq \underset{\sim}{b}$ and $-\underset{\sim}{A}\underset{\sim}{x} \leq -\underset{\sim}{b}$, and apply theorem 1, §12.1, to conclude that $\underset{\sim}{x}^0$ is a minimal solution iff there exist vectors $\bar{\underset{\sim}{u}}^0$, $\bar{\bar{\underset{\sim}{u}}}^0$, $\bar{\underset{\sim}{y}}^0$, $\bar{\bar{\underset{\sim}{y}}}^0 \in R^m$ and $\underset{\sim}{v}^0 \in R^n$ such that

$$\underset{\sim}{A}\underset{\sim}{x}^0 + \bar{\underset{\sim}{y}}^0 = \underset{\sim}{b}, \quad -\underset{\sim}{A}\underset{\sim}{x}^0 + \bar{\bar{\underset{\sim}{y}}}^0 = -\underset{\sim}{b},$$
$$\underset{\sim}{v}^0 - 2\underset{\sim}{C}\underset{\sim}{x}^0 - \underset{\sim}{A}'\bar{\underset{\sim}{u}}^0 + \underset{\sim}{A}'\bar{\bar{\underset{\sim}{u}}}^0 = \underset{\sim}{p},$$
$$\bar{\underset{\sim}{u}}^0 \geq \underset{\sim}{0}, \quad \bar{\bar{\underset{\sim}{u}}}^0 \geq \underset{\sim}{0}, \quad \underset{\sim}{v}^0 \geq \underset{\sim}{0}, \quad \bar{\underset{\sim}{y}}^0 \geq \underset{\sim}{0}, \quad \bar{\bar{\underset{\sim}{y}}}^0 \geq \underset{\sim}{0},$$
$$\underset{\sim}{x}^{0'}\underset{\sim}{v}^0 + \bar{\underset{\sim}{u}}^{0'}\bar{\underset{\sim}{y}}^0 + \bar{\bar{\underset{\sim}{u}}}^{0'}\bar{\bar{\underset{\sim}{y}}}^0 = 0.$$

These conditions are satisfied iff $\bar{\underset{\sim}{y}}^0 = \bar{\bar{\underset{\sim}{y}}}^0 = \underset{\sim}{0}$ and $\underset{\sim}{u}^0 = \bar{\underset{\sim}{u}}^0 - \bar{\bar{\underset{\sim}{u}}}^0$ satisfies conditions (12.6) and (12.7).

B. <u>Variables without Positivity Constraints</u>. For a problem of the form

$$Q(\underset{\sim}{x}) = \underset{\sim}{p}'\underset{\sim}{x} + \underset{\sim}{x}'\underset{\sim}{C}\underset{\sim}{x} = \text{Min!}, \quad \underset{\sim}{A}\underset{\sim}{x} \leq \underset{\sim}{b}, \tag{12.8}$$

where there are no positivity constraints on $\underset{\sim}{x}$, we have the

<u>Theorem 5</u>: A vector $\underset{\sim}{x}^0 \in R^n$ is a minimal solution of (12.8) iff there exist vectors $\underset{\sim}{u}^0$, $\underset{\sim}{y}^0 \in R^m$ such that

$$\begin{gathered} \underset{\sim}{A}\underset{\sim}{x}^0 + \underset{\sim}{y}^0 = \underset{\sim}{b}, \quad -2\underset{\sim}{C}\underset{\sim}{x}^0 - \underset{\sim}{A}'\underset{\sim}{u}^0 = \underset{\sim}{p}, \\ \underset{\sim}{u}^0 \geq \underset{\sim}{0}, \quad \underset{\sim}{y}^0 \geq \underset{\sim}{0}, \quad \underset{\sim}{u}^{0'}\underset{\sim}{y}^0 = 0. \end{gathered} \tag{12.9}$$

The proof of this theorem is similar to the proof of

theorem 1. The idea is to let $x = \bar{x} - \bar{\bar{x}}$, where $\bar{x} \geq 0$ and $\bar{\bar{x}} \geq 0$. The details we leave to the reader.

§13. Duality for Quadratic Optimization

For every quadratic optimization problem, one can find a dual, which is again a quadratic optimization. The behavior of the solutions of the two problems can be described by theorems which are similar to those in §5 on dual problems in linear optimization. However, in the present case, we no longer have the property that the second dual is identical to the original problem.

Nevertheless, theorem 5 contains a symmetric condition for duality. The duality theorem has not found as many applications in quadratic optimization, as the corresponding theorem for linear problems. Yet, like the theorem for linear problems, it provides a convenient means of finding upper and lower bounds for the extreme values, which makes it significant for numerical purposes. Although these have already been presented, in §12.2, they could also be derived from the duality theorem 3 below.

13.1. Formulating the Dual Problem

As in §11, we are given the problem

D^0: Find an $x \in R^n$ such that

$$Q(x) = p'x + x'Cx = \text{Min!}, \quad f(x) = Ax - b \leq 0.$$

The Lagrange function for this problem is

$$\Phi(x,u) = p'x + x'Cx + u'(Ax - b).$$

Let Φ_x be defined as in 12.1. The dual problem is

D^1: Find $w \in R^n$ and $u \in R^m$ (in brief: Find (w,u) ...) such that

$$\Phi(w,u) = p'w + w'Cw + u'(Aw-b) = \text{Max!},$$
$$\Phi_x(w,u) = 2Cw + A'u + p = 0, \quad u \geq 0.$$

For a maximum problem in quadratic optimization, we require the objective function to be concave, i.e., the negative of a convex function, so as to stay within the bounds of previously constructed theory. At first glance, it would appear that the objective function, $\Phi(w,u)$, in D^1 is not convave. But after considering the equivalent reformulation as problem $\tilde{D}^1$ in the proof of theorem 3, we see that the objective function of D^1 is at least concave on the linear submanifold defined by the constraints $\Phi_x(w,u) = 0$, even if not on all of R^{n+m}.

<u>Theorem 1</u>: Let D^0 and D^1 each have at least one feasible vector. If x is feasible for D^0 and (w,u) is feasible for D^1, then

$$Q(x) \geq \Phi(w,u).$$

(The analog of theorem 1, §5.1).

<u>Proof</u>: Let x and (w,u) be feasible for D^0 and D^1, respectively. $\Phi(t,u)$ is a convex function of $t \in R^n$. Therefore, $\Phi_x(w,u) = 0$ implies

$$\Phi(\underset{\sim}{w},\underset{\sim}{u}) = \underset{t \in R^n}{\text{Min}}\ \Phi(\underset{\sim}{t},\underset{\sim}{u}). \tag{13.1}$$

But then

$$\Phi(\underset{\sim}{w},\underset{\sim}{u}) \leq \Phi(\underset{\sim}{x},\underset{\sim}{u}) = Q(\underset{\sim}{x}) + \underset{\sim}{u}'(\underset{\sim}{A}\underset{\sim}{x} - \underset{\sim}{b}) \leq Q(\underset{\sim}{x}).$$

An immediate consequence of theorem 1 is

Theorem 2: Let $\underset{\sim}{x}^0$ and $(\underset{\sim}{w}^0,\underset{\sim}{u}^0)$ be feasible for D^0 and D^1 respectively. If

$$Q(\underset{\sim}{x}^0) = \Phi(\underset{\sim}{w}^0,\underset{\sim}{u}^0),$$

then $\underset{\sim}{x}^0$ is a solution for D^0 and $(\underset{\sim}{w}^0,\underset{\sim}{u}^0)$ is a solution for D^1 (and therefore the extreme values of the two dual problem are the same).

13.2. The Duality Theorem

The following theorem (like theorem 2, §5) contains the main result.

Theorem 3: D^0 has a finite minimal solution iff D^1 has a finite maximal solution; and these extreme values are equal, if they exist.

Proof: I. Let D^0 have a finite minimal solution, $\underset{\sim}{x}^0$. By theorem 5, §12.3, there exist vectors $\underset{\sim}{u}^0,\underset{\sim}{y}^0 \in R^m$ such that

$$\left.\begin{array}{r} \underset{\sim}{A}\underset{\sim}{x}^0 + \underset{\sim}{y}^0 = \underset{\sim}{b}, \quad 2\underset{\sim}{C}\underset{\sim}{x}^0 + \underset{\sim}{A}'\underset{\sim}{u}^0 + \underset{\sim}{p} = \underset{\sim}{0}, \\ \underset{\sim}{u}^0 \geq \underset{\sim}{0}, \quad \underset{\sim}{y}^0 \geq \underset{\sim}{0}, \end{array}\right\} \tag{13.2}$$

$$\underset{\sim}{u}^{0'}\underset{\sim}{y}^0 = 0. \tag{13.3}$$

If we set $\underset{\sim}{w}^0 = \underset{\sim}{x}^0$, it follows from (13.2) that

(w^0,u^0) is feasible for D^1. By (13.3) and $y^0 = b - Aw^0$,

$$\Phi(w^0,u^0) = Q(w^0) + u^{0'}(Aw^0-b) = Q(w^0) = Q(x^0). \quad (13.4)$$

It follows from this that (w^0,u^0) is a solution of D^1, by theorem 2.

II. Let D^1 have a finite maximal solution, (w^0,u^0). If (w,u) is feasible for D^1, then $A'u = -p - 2Cw$, and therefore,

$$\Phi(w,u) = p'w + w'Cw - p'w - 2w'Cw - u'b = -b'u - w'Cw.$$

D^1 can thus be replaced by the following, equivalent problem.

$\tilde{D}^1$: Find (w,u) such that

$$-\Phi(w,u) = b'u + w'Cw = \text{Min!},$$
$$2Cw + A'u + p = 0, \quad u \geq 0.$$

Set $w = w^+ - w^-$, where $w^+ \geq 0$ and $w^- \geq 0$. Then $\tilde{D}^1$ is equivalent to the following problem.

$\hat{D}^1$: Find $w^+, w^- \in R^n$ and $u \in R^m$ such that

$$\begin{pmatrix} 0 \\ 0 \\ b \end{pmatrix}' \begin{pmatrix} w^+ \\ w^- \\ u \end{pmatrix} + \begin{pmatrix} w^+ \\ w^- \\ u \end{pmatrix}' \begin{pmatrix} C & -C & 0 \\ -C & C & 0 \\ 0 & 0 & 0 \end{pmatrix} \begin{pmatrix} w^+ \\ w^- \\ u \end{pmatrix} = \text{Min!}$$

$$(2C \mid -2C \mid A') \begin{pmatrix} w^+ \\ w^- \\ u \end{pmatrix} = -p, \quad \begin{pmatrix} w^+ \\ w^- \\ u \end{pmatrix} \geq 0.$$

Here the first matrix consists of nine submatrices, where each 0 is a zero matrix of appropriate size.

Now $\hat{D}^1$ is a problem of the type considered in §12.3A. By theorem 4 of that section, the vector in R^{2n+m} given by

$$\begin{pmatrix} \underset{\sim}{w}^{+0} \\ \underset{\sim}{w}^{-0} \\ \underset{\sim}{u}^{0} \end{pmatrix} \geq \underset{\sim}{0} \tag{13.5}$$

is a minimal solution of D^1 iff there exist vectors

$$\underset{\sim}{z}^0 \in R^n, \begin{pmatrix} \underset{\sim}{v}^{+0} \\ \underset{\sim}{v}^{-0} \\ \underset{\sim}{y}^{0} \end{pmatrix} \in R^{2n+m} \quad (\underset{\sim}{v}^{+0}, \underset{\sim}{v}^{-0} \in R^n, \ \underset{\sim}{y}^0 \in R^m)$$

such that

$$(2\underset{\sim}{C}|-2\underset{\sim}{C}|\underset{\sim}{A}') \begin{pmatrix} \underset{\sim}{w}^{+0} \\ \underset{\sim}{w}^{-0} \\ \underset{\sim}{u}^{0} \end{pmatrix} = -\underset{\sim}{p}, \begin{pmatrix} \underset{\sim}{v}^{+0} \\ \underset{\sim}{v}^{-0} \\ \underset{\sim}{y}^{0} \end{pmatrix} \geq \underset{\sim}{0}, \tag{13.6}$$

$$\begin{pmatrix} \underset{\sim}{v}^{+0} \\ \underset{\sim}{v}^{-0} \\ \underset{\sim}{y}^{0} \end{pmatrix} - 2 \begin{pmatrix} \underset{\sim}{C} & -\underset{\sim}{C} & \underset{\sim}{0} \\ -\underset{\sim}{C} & \underset{\sim}{C} & \underset{\sim}{0} \\ \underset{\sim}{0} & \underset{\sim}{0} & \underset{\sim}{0} \end{pmatrix} \begin{pmatrix} \underset{\sim}{w}^{+0} \\ \underset{\sim}{w}^{-0} \\ \underset{\sim}{u}^{0} \end{pmatrix} - \begin{pmatrix} 2\underset{\sim}{C} \\ -2\underset{\sim}{C} \\ \underset{\sim}{A} \end{pmatrix} \underset{\sim}{z}^0 = \begin{pmatrix} \underset{\sim}{0} \\ \underset{\sim}{0} \\ \underset{\sim}{b} \end{pmatrix} \tag{13.7}$$

$$\underset{\sim}{w}^{+0'}\underset{\sim}{v}^{+0} + \underset{\sim}{w}^{-0'}\underset{\sim}{v}^{-0} + \underset{\sim}{u}^{0'}\underset{\sim}{y}^{0} = 0. \tag{13.8}$$

Write (13.7) as three equations (in $\underset{\sim}{v}^{+0}$, $\underset{\sim}{v}^{-0}$, and $\underset{\sim}{y}^0$) and add the equations in $\underset{\sim}{v}^{+0}$ and $\underset{\sim}{v}^{-0}$. This yields $\underset{\sim}{v}^{+0} + \underset{\sim}{v}^{-0} = \underset{\sim}{0}$. Together with $\underset{\sim}{v}^{+0} \geq 0$ and $\underset{\sim}{v}^{-0} \geq 0$, this implies $\underset{\sim}{v}^{+0} = \underset{\sim}{v}^{-0} = \underset{\sim}{0}$. Condition (13.8) is thus reduced to $\underset{\sim}{u}^{0'}\underset{\sim}{y}^0 = 0$.

Since $\tilde{D}^1$ and $\hat{D}^1$ are equivalent problems, the vector given by (13.5) is a solution of $\hat{D}^1$ iff the vector $(\underset{\sim}{w}^0, \underset{\sim}{u}^0)$, where $\underset{\sim}{w}^0 = \underset{\sim}{w}^{+0} - \underset{\sim}{w}^{-0}$, is a solution of $\tilde{D}^1$. Conditions (13.6) through (13.8) thus become

$$2\underset{\sim}{C}\underset{\sim}{w}^0 + \underset{\sim}{A}'\underset{\sim}{u}^0 + \underset{\sim}{p} = \underset{\sim}{0}, \quad \underset{\sim}{y}^0 \geq \underset{\sim}{0}, \tag{13.6'}$$

$$2\underset{\sim}{C}\underset{\sim}{w}^0 + 2\underset{\sim}{C}\underset{\sim}{z}^0 = \underset{\sim}{0}, \quad \underset{\sim}{A}\underset{\sim}{z}^0 - \underset{\sim}{y}^0 + \underset{\sim}{b} = \underset{\sim}{0}, \tag{13.7'}$$

$$\underset{\sim}{u}^{0'}\underset{\sim}{y}^0 = 0. \tag{13.8'}$$

Because of the equivalence of problems D^1, $\tilde{D}^1$, and $\hat{D}^1$, the assumption that $(\underset{\sim}{w}^0,\underset{\sim}{u}^0)$ is a maximal solution of D^1 implies the existence of vectors $\underset{\sim}{z}^0 \in R^n$ and $\underset{\sim}{y}^0 \in R^m$ for which (13.6') through (13.8') hold. Let $\underset{\sim}{x}^0 = -\underset{\sim}{z}^0$ and (13.7') implies that $\underset{\sim}{A}\underset{\sim}{x}^0 - \underset{\sim}{b} = -\underset{\sim}{y}^0 \leq \underset{\sim}{0}$. Thus $\underset{\sim}{x}^0$ is a feasible vector for D^0. It also follows from (13.7') that

$$\underset{\sim}{C}\underset{\sim}{w}^0 = -\underset{\sim}{C}\underset{\sim}{z}^0 = \underset{\sim}{C}\underset{\sim}{x}^0. \tag{13.9}$$

Therefore, by (13.6'),

$$\Phi_x(\underset{\sim}{x}^0,\underset{\sim}{u}^0) = 2\underset{\sim}{C}\underset{\sim}{x}^0 + \underset{\sim}{A}'\underset{\sim}{u}^0 + \underset{\sim}{p} = 2\underset{\sim}{C}\underset{\sim}{w}^0 + \underset{\sim}{A}'\underset{\sim}{u}^0 + \underset{\sim}{p} = 0$$

and $(\underset{\sim}{x}^0,\underset{\sim}{u}^0)$ is a feasible vector for D^1. Finally, by (13.7') and (13.8'),

$$\Phi(\underset{\sim}{x}^0,\underset{\sim}{u}^0) = Q(\underset{\sim}{x}^0) + \underset{\sim}{u}^{0'}(\underset{\sim}{A}\underset{\sim}{x}^0-\underset{\sim}{b}) = Q(\underset{\sim}{x}^0) - \underset{\sim}{u}^{0'}\underset{\sim}{y}^0 = Q(\underset{\sim}{x}^0).$$

By theorem 2, $\underset{\sim}{x}^0$ is a minimal solution of D^0 (and in addition to $(\underset{\sim}{w}^0,\underset{\sim}{u}^0)$, $(\underset{\sim}{x}^0,\underset{\sim}{u}^0)$ is also a maximal solution of D^1).

Corollary. Let $\underset{\sim}{C}$ be positive definite. If $(\underset{\sim}{w}^0,\underset{\sim}{u}^0)$ is a solution of D^1, then $\underset{\sim}{w}^0$ is a solution of D^0.

Proof: A positive definite matrix is non-singular. So the conclusion follows from (13.9).

Theorem 4: Suppose D^0 has no feasible vectors. Then there are two possible cases:

1) D^1 also has no feasible vectors; or

2) Φ is not bounded above on the set M^1 of feasible vectors, $(\underset{\sim}{w},\underset{\sim}{u})$, for D^1.

Proof: If M^1 is not empty and Φ is bounded above on M^1, we may conclude that Φ attains its maximum on M^1, by the theorem which we used in §12.2 and prove in the appendix. In that case, D^0 also has a solution, by theorem 3, and therefore, a feasible vector.

Remark. The duality properties of linear optimization problems may be derived by specializing the results obtained here. If we choose the zero matrix for C, problem D^0 and problem $\tilde{D}^1$, which is equivalent to D^1, become linear optimization problems which are dual by §5.1.

The converse of theorem 4 also holds. If D^1 has no feasible vectors, then either D^0 also has no feasible vectors, or the objective function $Q(\underset{\sim}{x})$ of D^0 is not bounded below on the set M^0 of feasible vectors for D^0.

13.3. A Symmetric Condition for Duality

The unsymmetric form of the duality theorem 3 was based on problems D^0 and D^1. In contrast, we can obtain a symmetric form as a consequence of Stoer, 1963, 1964, where duality for general convex optimization problems is also considered.

Theorem 5: The assertion of the duality theorem 3 is

equivalent to the assertion

$$\underset{\underset{\sim}{u} \geq 0}{\text{Max}} \ \underset{\underset{\sim}{x} \in R^n}{\text{Min}} \ \Phi(\underset{\sim}{x},\underset{\sim}{u}) = \underset{\underset{\sim}{x} \in R^n}{\text{Min}} \ \underset{\underset{\sim}{u} \geq 0}{\text{Max}} \ \Phi(\underset{\sim}{x},\underset{\sim}{u}).$$

Proof: I.   By (13.1), D^1 is equivalent to the problem: Find a $(\underset{\sim}{w}^0,\underset{\sim}{u}^0)$ such that

$$\Phi(\underset{\sim}{w}^0,\underset{\sim}{u}^0) = \underset{\underset{\sim}{u} \geq 0}{\text{Max}} \ \underset{\underset{\sim}{x} \in R^n}{\text{Min}} \ \Phi(\underset{\sim}{x},\underset{\sim}{u}).$$

II.   If $\underset{\sim}{x} \in R^n$ is a vector for which $\underset{\sim}{Ax} - \underset{\sim}{b} \leq \underset{\sim}{0}$
is false, that is, for which at least one component of
$\underset{\sim}{Ax} - \underset{\sim}{b}$ is positive, then, with the appropriate choice of
$\underset{\sim}{u} \geq \underset{\sim}{0}$, the expression $\underset{\sim}{u}'(\underset{\sim}{Ax}-\underset{\sim}{b})$ can be made arbitrarily
large.   If the set of feasible vectors for D^0 is not
empty, there must then be an $\underset{\sim}{x} \in R^n$ for which $\underset{\sim}{Ax} - \underset{\sim}{b} \leq \underset{\sim}{0}$.
Therefore, in considering

$$\underset{\underset{\sim}{x} \in R^n}{\text{Min}} \ \underset{\underset{\sim}{u} \geq 0}{\text{Max}} \ \Phi(\underset{\sim}{x},\underset{\sim}{u})$$

it suffices to take the minimum with respect to those $\underset{\sim}{x} \in$
R^n for which $\underset{\sim}{Ax} - \underset{\sim}{b} \leq \underset{\sim}{0}$ is valid.   But for such $\underset{\sim}{x}$, the
maximum with respect to $\underset{\sim}{u} \geq \underset{\sim}{0}$ is always attained at $\underset{\sim}{u} = \underset{\sim}{0}$;
therefore,

$$\underset{\underset{\sim}{x} \in R^n}{\text{Min}} \ \underset{\underset{\sim}{u} > 0}{\text{Max}} \ \Phi(\underset{\sim}{x},\underset{\sim}{u}) = \underset{\underset{\sim}{Ax}-\underset{\sim}{b} \leq 0}{\text{Min}} \ Q(\underset{\sim}{x}).$$

Thus D^0 is equivalent to the problem: Find $(\underset{\sim}{x}^0,\underset{\sim}{u}^0)$
such that

$$\Phi(\underset{\sim}{x}^0,\underset{\sim}{u}^0) = \underset{\underset{\sim}{x} \in R^n}{\text{Min}} \ \underset{\underset{\sim}{u} \geq 0}{\text{Max}} \ \Phi(\underset{\sim}{x},\underset{\sim}{u}).$$

§14. The Numerical Treatment of Quadratic Optimization Problems

We will present only a small selection from the multitude of numerical methods of quadratic optimization which various authors have proposed. A more extensive survey of such methods may be found in Künzi and Krelle, 1962, and in Wolfe, 1963. Here we present first the cutting plane method in Kelley, which we considered in §10 earlier, but now specialize to the case of quadratic optimization. In this case it will generally arrive at a solution in a finite number of steps. The second method we describe is due to Wolfe. A modification of the simplex method, it provides a solution for the Kuhn-Tucker conditions, and therefore for the quadratic optimization problem, if one exists at all. The process terminates after a finite number of steps in either case.

14.1. The Cutting Plane Method for Quadratic Optimization Problems

We presuppose the following problem type: Find an $\underset{\sim}{x} \in R^n$ such that

$$Q(\underset{\sim}{x}) = \underset{\sim}{p}'\underset{\sim}{x} + \underset{\sim}{x}'\underset{\sim}{C}\underset{\sim}{x} = \text{Min!}, \quad \underset{\sim}{A}\underset{\sim}{x} \leq \underset{\sim}{b}. \tag{14.1}$$

This is a quadratic optimization problem of the type considered in §12.3B, with inequalities as constraints, and without positivity constraints. The following condition is to be satisfied.

(A) The subset M of R^n defined by $\underset{\sim}{A}\underset{\sim}{x} \le \underset{\sim}{b}$ is bounded and is not empty (so we have an m-by-n matrix with $m > n$).

As in §10, we introduce a real variable, z, to transform the problem into one with a linear objective function:

$$z = \text{Min!}, \quad f(\underset{\sim}{x},z) = \underset{\sim}{p}'\underset{\sim}{x} + \underset{\sim}{x}'\underset{\sim}{C}\underset{\sim}{x} - z \le 0, \quad \underset{\sim}{A}\underset{\sim}{x} \le \underset{\sim}{b}. \tag{14.2}$$

The set of feasible points for this problem is not bounded. Within the limits of the constraints, z may assume arbitrarily large values. Yet, because the function $Q(\underset{\sim}{x}) = \underset{\sim}{p}'\underset{\sim}{x} + \underset{\sim}{x}'\underset{\sim}{C}\underset{\sim}{x}$ is bounded on the bounded set M, we may add the inequality $z \le \bar{z}$, with $\bar{z}$ sufficiently large, to the constraints of problem (14.2) without affecting the behavior of the solutions. For the initial set, S_0, of the cutting plane method, we could choose the subset of R^{n+1} of points $(\underset{\sim}{x},z)$ defined by $\underset{\sim}{A}\underset{\sim}{x} \le \underset{\sim}{b}$, $z \le \bar{z}$, and $z \ge \bar{\bar{z}}$, where $\bar{\bar{z}}$ is sufficiently small. However, the constraints $z \le \bar{z}$ and $z \ge \bar{\bar{z}}$ (if $\bar{\bar{z}}$ is chosen sufficiently small) never enter into the process at any stage; we may ignore these constraints in applying the cutting plane method. The convergence proof of §10 remains valid because these constraints are always satisfied.

The process is modified by choosing $(\underset{\sim}{x}^0,z_0)$ as follows. $\underset{\sim}{x}^0$ is to be a point for which $\underset{\sim}{A}\underset{\sim}{x}^0 \le \underset{\sim}{b}$ and $z_0 = \underset{\sim}{p}'\underset{\sim}{x}^0 + \underset{\sim}{x}^{0\prime}\underset{\sim}{C}\underset{\sim}{x}^0$. Then by rule (II) of §10, the set S_1 is defined by

$$\underset{\sim}{A}\underset{\sim}{x} \le \underset{\sim}{b},$$

$$f(\underset{\sim}{x}^0,z_0) + (\underset{\sim}{x}-\underset{\sim}{x}^0)'(\underset{\sim}{p} + 2\underset{\sim}{C}\underset{\sim}{x}^0) - (z-z_0) =$$

$$= (p + 2Cx^0)'x - z - x^{0'}Cx^0 \leq 0.$$

If we now continue the execution of the method, the only constraint in (14.2) which might be violated by a point (x^t, z_t) is the constraint $f(x,z) \leq 0$; for the remaining constraints, $Ax \leq b$, are satisfied by every point of every set S_t.

If for some $t > 0$ we ever have $f(x^t, z_t) \leq 0$, then (x^t, z_t) is a solution of problem (14.2), and therefore, x^t is a solution of problem (14.1). But if $f(x^t, z_t) > 0$, then rule (II) of §10 comes into play, with $f(x,z)$ in place of the original $f_k(z)$. In general, the set S_t, for $t = 1,2,\ldots,$ is defined by the inequalities

$$\begin{aligned} &Ax \leq b, \\ &(p+2Cx^\tau)'x - z \leq x^{\tau'}Cx^\tau \quad (\tau = 0,1,\ldots,t-1). \end{aligned} \qquad (14.3)$$

We must find a point (x^t, z_t) at which the objective function z attains its minimum subject to the constraints (14.3). Once again, it is advisable to solve the dual problem,

$$\left.\begin{aligned} b'u + \sum_{\tau=0}^{t-1} (x^{\tau'}Cx^\tau)v_\tau &= \text{Min!} \\ A'u + \sum_{\tau=0}^{t-1} (p + 2Cx^\tau)v_\tau &= 0, \\ \sum_{\tau=0}^{t-1} v_\tau &= 1, \\ v_\tau \geq 0, \quad u \geq 0. & \end{aligned}\right\} \qquad (14.4)$$

The convergence conclusion of §10 then applies. The sequence of the x^t contains a convergent subsequence whose limit is then a solution of problem (14.1). If the solution of (14.1) is uniquely determined, the sequence of x^t's converges to this solution.

Wolfe, 1961, proves that the cutting plane method for quadratic optimization solves the problem in a finite number of steps if certain additional conditions are met. For each $t = 1,2,\ldots$ use the numbers v_τ given by (14.4) to form a vector,

$$w^t = \sum_{\tau=0}^{t-1} v_\tau x^\tau. \tag{14.5}$$

Then there is a t such that x^t or w^t is a solution of (14.1) if the following conditions, in addition to (A), are satisfied.

(B) The matric C is positive definite (and there is exactly one solution, $\bar{x}$, therefore).

(C) The polyhedron defined by $Ax \le b$ has no degenerate vertices.

(D) If $\bar{x}$ is the uniquely determined solution of (14.1) and if $\bar{u}$ and $\bar{y}$ are vectors satisfying the Kuhn-Tucker conditions (denoted there by u^0 and y^0), then for $j = 1,\ldots,m$, it is never true that $\bar{u}_j = \bar{y}_j = 0$; hence it is true that either

$$u_j = 0,\ y_j > 0 \quad \text{or} \quad u_j > 0,\ y_j = 0.$$

The proof of this assertion can be found in Wolfe and will not be reproduced here.

14.2. An Example Using the Cutting Plane Method

We will solve the problem which served for a sample application of the inclusion theorem in §12.2. In the notation of (14.1), it reads

$$\begin{aligned} x_1 + x_2 &\le 8, \\ x_1 &\le 6, \\ x_1 + 3x_2 &\le 18, \\ -x_1 &\le 0, \\ -x_2 &\le 0, \end{aligned}$$

$$Q(\underset{\sim}{x}) = -48x_1 - 40x_2 + 2x_1^2 + x_2^2 = \text{Min!}$$

For the initial point, we may choose $\underset{\sim}{x}^0 = \binom{0}{0}$. Problem (14.4) for $t = 1$ becomes

$$\begin{aligned} 8u_1 + 6u_2 + 18u_3 \qquad\qquad - 0.v_0 &= \text{Min!} \\ u_1 + u_2 + u_3 - u_4 \qquad - 48v_0 &= 0 \\ u_1 + 3u_3 - u_5 - 40v_0 &= 0 \\ v_0 &= 1, \end{aligned}$$

$$u_i \ge 0 \quad (i = 1,2,\ldots,5).$$

If we choose the vectors for u_1, u_2, and v_0 as basis vectors, we obtain the following tableau (where blanks are to be understood as zeros).

	u_3	u_4	u_5	
u_1	3		-1	40
u_2	-2	-1	1	8
v_0				1
	-6	$\boxed{-6}$	$\boxed{-2}$	368
	6	8	3	-416

This already solves (14.4) for $t = 1$. Since $u_1 > 0$ and $u_2 > 0$, the solution of (14.3) satisfies the following constraints with an equality sign:

$$\begin{aligned} x_1 + x_2 &= 8 \\ x_1 \quad &= 6. \end{aligned}$$

Therefore, $\underset{\sim}{x}^1 = \begin{pmatrix} 6 \\ 2 \end{pmatrix}$, and consequently, $\underset{\sim}{x}^{1'} \underset{\sim}{C} \underset{\sim}{x}^1 = 76$ and $\underset{\sim}{p} + 2\underset{\sim}{C}\underset{\sim}{x}^1 = \begin{pmatrix} -24 \\ -36 \end{pmatrix}$.

Incidentally, by recalling §5.1, we can read off $\underset{\sim}{x}^1$ from the tableau immediately (the boxed-in numbers).

The above tableau needs one more column (problem (14.4) for $t = 3$):

	u_3	u_4	u_5	v_1		
u_1	3		-1	4	40	10
u_2	-2	-1	1	20	8	0.4
v_0				1	1	1
	-6	-6	-2	76	368	
	6	8	3	-100	-416	

After the next two simplex steps --

	u_3	u_4	u_5	u_2		
u_1	3.4	0.2	-1.2	-0.2	38.4	11.3
v_1	-0.1	-0.05	0.05	0.05	0.4	--
v_0	[0.1]	0.05	-0.05	-0.05	0.6	6
	1.6	-2.2	-5.8	-3.8	337.6	
	-4	3	8	5	-376	

	v_0	u_4	u_5	u_2	
u_1	-34	-1.5	0.5	1.5	18
v_1	1				1
u_3	10	0.5	-0.5	-0.5	6
	-16	[-3]	[-5]	-3	328
	40	5	6	3	-352

-- we obtain the solution $\underset{\sim}{x}^2 = \begin{pmatrix} 3 \\ 5 \end{pmatrix}$. Now we again add a column:

	v_0	u_4	u_5	u_2	v_2		
u_1	-34	-1.5	0.5	1.5	-21	18	--
v_1	1				[1]	1	1
u_3	10	0.5	-0.5	-0.5	9	6	2/3
	-16	-3	-5	-3	27	328	
	40	5	6	3	-15	-352	

	v_0	u_4	u_5	u_2	u_3	
u_1	-32/3	-1/3	-2/3	1/3	7/3	32
v_1	- 1/9	-1/18	1/18	1/18	-1/9	1/3
v_2	10/9	1/18	-1/18	-1/18	1/9	2/3
	-46	[-9/2]	[-7/2]	-3/2	-3	310
	170/3	35/6	31/6	13/6	5/3	-342

Thus, $\underset{\sim}{x}^3 = \begin{pmatrix} 9/2 \\ 7/2 \end{pmatrix}$. By (14.5), $\underset{\sim}{w}^3 = \frac{1}{3}\begin{pmatrix} 6 \\ 2 \end{pmatrix} + \frac{2}{3}\begin{pmatrix} 3 \\ 5 \end{pmatrix} =$ $\begin{pmatrix} 4 \\ 4 \end{pmatrix}$ and therefore $Q(\underset{\sim}{w}^3) = -304$. This is the solution (see §12.2) of the quadratic optimization problem.

14.3. Wolfe's Method

The method described in Wolfe, 1959, and presented here in a slightly altered form, presupposes the following problem type.

$$Q(\underset{\sim}{x}) = \underset{\sim}{p}'\underset{\sim}{x} + \underset{\sim}{x}'\underset{\sim}{C}\underset{\sim}{x} = \text{Min!}, \quad \underset{\sim}{A}\underset{\sim}{x} = \underset{\sim}{b}, \quad \underset{\sim}{x} \geq \underset{\sim}{0} \qquad (14.6)$$

where we evidently have equality constraints and positively constrained variables. The m-by-n matrix $\underset{\sim}{A}$ (where $m < n$) is to be of rank m. The n-by-n matrix $\underset{\sim}{C}$ is to be symmetric and positive definite (the case of a positive semi-definite matrix $\underset{\sim}{C}$ can also be handled by this method, but the proof of this requires some far-reaching hypotheses, in order to exclude degenerate cases, and in practice it can be very difficult to check that these are satisfied).

In §12.3 we derived the Kuhn-Tucker conditions for a problem of type (14.6). A vector $\underset{\sim}{x} \in R^n$ is a minimal solution of (14.6) iff there exist vectors $\underset{\sim}{u} \in R^m$ and $\underset{\sim}{v} \in R^n$ such that the conditions

$$\left.\begin{aligned} \underset{\sim}{A}\underset{\sim}{x} \qquad\qquad &= \underset{\sim}{b} \\ -2\underset{\sim}{C}\underset{\sim}{x} + \underset{\sim}{v} - \underset{\sim}{A}'\underset{\sim}{u} &= \underset{\sim}{p} \\ \underset{\sim}{x} \geq \underset{\sim}{0}, \quad \underset{\sim}{v} &\geq \underset{\sim}{0} \\ \underset{\sim}{x}'\underset{\sim}{v} = 0 \end{aligned}\right\} \tag{14.7}$$

are satisfied; here we omit the indices on $\underset{\sim}{x}$, $\underset{\sim}{u}$, and $\underset{\sim}{v}$ included in §12.3. Wolfe's method consists of a modification of the simplex method for determining a solution of (14.7). First we must find a feasible vector $\overline{\underset{\sim}{x}}$ for problem (14.6). i.e., a vector $\overline{\underset{\sim}{x}}$ for which $\underset{\sim}{A}\overline{\underset{\sim}{x}} = \underset{\sim}{b}$ and $\overline{\underset{\sim}{x}} \geq \underset{\sim}{0}$. This we can do by the process described in §3.4 and §4.4. If no such vector exists, there is no solution of (14.6). If one does exist, the process cited will find a vector $\overline{\underset{\sim}{x}}$ which is a vertex of the set of feasible vectors for problem (14.6), and thereby find a basis consisting of the m linearly independent column vectors $\underset{\sim}{a}^k$, $k \in Z$, of $\underset{\sim}{A}$. Since matrix $\underset{\sim}{C}$ is positive definite by assumption (and for any given bound M, the objective function $Q(\underset{\sim}{x}) > M$ outside a sufficiently large ball), there is a solution of (14.6) and hence, of (14.7).

In order to find one, we solve the problem

$$\text{(I)} \qquad \left.\begin{aligned} \underset{\sim}{A}\underset{\sim}{x} \qquad\qquad\qquad &= \underset{\sim}{b}, \\ -\underset{\sim}{C}\underset{\sim}{x} + \underset{\sim}{v} - \underset{\sim}{A}'\underset{\sim}{u} + \underset{\sim}{h}\zeta \quad &= \underset{\sim}{p}, \\ \underset{\sim}{x} \geq \underset{\sim}{0}, \quad \underset{\sim}{v} \geq \underset{\sim}{0}, \quad \zeta &\geq 0, \end{aligned}\right\} \tag{14.8}$$

$$x'v = 0, \tag{14.9}$$

$$\zeta = \text{Min!}, \tag{14.10}$$

where we set $h = p + 2C\bar{x}$. Because of the non-linear constraint (14.9), this is not a linear optimization problem. But it is possible to modify the simplex method with an additional rule, so that problem (I) then can be solved.

One point which satisfies constraints (14.8) and (14.9) is given by $x = \bar{x}$, $v = 0$, $u = 0$, and $\zeta = 1$. As a basis at this point, we must choose the appropriate system of $n + m$ linearly independent column vectors from the matrix

$$\begin{pmatrix} A & 0 & 0 & 0 \\ -2C & E & -A' & h \end{pmatrix}. \tag{14.11}$$

First we observe that the following $n + m$ column vectors are linearly independent.

1. The m vectors whose "upper" components are the m basis vectors a^k, $k \in Z$, of A, and whose "lower" components are the corresponding columns of $-2C$.

2. The $n - m$ column vectors of $\begin{pmatrix} 0 \\ E \end{pmatrix}$ corresponding to components v_i, $i \notin Z$.

3. All m column vectors of $\begin{pmatrix} 0 \\ -A' \end{pmatrix}$.

This system of $n + m$ column vectors is not yet a suitable basis, because $\zeta = 1$ at the initial point and the column vector $\begin{pmatrix} 0 \\ h \end{pmatrix}$ for the variable ζ must therefore be contained in the basis. We may suppose that $h \neq 0$; otherwise we would already have a solution of (14.7). But then $\begin{pmatrix} \underset{\sim}{0} \\ \underset{\sim}{h} \end{pmatrix}$ may be exchanged for one of the column vectors in 2. or 3. in such a way that we still have a system of $n + m$ linearly independent vectors; and then we do have a basis.

We now apply the simplex method to the problem determined by (14.8) and (14.10), all the while obeying the following, additional rule. In an exchange step where the column vector for the component x_i, $i = 1,\ldots,n$, remains in the basis, the column vector for the component v_i may not be added to the basis, and vice-versa.

Then all of the vectors which one obtains in the course of the simplex process will satisfy the constraint $\underset{\sim}{x}'\underset{\sim}{v} = 0$. With a positive definite matrix $\underset{\sim}{C}$, the simplex method, as modified by the additional rule, will lead to a solution of problem (I) with the minimal value 0 of the objective function. The assumption that the process terminates at a point $(\hat{\underset{\sim}{x}}, \hat{\underset{\sim}{v}}, \hat{\underset{\sim}{u}}, \hat{\zeta})$ where $\hat{\zeta} > 0$ leads to a contradiction. For then we would have a solution of the linear optimization problem

$$\text{(II)} \qquad \begin{aligned} \underset{\sim}{A}\underset{\sim}{x} \qquad\qquad\qquad\qquad &= \underset{\sim}{b}, \\ -2\underset{\sim}{C}\underset{\sim}{x} + \underset{\sim}{v} - \underset{\sim}{A}'\underset{\sim}{u} + \underset{\sim}{h}\zeta &= \underset{\sim}{p}, \\ \hat{\underset{\sim}{v}}'\underset{\sim}{x} + \hat{\underset{\sim}{x}}'\underset{\sim}{v} \qquad\qquad &= 0, \\ \underset{\sim}{x} \geq \underset{\sim}{0}, \quad \underset{\sim}{v} \geq \underset{\sim}{0}, \quad \zeta &\geq 0, \\ \zeta = \text{Min!} & \end{aligned}$$

where $\hat{\zeta} > 0$. But then the problem

$$\text{(II*)} \qquad \begin{aligned} A'y - 2Cw + \hat{v}\xi &\le 0, \\ w + \hat{x}\xi &\le 0, \\ Aw &= 0, \\ h'w &\le 1, \\ b'y + p'w &= \text{Max!}, \end{aligned}$$

which is dual to (II) by §5.3, would also have a solution, $\hat{y}$, $\hat{w}$, $\hat{\xi}$, such that

$$b'\hat{y} + p'\hat{w} = \hat{\zeta} > 0. \tag{14.12}$$

Those constraints in (II*) which correspond to positive components of the solution of (II), are satisfied with an equality sign. Since $\hat{\zeta} > 0$, it follows that

$$h'\hat{w} = 1. \tag{14.13}$$

For every i, $1 \le i \le n$, exactly one of the three following cases occurs.

$$\begin{aligned} &(\alpha) \quad \hat{x}_i > 0,\ \hat{v}_i = 0; \text{ then } (A'\hat{y} - 2C\hat{w})_i = 0 \\ &(\beta) \quad \hat{x}_i = 0,\ \hat{v}_i > 0; \text{ then } \hat{w}_i = 0 \\ &(\gamma) \quad \hat{x}_i = \hat{v}_i = 0; \text{ then } (A'\hat{y} - 2C\hat{w})_i \le 0 \\ &\qquad \text{and } \hat{w}_i \le 0. \end{aligned}$$

Therefore, $\hat{w}'(A'\hat{y} - 2C\hat{w}) \ge 0$. Since $A\hat{w} = 0$, this implies that $\hat{w}'C\hat{w} \le 0$. But since C is assumed positive definite, this makes $\hat{w} = 0$, in contradiction to (14.13).

(If we assume only that C is positive semi-definite, then $\hat{\zeta} > 0$ implies only that $\hat{w}'C\hat{w} = 0$, and hence that $C\hat{w} = 0$, by the remark on positive semi-definite matrices in §6.2. Then

$$1 = \underset{\sim}{h}'\hat{\underset{\sim}{w}} = (\underset{\sim}{p} + 2\underset{\sim}{C}\overline{\underset{\sim}{x}})'\hat{\underset{\sim}{w}} = \underset{\sim}{p}'\hat{\underset{\sim}{w}} + 2\overline{\underset{\sim}{x}}'\underset{\sim}{C}\hat{\underset{\sim}{w}} = \underset{\sim}{p}'\hat{\underset{\sim}{w}},$$

and also, $\underset{\sim}{b}'\hat{\underset{\sim}{y}} = \hat{\underset{\sim}{x}}'\underset{\sim}{A}'\hat{\underset{\sim}{y}} = 0$, since by ($\alpha$), $\hat{x}_i > 0$ now implies $(\underset{\sim}{A}'\hat{\underset{\sim}{y}})_i = 0$. It then follows from (14.12) that $\hat{\zeta} = 1$. Thus for a positive semi-definite matrix C, if we do not obtain a solution of (I) with $\hat{\zeta} = 0$, no diminution of ζ whatsoever is possible (the initial value is $\zeta = 1$). This case occurs when the objective function $Q(\underset{\sim}{x})$ in (14.6) is not bounded below on the set of feasible points.)

14.4. An Example Using Wolfe's Method

The problem which was already considered in §12.2 and §14.2 is changed by the introduction of slack variables x_3, x_4, and x_5, into the required form, (14.6):

$$\left.\begin{aligned} x_1 + x_2 + x_3 &= 8, \\ x_1 + x_4 &= 6, \\ x_1 + 3x_2 + x_5 &= 18, \\ x_i \geq 0 \quad (i &= 1,\ldots,5) \end{aligned}\right\} \tag{14.14}$$

$$2x_1^2 + x_2^2 - 48x_1 - 40x_2 = \text{Min!} \tag{14.15}$$

As an initial vertex, we choose $\overline{\underset{\sim}{x}}$ with $\overline{x}_1 = \overline{x}_2 = 0$, $\overline{x}_3 = 8$, $\overline{x}_4 = 6$, and $\overline{x}_5 = 18$. Then $\underset{\sim}{h} = \underset{\sim}{p} + 2\underset{\sim}{C}\overline{\underset{\sim}{x}} = (-48, -40, 0, 0, 0)'$. In addition to (14.14), problem (I) contains the constraints

$$\left.\begin{aligned} -4x_1 + v_1 - u_1 - u_2 - u_3 - 48\zeta &= -48 \\ -2x_2 + v_2 - u_1 - 3u_3 - 40\zeta &= -40 \end{aligned}\right\} \tag{14.16}$$

$$\left.\begin{array}{llll} v_3 - u_1 & & & = 0 \\ v_4 & - u_2 & & = 0 \\ v_5 & & - u_3 & = 0 \end{array}\right\} \qquad (14.17)$$

$$v_i \geq 0 \quad (i = 1,\ldots,5), \quad \zeta \geq 0. \qquad (14.18)$$

We can substitute for u_1, u_2, and u_3 in (14.16) with (14.17), so that we won't have to carry along the (positively unconstrained) variables u_i when applying Wolfe's method. Let us first use the above rules 1., 2., and 3. to choose as the linearly independent vectors those corresponding to the variables x_3, x_4, x_5, v_1, and v_2, which gives us a simplex tableau (where blanks again signify zeros):

	x_1	x_2	v_3	v_4	v_5	ζ	
x_3	1	1					8
x_4	1						6
x_5	1	3					18
v_1	-4		-1	-1	-1	-48	-48
v_2		-2	-1		-3	[-40]	-40
	0	0	0	0	0	-1	0
	2	-1	3	2	5	90	57

This tableau does not, however, correspond to any feasible point for problem (I). But with an exchange step at the (negative) pivot element -40, we can get the column vector for the variable ζ into the basis, obtaining the tableau

	x_1	x_2	v_3	v_4	v_5	v_2		
x_3	1	1					8	8
x_4	1						6	-
x_5	1	3					18	6
v_1	-4	[12/5]	1/5	-1	13/5	-6/5	0	0
ζ		1/20	1/40		3/40	-1/40	1	20
	0	1/20	1/40	0	3/40	-1/40	1	
	2	-11/2	3/4	2	-7/4	9/4	-33	

Since the vector for the variable v_2 is not in the basis, the vector for the variable x_2 may be taken into the basis through an exchange step with pivot 12/5. The remaining steps require no further explanation.

	x_1	v_1	v_3	v_4	v_5	v_2		
x_3	[8/3]	-5/12	-1/12	5/12	-13/12	1/2	8	3
x_4	1						6	6
x_5	6	-5/4	-1/4	15/12	-39/12	3/2	18	3
x_2	-5/3	5/12	1/12	-5/12	13/12	-1/2	0	-
ζ	1/12	-1/48	1/48	1/48	1/48	0	1	12
	1/12	-1/48	1/48	1/48	1/48	0	1	
	-43/6	55/24	29/24	-7/24	101/24	-1/2	-33	

	x_3	v_1	v_3	v_4	v_5	v_2		
x_1	3/8	-5/32	-1/32	5/32	-13/32	3/16	3	-
x_4	-3/8	5/32	1/32	-5/32	13/32	-3/16	3	96
x_5	-9/4	-5/16	-1/16	5/16	-13/16	3/8	0	-
x_2	5/8	5/32	1/32	-5/32	13/32	-3/16	5	160
ζ	-1/32	-1/128	3/128	1/128	7/128	-1/64	3/4	32
	-1/32	-1/128	3/128	1/128	7/128	-1/64	3/4	
	43/16	75/64	63/64	53/64	83/64	27/32	-23/2	

	x_3	v_1	ζ	v_4	v_5	v_2	
x_1			4/3				4
x_4			-4/3				2
x_5			8/3				2
x_2			-4/3				4
v_3	-4/3	-1/3	128/3	1/3	7/3	-2/3	32
	0	0	-1	0	0	0	0
			-42				-43

In the last tableau, all the ultimately extraneous numbers have been omitted, and with it, we have arrived at a solution, for $\zeta = 0$. We have $x_1 = 4$, $x_2 = 4$, and $Q_{min} = -304$.

IV. TCHEBYCHEV APPROXIMATION AND OPTIMIZATION

Three major areas of numerical analysis stand in close relationship: optimization, approximation, and boundary and initial value problems for differential equations. This chapter will present some of these relations, and show how methods developed for one of these areas can at times be applied profitably to another. For example, the methods previously described, for solving optimization problems, may be used successfully on computer installations to solve boundary value problems for ordinary and partial differential equations.

§15. Introduction

15.1. Approximation as Optimization

There is a close relationship between approximation problems and optimization problems. Approximation deals with problems of the following sort. We are given:

1. A continuous, real-valued function, $f(x)$, defined on a given point-set B (abbreviated, $f(x) \in C$ <B>), for example, a region of euclidean m-space, R^m, where $x = (x_1, x_2, \ldots, x_m)$;

2. A class, $V = \{g(x, a_1, \ldots, a_n)\}$, of real-valued functions, also continuous in x on B, and dependent on the real parameters, $a_1, \ldots, a_n$. We have $f \notin V$; when $f \in V$, we have a different kind of problem ("representation problem").

3. A metric, $\rho(f,g)$, for the distance between two functions, $f,g \in C$<B>. For a linear space such as C<B>, the notion of distance can be introduced through the definition of a norm, $||g||$. Then we have $\rho(f,g) = ||f-g||$. Various ways of defining a norm will be discussed in the sequel.

We are to find a function $g \in V$ for which the distance $\rho(f,g)$ is minimal. We introduce the infimum

$$\rho_0 = \inf_{g \in V} \rho(f,g),$$

where g ranges over the class V; ρ_0 is called the <u>minimal distance</u> and an element $h \in V$ is called a <u>minimal solution</u> if the distance equals the minimal distance:

$$\rho_0 = \rho(f,h).$$

While the minimal distance ρ_0 always exists, the same need not be true of a minimal solution; see Meinardus, 1967.

An approximation problem thus may be considered as an optimization problem with objective function $\rho(f,g)$.

When B is a finite set with m elements (and the discrete topology), the continuous, real-valued functions on B may be identified with the vectors $\underset{\sim}{v} \in R^m$. As a norm for $\underset{\sim}{v}$, we can take any <u>p-norm</u>,

$$||\underset{\sim}{v}||_p = [\sum_i |v_i|^p]^{1/p}, \tag{15.1}$$

with $1 \leq p < \infty$. For $p = 2$, this is the euclidean norm. As $p \to \infty$, $||\underset{\sim}{v}||_p$ becomes the <u>maximum norm</u> (or, sup norm),

$$||\underset{\sim}{v}||_\infty = \operatorname{Max}_i |v_i| \tag{15.2}$$

When B is an infinite set, in particular, a closed and bounded subset of R^q of positive Lebesgue measure, and again, $1 \leq p < \infty$, we may take

$$||g||_p = [\int_B |g(x)|^p dx]^{1/p} \tag{15.3}$$

as a norm. Now as $p \to \infty$, this becomes the <u>Tchebychev norm</u>,

$$||g||_\infty = \operatorname{Max}_{x \in B} |g(x)|. \tag{15.4}$$

15.2. Various Types of Approximation Problems

The approximation problems of especial importance are those for which distance is measured with one of the norms $||\underset{\sim}{v}||_2$, $||\underset{\sim}{v}||_\infty$, $||g||_2$, or $||g||_\infty$. For the cases $||\underset{\sim}{v}||_2$ and $||g||_2$, the terms <u>gaussian</u> approximation or <u>least-square</u> approximation are used. In the remaining cases, $||\underset{\sim}{v}||_\infty$ and $||g||_\infty$, one speaks of <u>Tchebychev</u> approximation. We will abbreviate this henceforth to T-approximation.

<u>Discrete</u> approximation refers to a finite set B and

the use of the norm (15.1) or (15.2), while <u>continuous</u> approximation refers to an infinite set B and the use of the norm (15.3) or (15.4).

When the set V, of functions with which f is to be approximated, is a finite dimensional linear subspace of C<B>, i.e., when $V = \{\sum_{i=1}^{n} a_i v_i(x)\}$, we have a <u>linear</u> approximation problem. We also frequently encounter <u>non-linear</u> approximation problems; for example, when V includes rational functions of the parameters a_i, or exponential functions:

$$V = \{a_1 e^{a_2 x_1 + a_3 x_2}\}. \tag{15.5}$$

In recent times, T-approximation has come to be of particular importance. For example, if a complicated function f(x) needs to be submitted to a machine, one often approximates it with an expression g(x) which the machine can compute easily, such as a polynomial. One would like the maximum error,

$$||\varepsilon||_\infty = ||g-f||_\infty,$$

on the interval of interest, $a \leq x \leq b$ (= the domain B), to remain within some prescribed bound, say 5×10^{-11}. In the following section we will consider classes of boundary value problems for which T-approximation is again preferable to gaussian approximation.

15.3. Boundary Value Problems for Elliptic Differential Equations and Tchebychev Approximation

Consider the boundary value problem for an unknown

function $u(x) = u(x_1,\ldots,x_n)$ satisfying the differential equation,

$$Lu \equiv -\sum_{j,k=1}^{n} a_{jk}\frac{\partial^2 u}{\partial x_j \partial x_k} - \sum_{j=1}^{n} b_j \frac{\partial u}{\partial x_j} + cu = r(x) \quad \text{on } D, \tag{15.6}$$

and the boundary condition,

$$Mu \equiv u - g\frac{\partial u}{\partial \sigma} = b(x) \quad \text{on } \Gamma, \tag{15.7}$$

where D is a given, bounded, open region in R^n with a piecewise smooth boundary Γ, and a_{jk}, b_j, c, r, g, and b are given continuous functions. Furthermore, the matrix (a_{jk}) is positive definite at every point of D, and $c \geq 0$ and $g \geq 0$. Let ν denote the inner normal, which need not be uniquely defined at every point of Γ, and let σ denote the so-called conormal. In the case of the potential equation $Lu = -\Delta u = r(x)$,

$$a_{jk} = \delta_{jk} = \begin{cases} 1 & \text{for } j = k \\ 0 & \text{for } j \neq k, \end{cases}$$

and the conormal coincides with the inner normal. For this and all other details, see Collatz, 1966, p. 122 and 385.

We try to find an approximate solution of the type

$$v = v_0 + \sum_{\nu=1}^{p} a_\nu v_\nu \tag{15.8}$$

with constants a_ν to be determined. In the simplest case, we choose the functions v_ν so that v satisfies the differential equation:

$$Lv_0 = r, \quad Lv_\nu = 0 \quad (\nu = 1,\ldots,p).$$

Then by the maximum principle, we have the error estimate

$$|v-u| \leq a \quad \text{on} \quad B \cup \Gamma, \tag{15.9}$$

whenever the defect, $d = Mv - b$, satisfies $|d| \leq a$ on all of Γ.

Using the known function $d_0 = Mv_0 - b$, we have to determine the constants a_ν so that the maximum modulus of

$$d = \sum_{\nu=1}^{p} a_\nu M v_\nu + d_0$$

on the boundary Γ is as small as possible. With the norm $||g||_\infty$, this becomes precisely a problem of linear Tchebychev approximation,

$$\rho(-d_0, \sum_{\nu=1}^{p} a_\nu M v_\nu) = \text{Min!} \tag{15.10}$$

Consider the best-known special case, the Dirichlet problem, the first boundary value problem for the potential equation,

$$\Delta u = 0 \quad \text{for} \quad r < 1,$$

for a disk in the plane (with polar coordinates (r,ϕ)) and given boundary values

$$u = f(\phi) \quad \text{for} \quad r = 1, \quad 0 \leq \phi \leq 2\pi$$
$$(f(0) = f(2\pi)).$$

If we should choose

$$v_0 = 0;\ v_1,\ldots,v_p = 1,\ r^n \cos n\phi,\ r^n \sin n\phi$$
$$(n = 1,2,\ldots,m), \text{ respectively,}$$

when $p = 2m + 1$, then we must find the best approximation of

$f(\phi)$, in the sense of Tchebychev, by a linear combination of the functions

$$1, \cos n\phi, \sin n\phi \quad (n = 1,\ldots,m),$$

i.e., we are faced with a trigonometric Tchebychev approximation.

Both here, and in the previously described, more general case, Tchebychev approximation is the more natural one, because it uses the norm $||g||_\infty$, and is preferable to others, such as least squares approximation, because it leads directly to the error estimate (15.9).

Similarly for parabolic equations T-approximation often leads immediately to an error estimate on the approximate solution; see Collatz, 1959, and Collatz, 1966, p. 393. Here we will consider only one example, which also leads to a trigonometric T-approximation. This is the heat conduction equation,

$$u_{xx} = ku_t \quad \text{for} \quad 0 < x < \pi,\ 0 < t, \tag{15.11}$$

with initial and boundary values

$$u(x,0) = f(x) \quad \text{for} \quad 0 \le x \le \pi,\ f(0) = f(\pi) = 0,$$
$$u(0,t) = u(\pi,t)t = 0 \quad \text{for} \quad 0 \le t.$$

For the approximate solution v we use

$$v(x,t) = \sum_{n=1}^{m} b_n \sin nx\, e^{-n^2t/k}.$$

The coefficients b_n must be chosen so that $v(x,0)$ is as close, in the Tchebychev sense, as possible to the function f(x).

For

$$|v(x,0) - f(x)| \le \delta \quad \text{for} \quad 0 \le x \le \pi$$

implies that the error estimate

$$|v-u| \le \delta \quad \text{for} \quad 0 < x < \pi, \quad 0 < t$$

holds on the whole domain. There are numerous other approximation problems which arise from boundary value problems for partial differential equations.

15.4. Contraction Mappings in Pseudometric Spaces and One-sided Tchebychev Approximation

We consider a non-linear boundary value problem, given by the quasilinear differential equation,

$$Lu = f(x_1,\ldots,x_n, u) \quad \text{on} \quad D, \tag{15.12}$$

and the boundary condition (15.7). Lu is again the linear differential expression given in (15.6); see Collatz, 1966, p. 251.

We assume that a non-negative Green function exists for the differential equation (15.6) and the homogeneous boundary condition $Mu = 0$. We will not repeat the theory of contraction mappings in pseudometric spaces here (see Schröder, 1956), but simply state the result of the theory suited for the present case.

Suppose we have found (perhaps by iteration) a pair of functions, u_0 and u_1, which satisfy

$$Lu_1 = f(x,u_0) \quad \text{on} \quad D, \quad Mu_1 = b(x) \quad \text{on} \quad \Gamma. \tag{15.13}$$

Suppose also that we have a bound, $N(x)$, for the partial derivative of f with respect to u,

$$\left|\frac{\partial f}{\partial u}\right| \leq N(x),$$

on a subset, F, of $(x_1,\ldots,x_n, u)$-space which contains u_0, u_1, and the solution u, and is convex with respect to u. Further, suppose there is a function $\sigma_0(x)$ such that

$$|u_1 - u_0| \leq \sigma_0(x).$$

For an error estimate on the function u_1,

$$|u-u_1| \leq \sigma(x),$$

we need a function, $\sigma(x)$, which satisfies the linear conditions,

$$\begin{aligned} \hat{L}\sigma \equiv L\sigma - N(x)\sigma &\geq N(x)\sigma_0 \quad \text{on} \quad D, \\ M\sigma &= 0 \quad \text{on} \quad \Gamma; \end{aligned} \tag{15.14}$$

and the bound on the error is particularly sharp if the inequality is "just barely" satisfied. That is, we want

$$L\sigma - N\sigma - N\sigma_0 \geq 0,$$

but with the left side "as small as possible". To this end, we let

$$\sigma = \sum_{\nu=1}^{p} a_\nu \phi_\nu(x), \quad M\phi_\nu = 0 \quad (\nu = 1,\ldots,p)$$

with functions ϕ_ν, which, as noted, satisfy the homogeneous boundary condition. The free parameters, a_ν, have to be determined so that the expression

$$\Phi = \sum_{\nu=1}^{p} a_\nu \hat{L}\phi_\nu$$

is close (in the Tchebychev sense) to the function $N\sigma_0$, but only from above, i.e., we want

$$\Phi - N\sigma_0 \geq 0$$

to be valid (one-sided Tchebychev approximation). The corresponding optimization problem is described in §16.3, and the reader should look there for the sequel.

15.5. Boundary Value Problems and Optimization

Linear boundary value problems frequently can be tied into an optimization problem, and thus the method of linear optimization can be used directly to obtain approximate solutions of the boundary value problems. We demonstrate this with an example of a problem of "monotone type" (Collatz, 1952). On a region D, as in §15.3, let there be given a linear differential equation for an unknown function $u(x) = u(x_1,\ldots,x_n)$,

$$Lu = r(x) \quad \text{on} \quad D,$$

and on the boundary Γ, again as in §15.3, let there be given a linear boundary condition,

$$Mu = \gamma(x) \quad \text{on} \quad \Gamma.$$

This problem is said to be of monotone type if, for every function $\varepsilon(x)$,

$$L\varepsilon \geq 0 \text{ on } D \text{ and } M\varepsilon \geq 0 \text{ on } \Gamma \text{ implies}$$
$$\varepsilon \geq 0 \quad \text{on} \quad D \cup \Gamma.$$

For these problems, there is a simple principle for estimation of the error. If one knows that a solution, u, of the boundary value problem exists, and if two approximations have been found, say $v(x)$ and $w(x)$, which satisfy

$$Lv \leq r(x) \leq Lw, \quad Mv \leq \gamma(x) \leq Mw, \tag{15.15}$$

then there is an inclusion, $v(x) \leq u(x) \leq w(x)$.

Large classes of boundary value problems for ordinary and partial differential equations are of monotone type (cf. Collatz, 1966, pp. 380-404). There are various means of attaining such approximations v and w; we will mention two ways.

A) Let

$$\left.\begin{aligned} v(x) &= \sum_{\nu=1}^{N} a_\nu v_\nu(x) \\ w(x) &= \sum_{\nu=1}^{N} b_\nu v_\nu(x) \end{aligned}\right\} \tag{15.16}$$

with fixed functions $v_\nu(x)$ and as yet undetermined constants a_ν, and b_ν; (sometimes one adds an element $v_0(x)$, causing either v or w to satisfy the inhomogeneous differential equation or the inhomogeneous boundary condition with arbitrary constants a_ν, b_ν, and then one set of inequalities drops out of the following formulas (15.7)). Next choose a number of discrete points, P_σ, $\sigma = 1,\ldots,s$, in the domain D, and boundary points, Q_τ, $\tau = 1,\ldots,t$, on Γ, and also several points Z_λ, $\lambda = 1,\ldots,\ell$, at which a particularly good error estimate is desired (in case of point symmetry it will often suffice to choose only one point, Z, say the center point of the domain). Letting the subscripts σ, τ, λ, indicate the value of the functions at the points P_σ, Q_τ, Z_λ, we obtain the linear optimization problem

$$\left.\begin{aligned} (Lv)_\sigma \le r_\sigma \le (Lw)_\sigma \qquad & (\sigma = 1,\ldots,s) \\ (Mv)_\tau \le \gamma_\tau \le (Mw)_\tau \qquad & (\tau = 1,\ldots,t) \end{aligned}\right\} \tag{15.17}$$

$$\left.\begin{aligned} -\phi \le (w-v)_\lambda \le \phi \qquad & (\lambda = 1,\ldots,\ell) \\ \phi = \text{Min!} \qquad & \end{aligned}\right\} \tag{15.18}$$

B) Replace (15.16) by the somewhat coarser approximation

$$\begin{aligned} v(x) &= \alpha_0 v_0(x) + \sum_{\nu=1}^{N} a_\nu v_\nu(x), \\ w(x) &= \alpha_1 v_0(x) + \sum_{\nu=1}^{N} a_\nu u_\nu(x) \end{aligned} \tag{15.19}$$

with the same constants a_ν for $v(x)$ and $w(x)$; only α_0 and α_1 differ. Here $v_0(x)$ is a fixed, chosen non-negative function on D, $v_0(x) \ge 0$. The optimization problem now simply reads

$$\alpha_1 - \alpha_0 = \text{Min!}$$

with constraints (15.17).

One should note that the process described in A) and B) does not exclude the possibility that the inequalities (15.15) are satisfied at the points P_σ and Q_τ, by (15.17), but not at every point of the domain, D, or the boundary, Γ. If such be the case, the functions $v(x)$ and $w(x)$ are not necessarily exact bounds for the solution, u(x). If the points P_σ and Q_τ are chosen sufficiently close together in D and Γ, one can usually be satisfied with the approximate bounds obtained. Or, one can proceed iteratively, by first applying A) or B), then finding those points which do the most violence to (15.15), next adding these to P_σ and Q_τ, and finally repeating the process, and doing all this several times if necessary, until (15.15) is satisfied up to the limits of computational precision, in all of D and Γ. In the case of B), a diminution of α_0 or an enlargement of α_1 is frequently sufficient.

§16. Discrete Linear Tchebychev Approximation

16.1. The Reduction to a Linear Optimization Problem

By §15.1, we are presented with a problem of discrete, linear T-approximation whenever a vector $\underset{\sim}{v}^0 = (v_{01},\ldots, v_{0m})' \in R^m$ is to be approximated by a linear combination,

$$\sum_{i=1}^{n} a_i \underset{\sim}{v}^i, \text{ of vectors } \underset{\sim}{v}^i = (v_{i1},\ldots,v_{im})' \in R^m$$

so that

$$\gamma = \max_{k=1,\ldots,m} \left|v_{0k} - \sum_{i=1}^{n} a_i v_{ik}\right| \text{ is minimal} \qquad (16.1)$$

We can also write this as

$$\left|\left|\underset{\sim}{v}^0 - \sum_{i=1}^{n} a_i \underset{\sim}{v}^i\right|\right|_\infty = \text{Min!}, \qquad (16.1')$$

The following conditions are to be met.

1. The vectors $\underset{\sim}{v}^1,\ldots,\underset{\sim}{v}^n$ are linearly independent

(since one could otherwise reduce the problem to one with a smaller n).

2. $\tilde{v}^0$ is not a linear combination of $\tilde{v}^1, \ldots, \tilde{v}^n$ (in argreement with the demand that $f \notin V$, made in 2. of §15.1).

Condition 1 requires that $n \leq m$, and the two conditions together even imply that

$$n < m. \qquad (16.2)$$

Problems of type (16.1) arise, among other places, in the treatment of overdetermined systems of linear equations, $v_{0k} - \sum_{i=1}^{n} a_i v_{ik} = 0$, in the unknowns a_i. If no solution exists, one can then consider the problem of determining those numbers a_i which "optimally" satisfy the system, in the sense of (16.1).

One often attempts to solve approximately a problem of continuous linear T-approximation,

$$\operatorname*{Max}_{B} \left| f(x) - \sum_{i=1}^{n} a_i v_i(x) \right| = \text{Min!},$$

by choosing m sample points, $x_1, \ldots, x_m \in B$, and, using the notation

$$f(x_k) = v_{0k}, \qquad v_i(x_k) = v_{ik},$$

then considering the discrete problem, (16.1).

Problem (16.1) admits a geometric interpretation. If there is no index k for which every $v_{ik} = 0$, $i = 1, \ldots, n$, we can assume that the v_{ik} are normalized, so that $\sum_{i=1}^{n} v_{ik}^2 = 1$ for $k = 1, \ldots, m$.

Then

$$v_{0k} - \sum_{i=1}^{n} a_i v_{ik} = 0 \quad (k = 1,\ldots,m) \tag{16.3}$$

are the equations of m hyperplanes in R^n consisting of the points $\tilde{a} = (a_1,\ldots,a_n)'$, and these equations are given in the "Hesse normal form". If $\tilde{a}$ is an arbitrary point of R^n, $|v_{0k} - \sum_{i=1}^{n} a_i v_{ik}|$ is the (euclidean) distance of this point from the kth hyperplane. The γ defined by (16.1) is the maximum of these distances. Problem (16.1) says: Find a point, $\tilde{a}$, for which this maximal distance is minimized. A point $\tilde{a}$ which solves this problem is called a Tchebychev point for the system of hyperplanes (16.3). In R^2 the Tchebychev point for three lines which form the sides of a triangle is the center of the inscribed circle of the triangle.

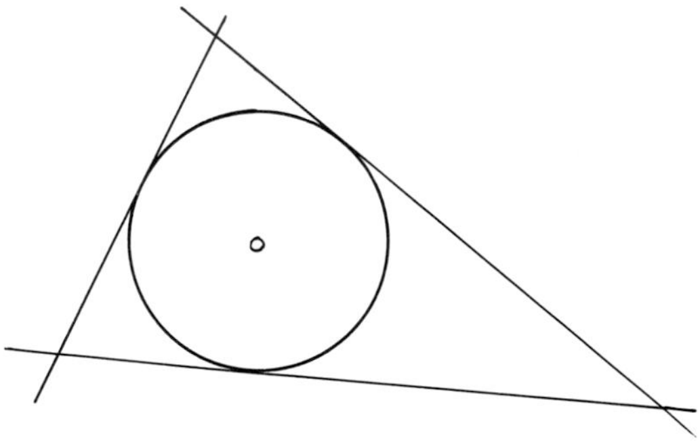

Figure 16.1

(16.1) may be written as a linear optimization problem:

$$\left.\begin{aligned} &\gamma + \sum_{i=1}^{n} a_i v_{ik} \geq v_{0k} \\ & \qquad\qquad\qquad\qquad (k = 1,\ldots,m) \\ &\gamma - \sum_{i=1}^{n} a_i v_{ik} \geq -v_{0k} \\ &\gamma = \text{Min!} \end{aligned}\right\} \tag{16.4}$$

This problem is one without positivity constraints on the variables γ, a_i; γ can be included because there is no explicit positivity constraint, although it is implicit in the constraints that $\gamma \geq 0$.

<u>Theorem 1</u>: Problem (16.4) has a minimal solution.

<u>Proof</u>: The set M of feasible points is not empty; for once $a_1,\ldots,a_n$ have been chosen arbitrarily, the constraints of (16.4) will be satisfied with sufficiently large γ. The objective function γ is bounded below on M. By theorem 16, §5.6 (which was formulated for positively constrained variables; (16.4) may be rewritten as such), a minimal solution exists.

By condition 2 above, the vector $\underset{\sim}{v}^0$ is not a linear combination of $\underset{\sim}{v}^1,\ldots,\underset{\sim}{v}^n$. This implies the

<u>Corollary</u>. For problem (16.4) the minimal value $\gamma > 0$.

We can therefore introduce new variables,

$$b_0 = \frac{1}{\gamma}, \quad b_i = -\frac{a_i}{\gamma} \qquad (i = 1,\ldots,n),$$

and transform (16.4) into the equivalent problem

$$\left.\begin{array}{ll} \sum_{i=0}^{n} b_i v_{ik} \leq 1 & \\ -\sum_{i=0}^{n} b_i v_{ik} \leq 1 & (k = 1,\ldots,m) \\ b_0 = \text{Max!} & \end{array}\right\} \qquad (16.5)$$

If we wish to solve this problem by the simplex method, as

described in §3 and §4, we introduce slack variables, and pick as the initial point the one for which the $b_i = 0$, $i = 1,\dots,n$, and the slack variables all have value 1. Note that the variables b_i are not positively constrained so that we must proceed as per §4.6.

We can even modify the simplex method so that the tableaux which we must construct need not have $2m$ rows, but merely m rows. To this end, we rewrite (16.5) in the form

$$\left.\begin{aligned} &\sum_{i=0}^{n} b_i v_{ik} + y_k = 0 \\ &-1 \le y_k \le 1 \qquad\qquad (k = 1,\dots,m) \\ &b_0 = \text{Max!} \end{aligned}\right\} \tag{16.6}$$

and for an exchange step, we replace rules 1 and 2 of §4.6, which are based on (3.4), by an appropriately constructed rule which guarantees that $-1 \le y_k \le 1$ for $k = 1,\dots,m$. We shall not go into the details here, and leave these to the reader to determine if needed.

16.2. Dualization

Problem (16.1) has been formulated in two ways as a linear optimization problem; once as the minimum problem (16.4), and again, as the equivalent maximum problem (16.5). We now will formulate the two dual problems, which are also equivalent to each other, and interpret these geometrically. By §5.1 and §5.3, respectively, the problem dual to (16.4) reads (with positively constrained variables, y_k^+, y_k^-):

$$\left.\begin{aligned}
&\sum_{k=1}^{m} (y_k^+ + y_k^-) = 1, \\
&\sum_{k=1}^{m} v_{ik}(y_k^+ - y_k^-) = 0, \quad (i = 1,\ldots,n) \\
&\quad y_k^+ \geq 0, \quad y_k^- \geq 0, \quad (k = 1,\ldots,m) \\
&\sum_{k=1}^{m} v_{0k}(y_k^+ - y_k^-) = \text{Max!}
\end{aligned}\right\} \quad (16.7)$$

The problem dual to (16.5), again with positively constrained variables, z_k^+, z_k^-, reads:

$$\left.\begin{aligned}
&\sum_{k=1}^{m} v_{0k}(z_k^+ - z_k^-) = 1, \\
&\sum_{k=1}^{m} v_{ik}(z_k^+ - z_k^-) = 0 \quad (i = 1,\ldots,n), \\
&\quad z_k^+ \geq 0, \quad z_k^- \geq 0 \quad (k = 1,\ldots,m), \\
&\sum_{k=1}^{m} (z_k^+ + z_k^-) = \text{Min!}
\end{aligned}\right\} \quad (16.8)$$

Since problems (16.4) and (16.5) have finite optimal solutions, by theorem 1 and the corollary, the dual problems (16.7) and (16.8) also have solutions, by theorem 2, §5.1.

If we have found a solution, z_k^+, z_k^-, $k = 1,\ldots,m$, of problem (16.8), then either $z_k^+ = 0$ or $z_k^- = 0$. For if $\delta = \text{Min}(z_k^+, z_k^-) > 0$ for some index k, then the constraints of (16.8) are also satisfied by $z_k^+ - \delta$ and $z_k^- - \delta$, thereby reducing the value of the objective function by 2δ. Thus we obtain a problem which is equivalent to (16.8) by adding the (non-linear) constraint

$$z_k^+ \cdot z_k^- = 0 \quad (k = 1,\ldots,m)$$

to the constraints of (16.8). If we then set $z_k = z_k^+ - z_k^-$, then $z_k^+ + z_k^- = |z_k|$. Problem (16.8) thus is equivalent to the following problem, which has no positivity constraints on its variables z_k.

$$\left.\begin{aligned} &\sum_{k=1}^{m} |z_k| = \text{Min!} \\ &\sum_{k=1}^{m} v_{0k} z_k = 1, \\ &\sum_{k=1}^{m} v_{ik} z_k = 0, \qquad (i = 1,\ldots,n). \end{aligned}\right\} \tag{16.9}$$

Here we no longer have a linear optimization problem. If we set

$$y_k = \frac{z_k}{\sum_{\ell=1}^{m} |z_\ell|}$$

(16.9) is transformed to an equivalent problem,

$$\left.\begin{aligned} &\sum_{k=1}^{m} v_{0k} y_k = \text{Max!} \\ &\sum_{k=1}^{m} v_{ik} y_k = 0 \qquad (i = 1,\ldots,n) \\ &\sum_{k=1}^{m} |y_k| = 1, \end{aligned}\right\} \tag{16.10}$$

which coincides with (16.7) if we make the substitutions $y_k = y_k^+ - y_k^-$ and $|y_k| = y_k^+ + y_k^-$.

Conditions 1 and 2 of §16.1 should also be satisfied here. Condition 1 says that the n-by-m matrix $\underset{\sim}{V}$ with row vectors $\underset{\sim}{v}^{1'},\ldots,\underset{\sim}{v}^{n'}$ is of rank n. In addition, we now add the (Haar) condition

3. Every n-rowed square submatrix of $\underset{\sim}{V}$ is nonsingular.

Theorem 2: If condition 3 is satisfied, then problems (16.7) and (16.8) have no degenerate vertices.

Proof: Consider the constraints in (16.7),

$$\sum_k v_{ik}(y_k^+ - y_k^-) = 0 \qquad (i = 1,\ldots,n),$$

where $y_k^+ \geq 0$ and $y_k^- \geq 0$. $y_k^+ - y_k^-$ cannot be zero for all k because $\sum_k (y_k^+ + y_k^-) = 1$. Since any n column vectors of matrix $\underset{\sim}{V}$ are linearly independent, the constraints therefore can be satisfied only if at least $n + 1$ of the numbers y_k^+, y_k^- are different from zero. Now the number of constraints is $n + 1$. At a degenerate vertex, fewer than $n + 1$ of the y_k^+, y_k^- would be non-zero. The theorem is proven similarly for (16.8).

One property of the solutions of problems (16.4) and (16.5) can be read off of theorem 2.

Theorem 3: If condition 3 is satisfied and a solution of problem (16.4) has been found, then at least $n + 1$ of the constraints for this problem are satisfied with equality.

Proof: Since every feasible point of the dual problem (16.7) has at least $n + 1$ positive components, the conclusion follows from theorem 5, §5.1.

This theorem, which is also valid, word for word, for problem (16.5), implies the following for problem (16.1).

If condition 3 is satisfied, and a solution of problem (16.1) is at hand, the maximal value, γ, will be attained for at least $n + 1$ indices k.

In 16.1 it was suggested that the approximation problem (16.1) could be attacked by solving the optimization problem (16.5) or (16.6) by the simplex method. Alternatively, we could apply the simplex method to one of the dual problems, (16.7) and (16.8), which we have considered here, or for that matter, we could apply a suitable modification of the simplex method to one of the problems (16.9) and (16.10). This approach is taken by Stiefel, 1959.

We now want to give a geometric interpretation to problem (16.7). Let conditions 1, 2, and 3 be satisfied. By theorems 6 and 7 of §2.2, it suffices to search for a solution of problem (16.7) among the vertices of the set M of feasible points. None of these vertices is degenerate, by theorem 2, and so exactly $n + 1$ of the numbers y_k^+, y_k^- are different from zero at each such vertex, and furthermore, for no index k can $y_k^+ > 0$ and simultaneously, $y_k^- > 0$ (for then there would not be a linearly independent system of basis vectors). Thus there are exactly $n + 1$ indices k for which the numbers $y_k = y_k^+ - y_k^-$ are non-zero. Let these indices define an index set S. Every vertex of M then has a unique index set of this sort. Conversely, every such set S of $n + 1$ indices k uniquely determines a pair of vertices of the set S. For by condition 3, the linear system of equations,

$$\sum_{k \varepsilon S} v_{ik} y_k = 0 \qquad (i = 1,\ldots,n),$$

determines the y_k uniquely, up to a common factor, and makes them all non-zero (except for the trivial solution).

Because of the additional condition

$$\sum_{k\varepsilon S} |y_k| = 1,$$

the y_k are determined up to a common factor of ± 1. The y_k^+ and y_k^- are obtained from y_k, $k \varepsilon S$, by setting $y_k^+ = y_k$ and $y_k^- = 0$ if $y_k > 0$, and setting $y_k^+ = 0$ and $y_k^- = -y_k$ if $y_k < 0$.

For a given index set S, we now consider the approximation problem

$$\gamma_S = \underset{k\varepsilon S}{\mathrm{Max}} \left| v_{0k} - \sum_{i=1}^{n} a_i v_{ik} \right| = \mathrm{Min!}. \tag{16.11}$$

The point $\underset{\sim}{a} = (a_1,\ldots,a_n)'$ which solves this problem is the center of the ball inscribed in the simplex bounded by the $n + 1$ hyperplanes in R^n,

$$v_{0k} - \sum_{i=1}^{n} a_i v_{ik} = 0 \quad (k \varepsilon S), \tag{16.12}$$

whenever these hyperplane equations are in Hesse normal form, which we assume here for purposes of geometric interpretation. The minimal value γ_S is the radius of this inscribed ball.

Just like the corresponding problem (16.1), problem (16.11) can be written as a linear optimization problem of the form (16.4), and this problem can be dualized, yielding the problem

$$\left.\begin{array}{l} \sum\limits_{k\in S}(y_k^+ + y_k^-) = 1, \quad \sum\limits_{k\in S} v_{ik}(y_k^+ - y_k^-) = 0 \quad (i = 1,\ldots,n) \\ \qquad y_k^+ \geq 0, \quad y_k^- \geq 0 \qquad (k \in S) \\ \sum\limits_{k\in S} v_{0k}(y_k^+ - y_k^-) = \text{Max!}. \end{array}\right\} \quad (16.13)$$

It follows from the above considerations that there are exactly two vertices for this problem, that the objective function is non-negative at one of these vertices and non-positive at the other, and that the absolute value of the objective function is the same at both vertices. The solution of problem (16.13) is provided by the vertex at which the objective function is non-negative. The maximal value of the objective function of (16.13) is equal to the minimal value of the objective function of the dual problem, and thus equal to the radius of the inscribed ball of the simplex bounded by the hyperplanes (16.12). The objective functions of (16.7) and (16.13) thus agree at the vertex under consideration. Therefore, (16.7) poses the following problem.

From among all systems S of $n + 1$ indices k, and thus from among all simplices bounded by $n + 1$ hyperplanes (16.12), find that one for which the radius of the inscribed ball is maximal.

Figure 16.2 illustrates the case $n = 2$ and $m = 4$. The inscribed circles of the four simplexes which appear are shown. The center of the largest inscribed circle is the Tchebychev point of the problem.

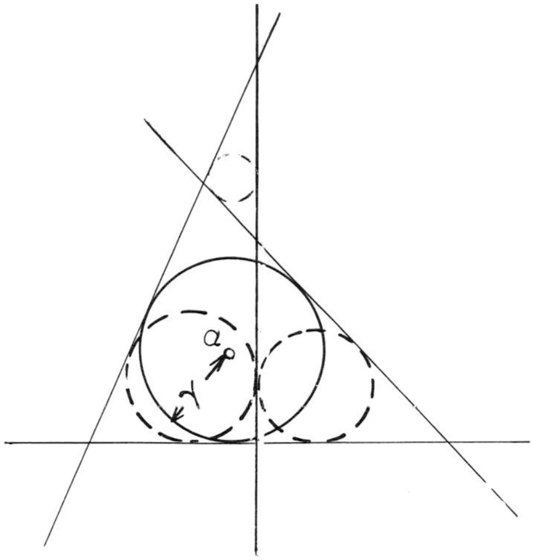

16.3. Further Problems of Discrete T-approximation

We list a few additional problems. These are so similar to those previously considered, that a brief formulation suffices.

A. Discrete linear T-approximation of several functions. Sometimes one is faced with the problem of simultaneously approximating several functions $f_\sigma(x)$, $\sigma = 1,\dots,s$, given on a domain B. Suppose we are to minimize

$$\max_{1 \le \sigma \le s} \left[\max_{x \in B} \left| f_\sigma(x) - \sum_{\nu=1}^{n} a_{\sigma\nu} v_{\sigma\nu}(x) \right| \right] = \tau, \tag{16.14}$$

where the $v_{\sigma\nu}(x)$ are given, fixed, continuous functions on B, and the constants $a_{\sigma\nu}$ may be chosen freely. With a choice of points $P_k \in B$, $k = 1,\dots,N$, we are lead to the linear optimization problem

$$-\gamma \le f_\sigma(P_k) - \sum_{\nu=1}^{n} a_{\sigma\nu} v_{\sigma\nu}(P_k) \le \gamma \ , \quad \begin{array}{l}(k = 1,\dots,N)\\ (\sigma = 1,\dots,s)\end{array} \tag{16.15}$$

$$\gamma = \text{Min!}$$

The situation is similar if a given function $f(x) = f_0(x)$ and certain of its derivatives are to be approximated simultaneously; for example, certain of the first partial derivatives, $f_j(x) = \partial f/\partial x_j$; where we again abbreviate $(x_1,\dots,x_n)$ by x.

B. Discrete one-sided T-approximation (cf. §15.4). In contrast to (16.1), the change now is that the second line of (16.4) has γ replaced by 0; thus we now have

$$\left.\begin{array}{l} 0 \le v_{0k} - \sum_{\nu=1}^{n} a_\nu v_{\nu k} \le \gamma, \quad (k = 1,\dots,m) \\ \gamma = \text{Min!} \end{array}\right\} \tag{16.16}$$

The same reformulations as in §16.1 and §16.2 are possible here, leading to other types of linear optimization problems.

C. Bounded least squares method. In this case, we would like to combine the advantages of the Gaussian and the Tchebychev approximation, by minimizing the root mean square deviation, $||f-v||_2$, while prescribing a maximum for the absolute deviation; see Krabs, 1964. Additionally, an approximation obtained in this way may be used as the initial estimate for an iterative process whenever we want to find the Tchebychev approximation. The prescribed bound on the deviation naturally must be larger than the minimal deviation for the Tchebychev approximation:

$$R > \inf_{v \varepsilon V} ||f - v||_\infty.$$

Thus the problem is to determine

$$\hat{\rho} = \inf_{v \varepsilon V_R} ||f - v||_\infty,$$

where V_R is the set defined by

$$V_R = \{v | v \ \varepsilon \ V, \quad ||f - v||_\infty \le R\}. \tag{16.17}$$

Again, we choose and fix points z_k, $k = 1,\ldots,N$, in B (which are no longer called x_k, so as to avoid confusion), and we again let $v_{k\nu} = v_\nu(z_k)$, $\nu = 1,\ldots,n$, $k = 1,\ldots,N$, $N > n$. The real matrix $\underset{\sim}{A} = (v_{k\nu})$ is to be of rank n.

We form a vector $\underset{\sim}{a} = (a_1,\ldots,a_n)'$ from the parameters a_ν, and also a vector $\underset{\sim}{f}$, from the functional values $f(z_k)$. Then we wish to determine $\underset{\sim}{a}$ so that

$$\Phi = ||\underset{\sim}{f} - \underset{\sim}{A}\underset{\sim}{a}||_2$$

is as small as possible.

We want

$$\Phi^2 = (\underset{\sim}{f}-\underset{\sim}{A}\underset{\sim}{a})'(\underset{\sim}{f}-\underset{\sim}{A}\underset{\sim}{a}) = ||\underset{\sim}{f}||_2^2 - 2\underset{\sim}{f}'\underset{\sim}{A}\underset{\sim}{a} + (\underset{\sim}{A}\underset{\sim}{a})'\underset{\sim}{A}\underset{\sim}{a} = \text{Min!}$$

or

$$Q(\underset{\sim}{a}) = \frac{1}{2}\,\underset{\sim}{a}'\underset{\sim}{C}\underset{\sim}{a} - \underset{\sim}{f}'\underset{\sim}{A}\underset{\sim}{a} = \text{Min!}. \tag{16.18}$$

Here $\underset{\sim}{C} = \underset{\sim}{A}'\underset{\sim}{A}$ is a positive definite matrix.

In addition, there are the constraints $||\underset{\sim}{f}-\underset{\sim}{A}\underset{\sim}{a}||_\infty \le R$. By using the vector $\underset{\sim}{e}$ whose components are all 1's, so

$\underset{\sim}{e} = (1, 1,\ldots,1)'$, we make these constraints read

$$\underset{\sim}{A}\underset{\sim}{a} \leq \underset{\sim}{f} + R\underset{\sim}{e}, \quad -\underset{\sim}{A}\underset{\sim}{a} \leq -\underset{\sim}{f} + R\underset{\sim}{e}. \tag{16.19}$$

(16.18) and (16.19) combine to form a problem of quadratic optimization, for finding the vector $\underset{\sim}{a}$, which satisfies the additional conditions (on the matrix, $\underset{\sim}{C}$, and on the linearity of the constraints) mentioned in §11.

§17. Further Types of Approximation Problems

17.1. Discrete Non-linear Tchebychev Approximation

Faced with the problem of approximating a function $f(x)$, in the Tchebychev sense on a finite set B of points x_k, $k = 1,\ldots,m$, by functions $g(x, a_1,\ldots,a_n)$ which are non-linear in the parameters $a_1,\ldots,a_n$, we may proceed by again formulating an optimization problem. If we set $f(x_k) = f_k$, the problem becomes

$$\left.\begin{aligned} &f_k - g(x_k,a_1,\ldots,a_n) - \gamma \leq 0, \quad (k = 1,\ldots,m) \\ &-f_k + g(x_k,a_1,\ldots,a_n) - \gamma \leq 0, \quad (k = 1,\ldots,m) \\ &\gamma = \text{Min!} \end{aligned}\right\} \tag{17.1}$$

This is a non-linear optimization problem in the variables $a_1,\ldots,a_n$, and γ. It is, however, not generally a convex optimization problem. For if g, as a function of the parameters $a_1,\ldots,a_n$, is not affine linear, g and $-g$ cannot be convex simultaneously.

17.2. Linear Continuous Tchebychev Approximation

Here we consider a problem already mentioned in §16.1, namely that of finding the minimum of the function

$$\Phi(\underset{\sim}{a}) = \Phi(a_1,\ldots,a_n) = \underset{B}{\mathrm{Max}}\left|f(x) - \sum_{i=1}^{n} a_i v_i(x)\right|,$$

where B is an infinite, closed, and bounded subset of R^q and $f(x)$, $v_i(x) \in C\langle B\rangle$. The function $\Phi(\underset{\sim}{a})$ is convex, since the functions

$$\phi(x,\underset{\sim}{a}) = f(x) - \sum_i a_i v_i(x)$$

are affine linear in $\underset{\sim}{a}$ for $x \in B$, so that $|\phi(x,\underset{\sim}{a})|$ is convex in $\underset{\sim}{a}$ for $x \in B$, because

$$\begin{aligned} |\phi(x,\lambda\underset{\sim}{a} + (1-\lambda)\underset{\sim}{b})| &= |\lambda\phi(x,\underset{\sim}{a}) + (1-\lambda)\phi(x,\underset{\sim}{b})| \\ &\le \lambda|\phi(x,\underset{\sim}{a})| + (1-\lambda)|\phi(x,\underset{\sim}{b})| \end{aligned}$$

for $0 \le \lambda \le 1$; finally, $\underset{x\in B}{\mathrm{Max}}|\phi(x,\underset{\sim}{a})|$ is convex if $|\phi(x,\underset{\sim}{a})|$ is convex for all $x \in B$. The problem of linear continuous T-approximation,

$$\Phi(\underset{\sim}{a}) = \mathrm{Min!}, \tag{17.2}$$

thus can be regarded as a problem of convex optimization (without constraints).

One may also write problem (17.2) in the form (cf. (16.4))

$$\left.\begin{aligned} &\left.\begin{aligned} \gamma + \sum_{i=1}^{n} a_i v_i(x) &\ge f(x), \\ \gamma - \sum_{i=1}^{n} a_i v_i(x) &\ge -f(x), \end{aligned}\right\} x \in B \\ &\gamma = \mathrm{Min!}. \end{aligned}\right\} \tag{17.3}$$

This is a linear optimization problem with infinitely many constraints. Investigations into treating such problems with the iterative Newton's method may be found in Cheney and Goldstein, 1959.

17.3. Non-linear Approximation which Leads to Non-convex Optimization Problems

Although the linear continuous T-approximation of §17.2 lead to convex optimization problems, such is not necessarily the case with non-linear T-approximation; this will be demonstrated with several examples.

A. Exponential Approximation. For this, a given function $f(x)$, continuous on a real interval $J = [a,b]$, is to be approximated as closely as possible in the Tchebychev sense, by an expression of the form

$$v = \sum_{\nu=1}^{n} a_\nu e^{b_\nu x} \tag{17.4}$$

with a suitable choice of real parameters a_ν and b_ν. If we set

$$\Phi = \Phi(a_\nu, b_\nu) = ||f-v||_\infty = \operatorname*{Max}_{x\varepsilon J} |f(x)-v(x)|, \tag{17.5}$$

the objective of minimizing the value of Φ leads in general to a non-convex function Φ, and thus to a non-convex optimization problem. To show non-convexity, it suffices to give a single counter example in one parameter. Suppose the function $f(x) = e^x$ is to be approximated as closely as possible by a function $v(x,b) = e^{bx}$ on the interval $J = [0,1]$.

In this case, the maximal deviation will be at $x = 1$; see Figure 17.1. Thus $\Phi = \Phi(b) = |e - e^b|$, and this function has the appearance of the solid curve in Figure 17.2 and is not convex.

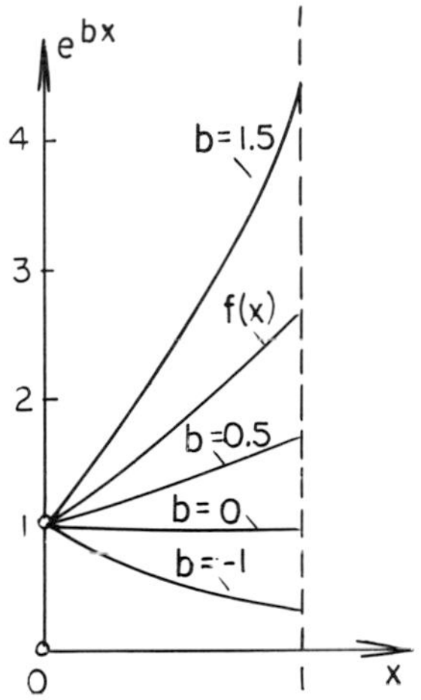

Figure 17.1

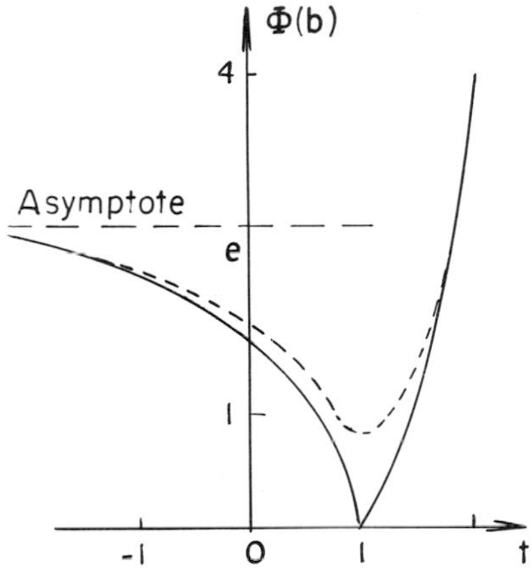

Figure 17.2

The objection that this is a problem of representation (cf. §15.1) and not of approximation because $f(x)$ itself belongs to the class of functions e^{bx}, is met by observing that the function $f(x) = e^x + \varepsilon x$ (where ε is small, e.g., $\varepsilon = 0.01$) is qualitatively almost the same and allows the same considerations; the function $\Phi(b)$ now has the appearance of the dashed curve in Figure 17.2 and is also not convex. This phenomenon is not limited to T-approximation, but appears in the same way in Gaussian or least mean square approximation.

B. Trigonometric Approximation; see §15.3. In general, this too does not lead to a convex optimization problem. For example, approximate the function $f(x) = 1 - x$ by functions $v(x,a) = \cos(ax)$ on the interval $J = [0,1]$. Again, we assume that this is a T-approximation (for Gaussian approximation, the phenomenon is the same). See Figure 17.3.

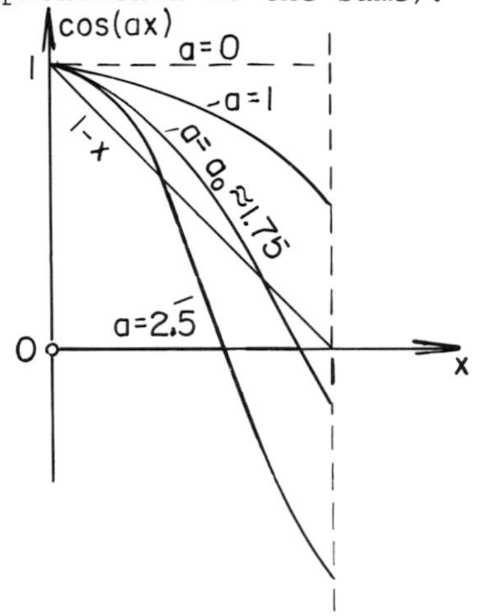

Figure 17.3

By (17.5), the function $\Phi = \Phi(a)$ has the appearance represented in Figure 17.4 and is not convex.

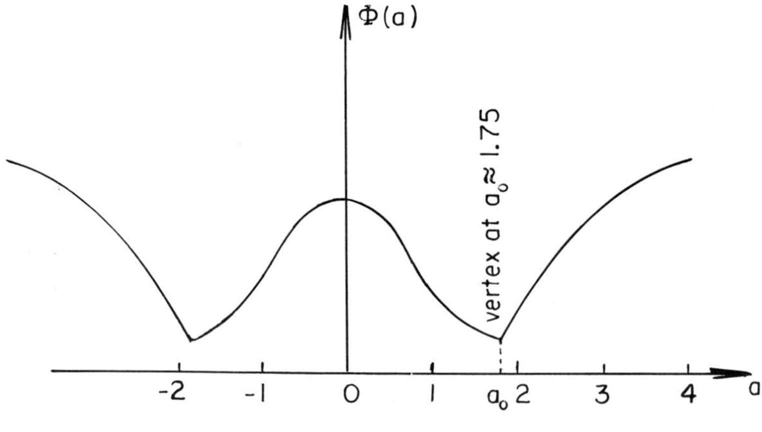

Figure 17.4

C. Rational Approximation. To see that in this case in general there also is no convexity, consider the example of the function $f(x) = \frac{1}{1+x}$ which is to be T-approximated on $[0,1]$ by functions $v(x,a) = \frac{1}{a+x}$ (where $a > 0$); see Figure 17.5.

By (17.5), the function $\Phi = \Phi(a)$ is the one represented by the solid curve in Figure 17.6 and is not convex. If we use the function $f(x) = \frac{1}{1+x} + \varepsilon e^{-x}$ (where again ε is small, say $\varepsilon = 0.01$), the vertex is smoothed away (the dashed curve in Figure 17.6), but the corresponding function $\Phi(a)$ remains non-convex.

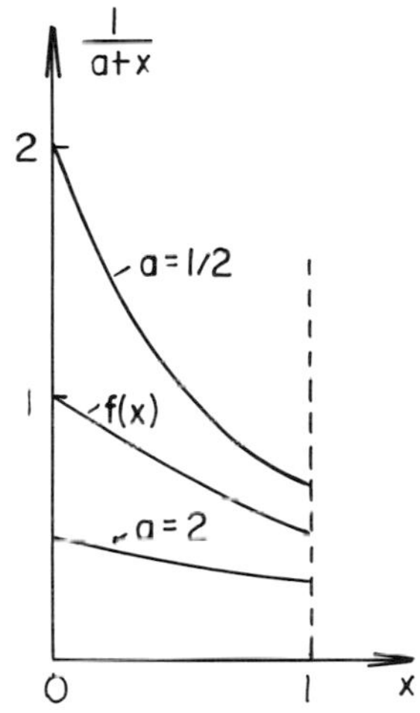

Figure 17.5

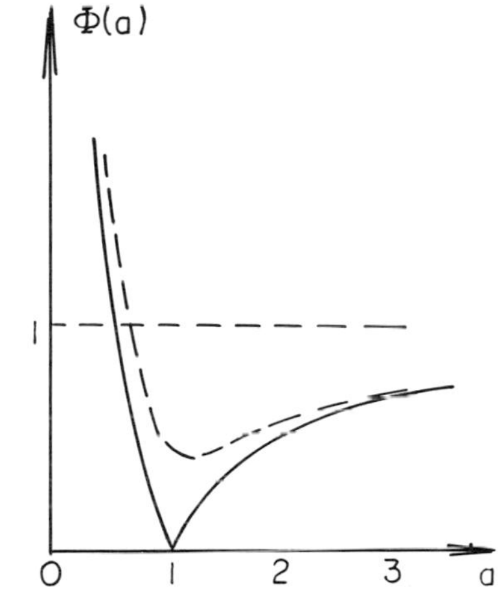

Figure 17.6

17.4. Separation Problems and Optimization

Separation problems appear in a certain counterpoint to approximation problems, although the two are mathematically equivalent. We again have exactly the same situation as in §15.1. A fixed function $f(x)$ and a class of functions $V = \{g(x, a_1,...,a_n)\}$, all in the space $C\langle B\rangle$, are given. Now, for fixed values of the parameters $a_1,...,a_n$ we set

$$\phi = \phi(a_\nu) = \underset{x\varepsilon B}{\text{Min}}|f(x) - g(x,a_\nu)|$$

and

$$\Phi = \Phi(a_\nu) = \underset{x\varepsilon B}{\text{Max}} |f(x) - g(x,a_\nu)|.$$

For T-approximation one asks for the minimum of $\Phi(a_\nu)$, and for separation, for the maximum of $\phi(a_\nu)$. Since the two problems are formally equivalent, separation problems naturally lead also to optimization problems. Nevertheless, they are presented here especially because in applications they often appear in a form where the separation is to be kept as large as possible.

Example I (from machine design). Certain machine parts with a periodic motion are to be arranged so that they not only do not come in contact when in motion, but actually remain separated by the greatest possible distance. This might be for reasons of safety, or, in the case of electrically charged parts, to avoid a spark or arcing over the gap. For example, if the movement of a machine part, or at least of a point on this part, is given by $f(t) = \sin(t) + (1/2)\sin(2t)$, and the movement of a point on another machine part is given by $g(t,t_0) = -3/2 + \sin(2(t-t_0))$, the problem would be to find t_0 so that the minimal separation,

$$\underset{t\varepsilon[0,2\pi]}{\text{Min}} |f(t)-g(t,t_0)| = \phi(t_0),$$ becomes as large as possible;

see Figure 17.7.

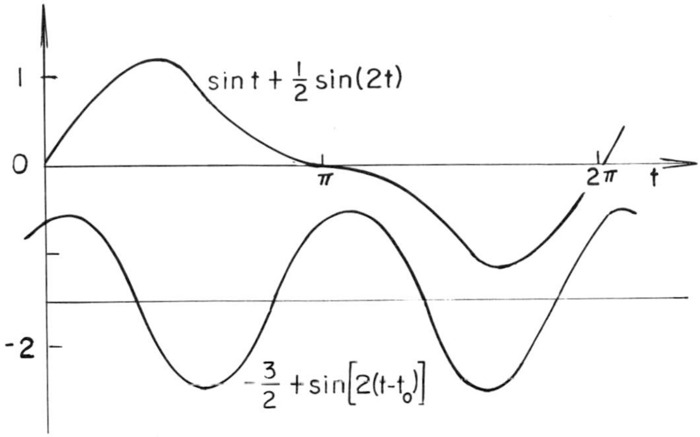

Figure 17.7

Example II (the unfriendly brothers). This is an example of a non-linear separation problem. We are to locate n antagonistic brothers in a region (e.g., a rectangle with sides of length 1 and a, representing the land on which they will build their houses), so that the minimal distance, ρ, between any two of the brothers is as large as possible. Figure 17.8 shows the solution for a square ($a = 1$) and $n = 6$. Another interpretation: n students are

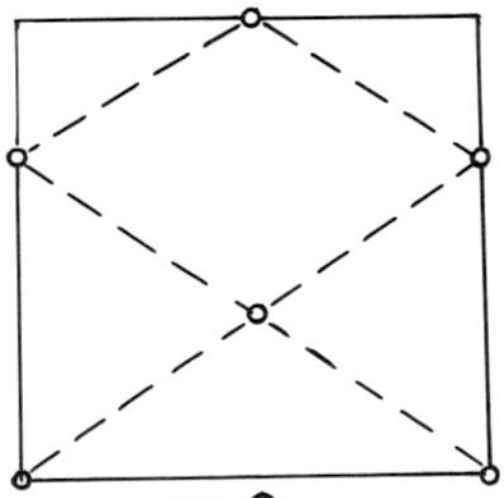

Figure 17.8

to be seated in a rectangular examination room so as to minimize the possibility of interaction. To formulate the optimization problem, we let (x_j, y_j) be the coordinates of person P_j, $j = 1,\ldots,n$, and then we have

$$(x_j - x_k)^2 + (y_j - y_k)^2 \geq \gamma \quad \text{for} \quad 1 \leq j < k \leq n$$
$$0 \leq x_j \leq a, \quad 0 \leq y_j \leq 1 \quad \text{for} \quad j = 1,\ldots,n$$
$$\gamma = \text{Max!}.$$

For the case $n = 5$, Figure 17.9 shows the shifts in the solution which occur as a increases.

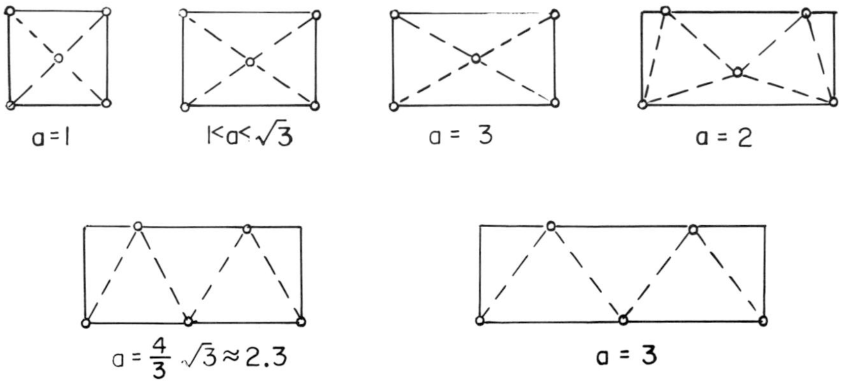

Figure 17.9

17.5. Linear T-approximation for Complex Variables

Let z be a complex variable, and let $f(z)$ and $w_\nu(z)$, $\nu = 1,\ldots,p$, be given holomorphic functions defined on some region B of the complex plane. We want to find complex parameters, a_ν, so that the error,

$$\varepsilon(z,a_\nu) = \sum_{\nu=1}^{p} a_\nu w_\nu(z) - f(z), \tag{17.6}$$

is in some sense as small as possible on B. In the discrete case, we select and fix points z_j, $j = 1,\ldots,m$, in B and demand that the function

$$\Phi(a_\nu) = \operatorname*{Max}_j |\varepsilon(z_j,a_\nu)| \tag{17.7}$$

be as small as possible. In the continuous case, however, we want the minimum of the function

$$\hat{\Phi}(a_\nu) = \operatorname*{Max}_{z\varepsilon B} |\varepsilon(z,a_\nu)|. \tag{17.8}$$

I. Discrete T-approximation.

In place of (17.7), we may put

$$\psi(a_\nu) = \operatorname*{Max}_j |\varepsilon^2(z_j,a_\nu)| \tag{17.9}$$

and then look for the minimum of this function. With the usual abbreviations and separation into real and imaginary parts, we have

$$\left.\begin{aligned} f_j &= f(z_j) = g_j + ih_j \\ w_{\nu j} &= w_\nu(z_j) = u_{\nu j} + iv_{\nu j} \\ a_\nu &= b_\nu + ic_\nu. \end{aligned}\right\} \quad \begin{aligned} &(j = 1,\ldots,m) \\ &(\nu = 1,\ldots,p) \end{aligned}$$

Letting

$$S_j = [\sum_\nu (b_\nu u_{\nu j} - c_\nu v_{\nu j}) - g_j]^2 + \\ + [\sum_\nu (b_\nu v_{\nu j} + c_\nu u_{\nu j}) - h_j]^2,$$

(17.9) becomes

$$\psi(a_\nu) = \max_j S_j.$$

The attempt to minimize this function leads to the convex optimization problem

$$S_j - \delta \leq 0, \quad \delta = \text{Min}; \tag{17.10}$$

the variables, δ, b_ν, c_ν, are real.

Discrete linear T-approximation for real variables led to linear optimization in §16.1; here, for complex variables, it leads to convex optimization.

II. Continuous T-approximation

Here we obtain a convex optimization problem, just as in the real case in 17.2, because the proof that the function $|\phi(x,\underset{\sim}{a})|$ of that section is convex in $\underset{\sim}{a}$ for fixed x remains unchanged if x and $\underset{\sim}{a}$ are complex. And the conclusion that the maximum of this function is again convex for $x \varepsilon B$ also remains valid in the complex case.

III. Numerical Example

The function

$$f(z) = \frac{3}{2 + z^2}$$

is to be approximated by a linear function of the form $w = a + bz$, where this is to be a discrete T-approximation at the points $P_j = (i)^{j-1}$, $j = 1, 2, 3, 4$. The solution is the constant function, $w = 2$, with minimal deviation $\rho_0 = 1$.

V. ELEMENTS OF GAME THEORY

Game theory contains one important application of the theory of linear optimization, including the duality theorems. In recent times, game theory has come to be of great significance with the realization that many economic and behavioral questions may be formulated as game-theoretic problems and that statistical decision theory is closely connected to game theory (Ferguson, 1967).

The main theorem in the theory of matrix games follows immediately from the theorems of §5.

§18. Matrix Games (Two Person Zero Sum Games)

18.1. Definitions and Examples

We consider a game with two players, P_1 and P_2, each of whom has available a certain set of possible courses of action, called moves for short. The set of moves for player P_i will be denoted by Σ_i, $i = 1, 2$, and the elements of

Σ_i, that is, the moves which player P_i may make, by σ^i.

To begin with, the definition will be tightened to allow only finite sets of moves:

$$\Sigma_1 = \{\sigma_1^1, \ldots, \sigma_m^1\}; \quad \Sigma_2 = \{\sigma_1^2, \ldots, \sigma_n^2\}.$$

The game proceeds as follows. P_1 and P_2 each simultaneously decide on a move; then one will have to pay another an amount which depends on the moves chosen by each. This completes the game. Before the game begins, then, it is once and for all made clear to both of the players that if P_1 chooses move σ_j^1 and P_2 simultaneously chooses move σ_k^2, then P_2 will have to pay an amount of a_{jk} to P_1. All of these previously agreed upon, possible pay-off amounts are collected in the obvious way into a pay-off matrix, $\underset{\sim}{A} = (a_{jk})$:

P_1 \ P_2	σ_1^2	σ_2^2	$\cdots$	σ_n^2
σ_1^1	a_{11}	a_{12}	$\cdots$	a_{1n}
$\vdots$	$\vdots$	$\vdots$		$\vdots$
σ_m^1	a_{m1}	a_{m2}	$\cdots$	a_{mn}

Pay-offs by P_2 to P_1

$\underset{\sim}{A}$ is the table of "winnings" for P_1 (which of course includes negative winnings, i.e., losses). Such a matrix game is totally determined by Σ_1, Σ_2, and A. In describing the elements, a_{jk}, of the pay-off matrix as amounts, we used the word "amount" in the broad sense. It may indeed refer to amounts of money, but may just as well refer simply to the bare fact of winning or losing. However, the matrix elements are required to be real numbers, always.

In terms of the definition given in §19 below, matrix games are two person zero sum games. A matrix game is called "fair" if the players have equal chances of winning; otherwise it is called "unfair". Just what "equal chances" might mean will be made precise in §18.3.

Examples.

1. Stone-Paper-Scissors Game. Both players have the three options "stone", "paper", and "scissors". The rules are that stone beats scissors, scissors beats paper, and paper beats stone. Letting 1 stand for winning, -1 for losing, and 0 for undecided, we have the pay-off matrix

P_1 \ P_2	Stone	Paper	Scissors
Stone	0	-1	1
Paper	1	0	-1
Scissors	-1	1	0

It is evident that the game is fair.

2. Skin Game (due to Kuhn, 1957). P_1 and P_2 each have a three card hand; P_1 has an ace of diamonds, an ace of spades, and a deuce of diamonds; P_2 has an ace of diamonds, an ace of spades, and a deuce of spades. The players simultaneously play one of their cards. P_1 wins if the

played cards are the same color; otherwise P_2 wins. An ace scores one point, and a spade, two. The amount won is equal to the number of points on the winning card. The game has the pay-off matrix

P_1 \ P_2	♢	♠	♠ ♠
♢	1	-1	-2
♠	-1	1	1
♢ ♢	2	-1	-2

This pay-off matrix contains 5 negative elements, as opposed to 4 positive elements, leaving the impression that it is unfair. Thus we are motivated to formulate the

<u>Additional Rule</u>: When a deuce is played by both, there will be no pay-off, i.e., the lower right-hand element of the pay-off matrix will be changed from -2 to 0.

When the additional rule is in effect, the sum of all the matrix elements is zero. This might lead one to believe that the game is now fair. In §18.2 we will see that the opposite is true: the skin game without the additional rule is the fair game, and the skin game with the additional rule is the unfair game.

3. <u>General Blotto Game</u>. The enemy threatens town S with four divisions. There are two approaches, w_1 and w_2, to S from the town P where the enemy, with all four divisions, is located. The enemy has the option of dividing

his troops, though not of subdividing a division, and sending some along w_1 and the rest along w_2.

In order to defend the town, General Blotto must use his five divisions to deny the enemy access. He, too, may split his troops between w_1 and w_2 as he wishes, so long as he does not subdivide any division. If one of the approaches has a superior number of enemy divisions on it, the enemy will break though and conquer the town. In the pay-off

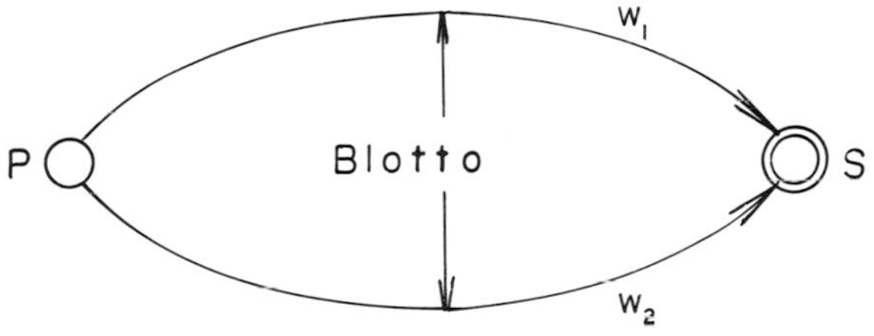

Figure 18.1

matrix below, 1 denotes a victory for the attacker and -1 a victory for General Blotto.

		Blotto on w_1	5	4	3	2	1	0
Enemy on		on w_2	0	1	2	3	4	5
w_1	w_2							
4	0		-1	-1	1	1	1	1
3	1		1	-1	-1	1	1	1
2	2		1	1	-1	-1	1	1
1	3		1	1	1	-1	-1	1
0	4		1	1	1	1	-1	-1

18.2. Strategies

The matrix game with the sets of moves $\Sigma_1 = \{\sigma_1^1\ldots, \sigma_m^1\}$ for player P_1 and $\Sigma_2 = \{\sigma_1^2,\ldots,\sigma_n^2\}$ for player P_2 is to be repeated several times in succession. Player P_1 chooses move σ_j^1 with a relative frequency x_j, $j = 1,\ldots,m$, letting some random mechanism with the appropriate probabilities decide which move to make next.

We have

$$0 \leq x_j \leq 1 \ (j = 1,\ldots,m); \quad \sum_{j=1}^{m} x_j = 1. \qquad (18.1)$$

Definition. The vector $\underset{\sim}{x} = (x_1,\ldots,x_m)'$ is called a strategy for P_1. If one component of $\underset{\sim}{x}$ is equal to 1 and the rest are equal to 0, $\underset{\sim}{x}$ is called a pure strategy; otherwise, $\underset{\sim}{x}$ is a mixed strategy.

Every vector $\underset{\sim}{x} = (x_1,\ldots,x_m)'$ such that $\underset{\sim}{x} \geq \underset{\sim}{0}$ and $\sum_j x_j = 1$ is a possible strategy for P_1.

Correspondingly, every vector $\underset{\sim}{y} = (y_1,\ldots,y_n)'$ such that $\underset{\sim}{y} \geq \underset{\sim}{0}$ and $\sum_k y_k = 1$ is a possible strategy for P_2.

Examples. 1) Suppose the pay-off matrix of a game (where $m \geq 2$ and $n = 4$) is

P_1 \ P_2	σ_1^2	σ_2^2	σ_3^2	σ_4^2
σ_1^1	2	-3	-1	1
σ_2^1	0	2	1	2

Assume for the moment that P_2 always chooses move σ_k^2, thus pursuing a pure strategy. Then the expected value of the pay-off to P_1 is

$$E = x_1 a_{1k} + (1 - x_1) a_{2k} \tag{18.2}$$

per game, if P_1 follows the strategy $(x_1, 1-x_1)'$. In Figure 18.2, the expected values of the pay-offs to P_1 per game corresponding to the four possible pure strategies of P_2 are shown as functions of x_1.

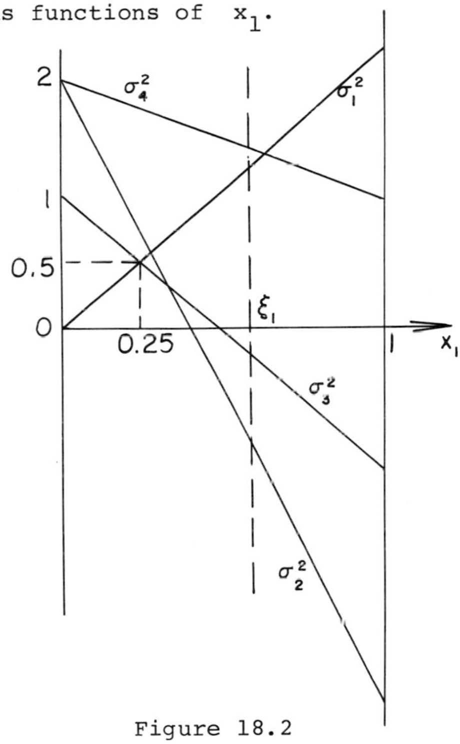

Figure 18.2

The move which P_2 always makes for any given pure strategy is indicated next to the corresponding line segment. Should P_1 pursue the strategy $(\xi_1, 1-\xi_1)'$, P_2 will be able to win on the average if he always makes move σ_2^2.

It is easy to see that there is one and only one optimal strategy for P_1, namely $\underset{\sim}{x} = (0.25, 0.75)'$. With this strategy, the average minimal win for P_1 is $v = 0.5$ per game. v is called the *value of the game*. Since $v > 0$, the chances are unequal, and the game is unfair in favor of P_1.

The optimal strategy for P_2 is also determined uniquely. One can see from Figure 18.2 that P_2 must never choose either of the moves σ_2^2 or σ_4^2, i.e., P_2's strategy must be of the form $\underset{\sim}{y} = (y_1, 0, 1-y_1, 0)'$, where no conditions are placed on y_1 for the moment. The expected value of the pay-off to P_1 per game is

$$\left.\begin{array}{ll} 2y_1 - (1-y_1), & \text{if } P_1 \text{ always chooses } \sigma_1^1, \\ 0\cdot y_1 + (1-y_1), & \text{if } P_1 \text{ always chooses } \sigma_2^1. \end{array}\right\} \quad (18.3)$$

The condition

$$2y_1 - (1-y_1) \le 0.5; \quad 1 - y_1 \le 0.5 \quad (18.4)$$

implies that $y_1 = 0.5$. Then we obtain uniquely determined optimal strategies for P_1 and P_2:

$$\left.\begin{array}{l} \underset{\sim}{x} = (1/4, 3/4)'; \quad \underset{\sim}{y} = (1/2, 0, 1/2, 0)'; \\ \text{value of the game: } v = 1/2. \end{array}\right\} \quad (18.5)$$

2a) *Skin game with additional rule.* (See §18.1, example 2.) We introduce the following notation for the moves of P_1 and P_2.

$$\sigma_1^1 = \diamondsuit, \quad \sigma_2^1 = \spadesuit, \quad \sigma_3^1 = \diamondsuit\,\diamondsuit;$$

$$\sigma_1^2 = \diamondsuit, \quad \sigma_2^2 = \spadesuit, \quad \sigma_3^2 = \spadesuit\,\spadesuit.$$

The pay-off matrix then is

P_1 \ P_2	σ_1^2	σ_2^2	σ_3^2
σ_1^1	1	-1	-2
σ_2^1	-1	1	1
σ_3^1	2	-1	0

We again begin by assuming that P_2 always makes moves σ_k^2. If $(x_1, x_2, x_3)'$ is the strategy for P_1, the expected value of the pay-off to P_1 per game is

$$E = x_1a_{1k} + x_2a_{2k} + x_3a_{3k}. \tag{18.6}$$

As in Example 1), we can graph the three planes corresponding to the three pure strategies available to P_2. In Figure 18.3, the triangle in the (x_1, x_2)-plane with the bold border contains all possible strategies for P_1. We see from the illustration that there is one and only one optimal strategy for P_1, namely $\underset{\sim}{x} = (0, 0.6, 0.4)'$. For this strategy, the average minimum win for P_1 per game is $v = 0.2 > 0$. Since the value v of the game is positive, the game is unfair in favor of P_1. The illustration also shows that P_2 must never choose the move σ_3^2, i.e., P_2 must pursue a strategy of the form $\underset{\sim}{y} = (y_1, 1-y_1, 0)'$. As in Example 1) where (18.4) was required, we here place the following conditions on y_1:

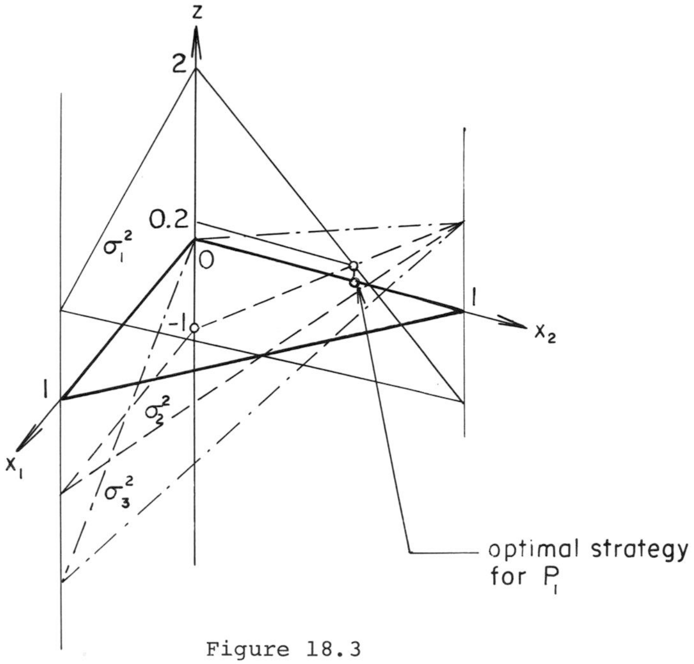

Figure 18.3

$$y_1 - (1-y_1) \leq 0.2; \quad -y_1 + (1-y_1) \leq 0.2;$$
$$2y_1 - (1-y_1) \leq 0.2.$$

This implies that $y_1 = 0.4$. Thus the uniquely determined optimal strategies for P_1 and P_2 are

$$\left.\begin{array}{l} \underset{\sim}{x} = (0,\ 0.6,\ 0.4)'; \quad \underset{\sim}{y} = (0.4,\ 0.6,\ 0)'; \\ \text{value of the game:} \quad v = 0.2. \end{array}\right\} \quad (18.7)$$

2b) Skin game without additional rule. Dropping the additional rule alters the position of the plane corresponding to σ_3^2 in Figure 18.3. This plane no longer passes through the origin, but rather through the point $(x_1, x_2, z) = (0, 0, -2)$. It is easy to see from the illustration that the optimal strategy for P_1 now is $\underset{\sim}{x} = (0, 2/3, 1/3)'$. With this strategy, the average minimum win for P_1 per game is $v = 0$. Thus the game is not unfair in favor of P_1, but fair.

Similarly one can easily see that P_2 must never choose move σ_2^2 now, i.e., P_2 must pursue a strategy of the form $(y_1, 0, 1-y_1)'$. As in Example 2a), we place the following conditions on y_1:

$$y_1 - 2(1-y_1) \leq 0; \qquad -y_1 + (1-y_1) \leq 0;$$
$$2y_1 - 2(1-y_1) \leq 0;$$

this implies that $y_1 = 0.5$. Then the optimal strategies are

$$\left.\begin{array}{l} \underset{\sim}{x} = (0, 2/3, 1/3)', \quad \underset{\sim}{y} = (1/2, 0, 1/2)'; \\ \text{value of the game:} \quad v = 0. \end{array}\right\} \tag{18.8}$$

18.3. Attainable Wins and Saddlepoint Games

Let the set of moves for player P_1, and P_2 respectively, be

$$\Sigma_1 = \{\sigma_1^1, \ldots, \sigma_m^1\}, \quad \Sigma_2 = \{\sigma_1^2, \ldots, \sigma_n^2\}.$$

Let $\underset{\sim}{A} = (a_{jk})$ be the pay-off matrix for the pay-offs from P_2 to P_1. The strategies for P_1 will be denoted by $\underset{\sim}{x} = (x_1, \ldots, x_m)'$ and the strategies for P_2, by $\underset{\sim}{y} = (y_1, \ldots, y_n)'$. The set of strategies for P_1 is the following set of vectors in R^m:

$$D_1 = \{\underset{\sim}{x} \in R^m \mid \underset{\sim}{x} \geq \underset{\sim}{0} \text{ and } \sum_{j=1}^{m} x_j = 1\}.$$

Correspondingly,

$$D_2 = \{\underset{\sim}{y} \in R^n | \underset{\sim}{y} \geq \underset{\sim}{0} \quad \text{and} \quad \sum_{k=1}^{n} y_k = 1\}$$

is the set of strategies for P_2. The sets D_1 and D_2 are closed and bounded; a continuous function attains its maximum and minimum on such sets.

If P_1 pursues strategy $\underset{\sim}{x}$ and P_2 pursues strategy $\underset{\sim}{y}$, stochastically independently from P_1, the expected value of the pay-off to P_1 per game is

$$W = \sum_{j,k} x_j a_{jk} y_k = \underset{\sim}{x}' \underset{\sim}{A} \underset{\sim}{y}. \tag{18.9}$$

(a) We begin by asking which strategies P_1 must choose in order to maximize W. Because P_2 conversely will attempt to minimize W, P_1 must choose an $\underset{\sim}{x} = \underset{\sim}{\tilde{x}} \in D_1$ so that $\underset{\underset{\sim}{y} \in D_2}{\text{Min}} \underset{\sim}{x}' \underset{\sim}{A} \underset{\sim}{y}$ is as large as possible. Since every $\underset{\sim}{y} \in D_2$ is a convex combination of the n pure strategies for P_2, we have

$$\phi(\underset{\sim}{x}) = \underset{\underset{\sim}{y} \in D_2}{\text{Min}} \underset{\sim}{x}' \underset{\sim}{A} \underset{\sim}{y} = \underset{k}{\text{Min}} \sum_j x_j a_{jk} \tag{18.10}$$

for all $\underset{\sim}{x} \in D_1$. P_1 therefore can restrict his attention to the pure strategies of P_2, only. Thus P_1 must choose a strategy $\underset{\sim}{\tilde{x}} \in D_1$ so as to maximize $\phi(\underset{\sim}{\tilde{x}})$. Let

$$v_1 = \underset{\underset{\sim}{x} \in D_1}{\text{Max}} \phi(\underset{\sim}{x}) = \underset{\underset{\sim}{x} \in D_1}{\text{Max}} \underset{k}{\text{Min}} \sum_j x_j a_{jk}. \tag{18.11}$$

(b) Now we ask which strategy P_2 must choose in order to minimize W. As in (a), we find that $\underset{\sim}{\tilde{y}} \in D_2$ must be chosen so that

$$\psi(\underset{\sim}{y}) = \underset{\underset{\sim}{x} \in D_1}{\text{Max}} \underset{\sim}{x}' \underset{\sim}{A} \underset{\sim}{y} = \underset{j}{\text{Max}} \sum_k a_{jk} y_k \tag{18.12}$$

is as small as possible. Let

$$v_2 = \min_{\underset{\sim}{y}\varepsilon D_2} \psi(\underset{\sim}{y}) = \min_{\underset{\sim}{y}\varepsilon D_2} \max_j \sum_k a_{jk}y_k. \tag{18.13}$$

As previously noted, D_1 and D_2 are closed and bounded, so that the continuity of the functions $\phi(\underset{\sim}{x})$ and $\psi(\underset{\sim}{y})$ implies that there exist vectors, $\underset{\sim}{\tilde{x}} \varepsilon D_1$ and $\underset{\sim}{\tilde{y}} \varepsilon D_2$, such that

$$v_1 = \phi(\underset{\sim}{\tilde{x}}), \quad v_2 = \psi(\tilde{y}).$$

Since $\underset{\sim}{x} \geq \underset{\sim}{0}$, $\underset{\sim}{y} \geq \underset{\sim}{0}$, and the sum of the components is always 1, we have the estimates

$$\left.\begin{aligned} v_1 &= \phi(\underset{\sim}{\tilde{x}}) = \min_k \sum_j \tilde{x}_j a_{jk} \leq \sum_{j,k} \tilde{x}_j a_{jk} \tilde{y}_k \leq \\ &\leq \max_j \sum_k a_{jk}\tilde{y}_k = \psi(\tilde{y}) = v_2, \end{aligned}\right\} \tag{18.14}$$

and therefore

$$v_1 \leq v_2. \tag{18.15}$$

<u>Remark</u>. If player P_1 pursues pure strategies only, he can at least attain an average win of

$$w_1 = \max_j \min_k a_{jk}.$$

Similarly, if player P_2 uses only pure strategies, he can at most suffer an average loss of

$$w_2 = \min_k \max_j a_{jk}.$$

Naturally

$$v_1 \geq w_1, \quad v_2 \leq w_2 \tag{18.16}$$

since we reduced the number of elements to be considered in forming the maximum and minimum, respectively. Because of (18.15) this implies that

$$w_1 \le v_1 \le v_2 \le w_2. \tag{18.17}$$

<u>Definition</u>. If $w_1 = w_2$, the game is called a <u>saddle point game</u>. By (18.17), $w_1 = w_2$ implies that

$$w_1 = v_1 = v_2 = w_2. \tag{18.18}$$

Since v_1 is the highest win and v_2 is the lowest loss which P_1 and P_2 respectively can attain in the game, it follows from (18.18) that for saddle point games there is a pure strategy for both players which is already optimal. These optimal strategies for P_1 and P_2 can be read directly off of the pay-off matrix. For this reason, saddle point games constitute a trivial special case of matrix games.

<u>Examples</u>.

1. <u>Skin game with additional rule</u>. The pay-off matrix is

$$\begin{pmatrix} 1 & -1 & -2 \\ -1 & 1 & 1 \\ 2 & -1 & 0 \end{pmatrix}.$$

We have

$$w_1 = \operatorname*{Max}_j \operatorname*{Min}_k a_{jk} = \operatorname{Max} \; (-2,\ -1,\ -1) = -1;$$

$$w_2 = \operatorname{Min} \operatorname{Max} a_{jk} = \operatorname{Min} \; (2,1,1) = 1 > -1 = w_1.$$

Therefore this game is not a saddle point game.

2. A game with pay-off matrix

$$\begin{pmatrix} 3 & \boxed{1} & 4 \\ 2 & 0 & 1 \\ -1 & -2 & -2 \end{pmatrix}.$$

In this case,

$$w_1 = \operatorname*{Max}_j \operatorname*{Min}_k a_{jk} = \operatorname{Max} \ (1,0,-2) = 1;$$

$$w_2 = \operatorname*{Min}_k \operatorname*{Max}_j a_{jk} = \operatorname{Min} \ (3,1,4) = 1 = w_1.$$

Thus the game is a saddle point game and the optimal strategies for P_1 and P_2 are

$$\underset{\sim}{x} = (1, 0, 0)' \quad \text{and} \quad \underset{\sim}{y} = (0, 1, 0)',$$

respectively.

18.4. The Minimax Theorem

We will show that the condition $v_1 = v_2$ is satisfied in general for matrix games, and not just for saddle point games.

Theorem 1 (The Minimax Theorem): Let Σ_1 and Σ_2 be finite sets of moves and let $\underset{\sim}{A}$ be a pay-off matrix for a game. If v_1 and v_2 are defined as in (18.11) and (18.13), then

$$v_1 = v_2. \tag{18.19}$$

Proof: Interpreting (18.11) as a rule for determining $\underset{\sim}{\tilde{x}} \in D_1$ allows us to write it as a linear optimization problem:

$$\left.\begin{array}{l} \sum_{j=1}^{m} x_j = 1 \\ \sum_{j=1}^{m} x_j a_{jk} + \xi \geq 0 \quad (k = 1,\ldots,n) \\ x_j \geq 0 \quad (j = 1,\ldots,m), \\ \xi \text{ not positively constrained} \\ \xi = \text{Min!} \ . \end{array}\right\} \tag{18.20}$$

We have already shown that this problem has a solution, given by $\underset{\sim}{x} = \tilde{\underset{\sim}{x}}$ and $\xi = -v_1$. Similarly we can write (18.13) as a linear optimization problem:

$$\left.\begin{array}{l} \eta + \sum_{k=1}^{n} a_{jk} y_k \leq 0 \quad (j = 1,\ldots,m) \\ \qquad \sum_{k=1}^{n} y_k = 1 \\ \eta \text{ not sign constrained, } y_k \geq 0 \ (k = 1,\ldots,n) \\ \eta = \text{Max!} \ . \end{array}\right\} \tag{18.21}$$

The solution of this problem is given by $\underset{\sim}{y} = \tilde{\underset{\sim}{y}}$ and $\eta = -v_2$. By §5.3, (18.20) and (18.21) are a pair of dual linear optimization problems. Both have solutions. Therefore the extreme values of the objective functions, $-v_1$ and $-v_2$, must be equal.

<u>Definition</u>. The quantity $v = v_1 = v_2$ is called the <u>value</u> of the matrix game.

In view of equations (18.10) through (18.13), the Minimax Theorem implies that every matrix game has a value v, namely

$$v = \underset{\underset{\sim}{x}\in D_1}{\text{Max}} \ \underset{\underset{\sim}{y}\in D_2}{\text{Min}} \ \underset{\sim}{x}'A\underset{\sim}{y} = \underset{\underset{\sim}{y}\in D_2}{\text{Min}} \ \underset{\underset{\sim}{x}\in D_1}{\text{Max}} \ \underset{\sim}{x}'A\underset{\sim}{y}. \tag{18.22}$$

If $\tilde{x} \in D_1$ and $\tilde{y} \in D_2$ are vectors such that $v = \phi(\tilde{x}) = \psi(\tilde{y})$, then it follows from (18.14), with the help of (18.10) and (18.12), that

$$x'A\tilde{y} \leq v = \tilde{x}'A\tilde{y} \leq \tilde{x}'Ay \tag{18.23}$$

for all $x \in D_1$ and all $y \in D_2$. This implies that $(\tilde{x},\tilde{y})$ is a saddle point of the bilinear form $x'Ay$.

$\tilde{x}$ and $\tilde{y}$ are the optimal strategies for P_1 and P_2 respectively. If P_1 pursues strategy $\tilde{x}$, the expected value of the pay-off per game to him is v, regardless of the strategy y which player P_2 might use. The corresponding comment holds for the corresponding situation where P_2 pursues strategy $\tilde{y}$.

The concept of a "fair" game was already mentioned in the discussion of the examples.

<u>Definition</u>. A matrix game is called <u>fair</u> if it has a value $v = 0$.

The Stone-Paper-Scissors game (Example 1 of §18.1) is one example of a fair game. It is also an example of a symmetric game.

<u>Definition</u>. A matrix game is <u>symmetric</u> if $\Sigma_1 = \Sigma_2$ and $A = -A'$.

Since the sets of moves are the same for both players in a symmetric game, the sets of strategies are also the same, so $D_1 = D_2 = D$.

<u>Theorem 2</u>: A symmetric game has a value $v = 0$, and both players can use the same optimal strategy.

<u>Proof</u>: Since $\underset{\sim}{A}$ is skew-symmetric,

$$\underset{\sim}{x}'\underset{\sim}{A}\underset{\sim}{x} = -\underset{\sim}{x}'\underset{\sim}{A}\underset{\sim}{x} = 0 \tag{18.24}$$

for all $\underset{\sim}{x} \in D$. Let $\underset{\sim}{\tilde{x}}$ and $\underset{\sim}{\tilde{y}}$ in D be optimal strategies for P_1 and P_2 respectively. In (18.23) let $\underset{\sim}{x} = \underset{\sim}{\tilde{y}}$ and $\underset{\sim}{y} = \underset{\sim}{\tilde{x}}$, and apply (18.24); then

$$0 = \underset{\sim}{\tilde{y}}'\underset{\sim}{A}\underset{\sim}{\tilde{y}} \leq v \leq \underset{\sim}{\tilde{x}}'\underset{\sim}{A}\underset{\sim}{\tilde{x}} = 0 \tag{18.25}$$

so that $v = 0$. Both players may use $\underset{\sim}{\tilde{x}}$ as an optimal strategy because

$$\underset{\sim}{x}'\underset{\sim}{A}\underset{\sim}{\tilde{x}} \leq 0 = \underset{\sim}{\tilde{x}}'\underset{\sim}{A}\underset{\sim}{\tilde{x}} \leq \underset{\sim}{\tilde{x}}'\underset{\sim}{A}\underset{\sim}{x} \tag{18.26}$$

for all $\underset{\sim}{x} \in D$. The right-hand inequality in (18.26) follows from (18.23); the left-hand one follows in turn from the right-hand one because of the skew-symmetry of $\underset{\sim}{A}$.

Every symmetric game is therefore fair. Intuitively, this is essentially obvious anyway. However, it is not true that conversely every fair game is symmetric (e.g., the skin-game without the additional rule, Example 2 of §18.1). For every matrix game, we can find an equivalent game, in a trivial way, which is fair. Suppose a matrix game has pay-off matrix $\underset{\sim}{A} = (a_{jk})$ and value v. The matrix game with the same sets of moves but with pay-offs of $\hat{a}_{jk} = a_{jk} + \lambda$, where λ is a constant, then has value $\hat{v} = v + \lambda$ (because $\lambda \sum_j \tilde{x}_j \sum_k \tilde{y}_k = \lambda$). Specializing to $\lambda = -v$ gives a game with value $\hat{v} = 0$.

18.5. Matrix Games and Linear Optimization Problems

One connection between linear optimization and the theory of matrix games was already established in the course of the proof of the minimax theorem in the previous section. The value of a game and the optimal strategies for both players can be determined numerically by solving the linear optimization problems (18.20) and (18.21). However, in that case the constraints were given in part as equations and otherwise as inequalities, and some of the variables were positively constrained and the rest not, so that it is advisable to reformulate those problems for numerical purposes.

Let $\underset{\sim}{e}^m$, respectively $\underset{\sim}{e}^n$, denote the vector in R^m, respectively R^n, whose components are all equal to 1. Those two problems may now be written in the form

$$\left.\begin{array}{l} \underset{\sim}{A}'\underset{\sim}{x} - (-\xi)\underset{\sim}{e}^n \geq \underset{\sim}{0}, \quad \underset{\sim}{x}'\underset{\sim}{e}^m = 1, \quad \underset{\sim}{x} \geq \underset{\sim}{0} \\ \qquad -\xi = \text{Max!} \end{array}\right\} \qquad (18.20a)$$

$$\left.\begin{array}{l} \underset{\sim}{A}\underset{\sim}{y} - (-\eta)\underset{\sim}{e}^m \leq \underset{\sim}{0}, \quad \underset{\sim}{y}'\underset{\sim}{e}^n = 1, \quad \underset{\sim}{y} \geq \underset{\sim}{0}, \\ \qquad -\eta = \text{Min!}\ . \end{array}\right\} \qquad (18.21a)$$

The maximal value of $-\xi$ as well as the minimal value of $-\eta$ are both equal to the value v of the game. Solutions of (18.20a) and (18.21a) are the optimal strategies $\tilde{\underset{\sim}{x}}$ and $\tilde{\underset{\sim}{y}}$. In the following reformulation we assume that the value of the game, v, is positive; this can always be arranged by adding an appropriate constant, λ, to each of the elements of the pay-off matrix (cf. the last sentence of §18.4). It suffices to choose λ so large that every $a_{jk} + \lambda > 0$. Then we can restrict $-\xi$ and $-\eta$ to positive values to begin with, and change over to the new variables

$$\underset{\sim}{w} = \frac{1}{-\xi}\cdot\underset{\sim}{x}, \quad \underset{\sim}{z} = \frac{1}{-\eta}\cdot\underset{\sim}{y}. \tag{18.27}$$

Then we obtain linear optimization problems

$$\underset{\sim}{A}'\underset{\sim}{w} \geq \underset{\sim}{e}^n, \quad \underset{\sim}{w} \geq \underset{\sim}{0}, \quad \underset{\sim}{w}'\underset{\sim}{e}^m = \text{Min!}, \tag{18.20b}$$

$$\underset{\sim}{A}\underset{\sim}{z} \leq \underset{\sim}{e}^m, \quad \underset{\sim}{z} \geq \underset{\sim}{0}, \quad \underset{\sim}{z}'\underset{\sim}{e}^n = \text{Max!} . \tag{18.21b}$$

which are equivalent to (18.20a) and (18.21a).

Once again we have a pair of dual problems (cf. §5.2). The optimal value of the two problems is $1/v$. If $\tilde{\underset{\sim}{w}}$ and $\tilde{\underset{\sim}{z}}$ are solutions to these problems, then

$$\tilde{\underset{\sim}{x}} = v\tilde{\underset{\sim}{w}} \quad \text{and} \quad \tilde{\underset{\sim}{y}} = v\tilde{\underset{\sim}{z}} \tag{18.28}$$

are a pair of optimal strategies of the game with pay-off matrix $\underset{\sim}{A}$. Solutions for problems (18.20b) and (18.21b) may be found with the simplex method, as described in §3 and §4. It suffices to solve one of these problems. The solution of the dual problem is then easily obtained via the remark at the end of §5.1.

18.6. Computational Examples for Matrix Games Using the Simplex Method

1. Example 1 of §18.2.

$$\text{Pay-off matrix} \quad \underset{\sim}{A} = \begin{pmatrix} 2 & -3 & -1 & 1 \\ 0 & 2 & 1 & 2 \end{pmatrix}.$$

We solve problem (18.21b) with the simplex method. t_1 and t_2 are slack variables. The objective function is $-\underset{\sim}{z}'\underset{\sim}{e}^4 = \text{Min!}$

	z_1	z_2	z_3	z_4		
t_1	[2]	-3	-1	1	1	1/2
t_2	0	2	1	2	1	-
	1	1	1	1	0	
	-2	1	0	-3	-1	

	t_1	z_2	z_3	z_4		
z_1	1/2	-3/2	-1/2	1/2	1/2	-
t_2	0	2	[1]	2	1	1
	-1/2	5/2	3/2	1/2	-1/2	
	1	-2	-1	-2	0	

	t_1	z_2	t_2	z_4		
z_1	1/2	-1/2	1/2	3/2	1	-
z_3	0	2	1	2	1	-
	[-1/2]	-1/2	[-3/2]	-5/2	-2	
	1	0	1	0	1	

Value of the game: $v = 1/2$.

Optimal strategies for P_2: $\tilde{y} = (1/2)\cdot(1, 0, 1, 0)' = (1/2, 0, 1/2, 0)'$.

Optimal strategy for P_1 (can be read off the boxed-in numbers in the last tableau): $\tilde{x} = (1/2)\cdot(1/2, 3/2)' = (1/4, 3/4)'$. See (18.5).

2. Example 2a of §18.2 (skin game with additional rule).

$$\text{Pay-off matrix } A = \begin{pmatrix} 1 & -1 & -2 \\ -1 & 1 & 1 \\ 2 & -1 & 0 \end{pmatrix}.$$

We solve problem (18.21b) with the simplex method. t_1, t_2, and t_3 are slack variables. The objective function is $-z'e^3 = \text{Min!}$

	z_1	z_2	z_3		
t_1	1	-1	-2	1	-
t_2	-1	[1]	1	1	1
t_3	2	-1	0	1	-
	1	1	1	0	
	-2	1	1	-2	

	z_1	t_2	z_3		
t_1	0	1	-1	2	-
z_2	-1	1	1	1	-
t_3	[1]	1	1	2	2
	2	-1	0	-1	
	-1	-1	0	-3	

	t_3	t_2	z_3		
t_1	0	1	-1	2	-
z_2	1	2	2	3	-
z_1	1	1	1	2	-
	$\boxed{-2}$	$\boxed{-3}$	-2	-5	
	1	0	1	-1	

Value of the game: $v = 1/5$.

Optimal strategy for P_2: $\tilde{y} = (1/5)\cdot(2, 3, 0)' = (2/5, 3/5, 0)'$.

Optimal strategy for P_1: $\tilde{x} = (1/5)\cdot(0, 3, 2)' = (0, 3/5, 2/5)'$. See (18.7).

§19. n-Person Games

As a sequel to the discussion of matrix games, we will present a few selections from the extensive theory of n-person games. For non-cooperative games, we will prove the theorem on the existence of an equilibrium point, which represents a generalization of the minimax theorem. For cooperative games, we will introduce and discuss characteristic functions. The far-ranging and as yet incomplete investigations into the concept of value for cooperative games lie beyond the scope of this short presentation.

19.1. Introduction

n persons, $P_1, P_2, \ldots, P_n$, take part in a game. Each player has a certain set of possible courses of action; let P_i have the set Σ_i, $i = 1,\ldots,n$. We call Σ_i the set of moves and the elements $\sigma^i \in \Sigma_i$, the <u>moves</u>. Let $A_1(\sigma^1,\ldots,\sigma^n),\ldots,A_n(\sigma^1,\ldots,\sigma^n)$ be real valued functions of the $\sigma^i \in \Sigma_i$, $i = 1,\ldots,n$.

Suppose each player has chosen a certain move; player P_i has chosen move $\sigma^i \in \Sigma_i$, $i = 1,\ldots,n$. Then player P_i is paid an amount of

$$A_i(\sigma^1,\ldots,\sigma^n) \quad (i = 1,\ldots,n).$$

$A_1,\ldots,A_n$ are the <u>pay-off functions</u> of the game. If

$$\sum_{i=1}^{n} A_i(\sigma^1,\ldots,\sigma^n) = c \quad \text{for all} \quad \sigma^1 \in \Sigma_1,\ldots,\sigma^n \in \Sigma_n, \tag{19.1}$$

then the game is called a <u>constant sum game</u>. If $c = 0$, the game is a <u>zero sum game.</u>

According to this definition, the matrix games of §18 are two person zero sum games. If all sets of moves are finite sets, the game is called <u>finite</u>. If all sets of moves consist of bounded intervals of the real number line, the game is called <u>continuous</u>.

<u>Definition</u>. An n-person game is <u>non-cooperative</u> if no collusion between the players, either with regard to the conduct of the game, or with regard to the division of the pay-offs is allowed; otherwise, the game is <u>cooperative</u>.

Most social games are non-cooperative, as are the two person zero sum games, by definition. In contrast, many economic and political systems are of a cooperative type.

For example, monopolies, cartels, and multi-party coalitions constitute groupings of players which carry on a cooperative game (against others). The cooperative games are the more important ones in applications to the real world.

19.2. Non-cooperative Games

We will only consider the simple type of game, which has a finite tree and complete information. A complete discussion may be found, e.g., in E. Burger, 1959.

Definition. A plane figure consisting of finitely many nodes and finitely many edges connecting these nodes is called a finite tree if the following conditions are met. The nodes are arranged in a number of levels. There is exactly one node A (the initial point) on the lowest level. Edges run only between nodes on adjacent levels. Every node other than A is connected to exactly one node on the next lower level.

Thus the edges branch out from A as in the tree illustrated in Figure 19.1a. The object in Figure 19.1b is not a tree. Those nodes of the tree which are not connected to any nodes on the next higher level are called endpoints of the tree. The height of a tree equals the number of levels. The tree in Figure 19.1a has height 4.

A game in which every player makes finitely many moves and all moves are made openly and whose initial description is known to all players may be represented by a finite tree.

The initial point A of the tree corresponds to the initial state of the game, and the branching at a node, to

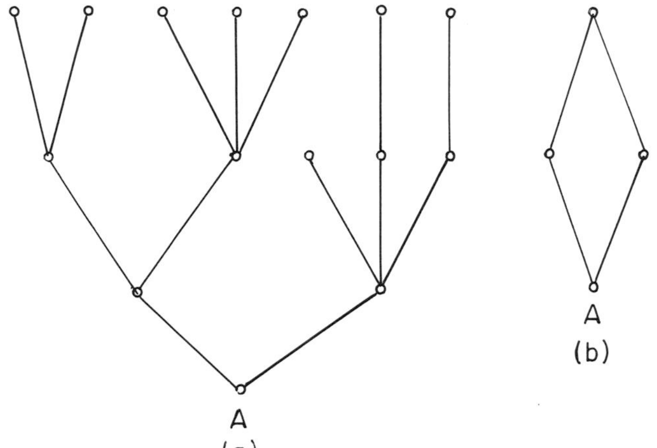

Figure 19.1

the possible moves at that state of the game. Chess is a game with a finite tree if we follow any of the conventions which limit the number of moves.

Let one of the numbers $1,2,\ldots,n$ be assigned to each node of the tree which is not an endpoint. This will signify that player P_i is to make the next move whenever the game has reached a node with number i. Let real-valued functions f_i, $i = 1,\ldots,n$, be defined on the set of endpoints as follows: if the game ends at a certain endpoint, the value of f_i at this endpoint equals the pay-off to player P_i. A move by player P_i, $i = 1,\ldots,n$, consists of a rule which picks out one of the upward leading edges at each node of the tree which has number i assigned to it. Let Σ_i be the set of all moves for player P_i. Clearly Σ_i is finite. If every player has decided on one of his moves, say player P_i on move $\sigma^i \in \Sigma_i$, then the course of the game is uniquely determined, i.e., an endpoint $E(\sigma^1,\ldots,\sigma^n)$ has been reached.

The pay-off functions A_i of the game are given by

$$A_i(\sigma^1,\ldots,\sigma^n) = f_i(E(\sigma^1,\ldots,\sigma^n)) \quad (i = 1,\ldots,n). \quad (19.2)$$

Inclusion of observable random moves: Let some of the nodes which are not endpoints be assigned the number 0, instead of one of the numbers $1,\ldots,n$. The (fictitious) player P_0 consists of a randomizing mechanism which chooses each edge leading upwards from a node assigned a 0 with a certain probability. If there are m edges leading up from such a node, let the probabilities be $p_1,\ldots,p_m$, where $p_1 + \cdots + p_m = 1$. Moves made by "player" P_0 are to be made openly. When all the probabilities with which player P_0 chooses the various edges are known, the game is a _game with complete information_.

The inclusion of random moves in no way alters the previous description of a move.

Let the endpoints of the tree be denoted by $E_1,\ldots,E_N$. Suppose each player has chosen a certain move σ^i, $i = 1,\ldots,n$. Because of the random moves, no longer is any endpoint reached with certainty, but rather with some probability which depends on $\sigma^1,\ldots,\sigma^n$ and which we denote by $w_\nu(\sigma^1,\ldots,\sigma^n)$, $\nu = 1,\ldots,N$. The pay-off functions A_i, $i = 1,\ldots,n$, are given by the expected values,

$$A_i(\sigma^1,\ldots,\sigma^n) = \sum_{\nu=1}^{N} w_\nu(\sigma^1,\ldots,\sigma^n)\cdot f_i(E_\nu). \quad (19.3)$$

Definition. An n-tuple of moves, $(\hat{\sigma}^1,\ldots,\hat{\sigma}^n)$ is called an <u>equilibrium point</u> of the game if, for $1 \leq i \leq n$ and for all $\sigma^i \in \Sigma_i$, it is true that

$$A_i(\hat{\sigma}^1,\ldots,\hat{\sigma}^{i-1},\sigma^i,\hat{\sigma}^{i+1},\ldots,\hat{\sigma}^n) \leq A_i(\hat{\sigma}^1,\ldots,\hat{\sigma}^n). \tag{19.4}$$

If $(\hat{\sigma}^1,\ldots,\hat{\sigma}^n)$ is an equilibrium point and if all of the players P_j, $j \neq i$, stick with move $\hat{\sigma}^j$, then player P_i is best off in also sticking with move $\hat{\sigma}^i$. For then a deviation from $\hat{\sigma}^i$ can not increase the pay-off to him, but generally only decrease it. If two players should collude, then they might be able to achieve a greater total pay-off by deviating from $(\hat{\sigma}^1,\ldots,\hat{\sigma}^n)$. But side agreements and the like are excluded from consideration for non-cooperative games.

<u>Theorem</u>: Every game with a finite tree and complete information has at least one equilibrium point.

<u>Proof</u>: (By induction on the height, λ, of the tree.)

The case $\lambda = 0$ is trivial; then the game consists of the agreed upon pay-offs only. So let $\lambda \geq 1$. Edges $s_1,\ldots,s_m$ lead upwards from the initial point A. Each of these edges determines a subtree, B_μ, $\mu = 1,\ldots,m$, whose height is less than λ. The pay-offs f_i at the endpoints of tree B_μ define pay-off functions, $A_i^{(\mu)} = A_i^{(\mu)}(\sigma_\mu^1,\ldots,\sigma_\mu^n)$, where the σ_μ^i are the moves of the player P_i in the game B_μ. By the induction hypothesis, every subtree B_μ has an equilibrium point with the moves $\hat{\sigma}_\mu^j$, $j = 1,\ldots,n$, i.e.,

$$A_i^{(\mu)}(\hat\sigma_\mu^1,\ldots,\hat\sigma_\mu^{i-1},\sigma_\mu^i,\hat\sigma_\mu^{i+1},\ldots,\hat\sigma_\mu^n) \le \le A_i^{(\mu)}(\hat\sigma_\mu^1,\ldots,\hat\sigma_\mu^n) \tag{19.5}$$

for $i = 1,\ldots,n$, and for all σ_μ^i.

Case I. The selection of an s_μ at the point A is made by a randomizing mechanism with probabilities p_μ, $\mu = 1,\ldots,m$, where $p_\mu \ge 0$ and $\Sigma p_\mu = 1$.

For each player P_i, the moves $\hat\sigma_\mu^i$ in all the subtrees determine a move $\hat\sigma^i$ for the complete game, and $(\hat\sigma^1,\ldots,\hat\sigma^n)$ is an equilibrium point. For

$$A_i(\sigma^1,\ldots,\sigma^n) = \sum_{\mu=1}^{m} p_\mu A_i^{(\mu)}(\sigma_\mu^1,\ldots,\sigma_\mu^n) \tag{19.6}$$

for $i = 1,\ldots,n$. Together with (19.5), this implies (19.4).

Case II. It is player P_k's turn at the initial point A. For all players P_i, $i \ne k$, the moves $\hat\sigma_\mu^i$, $\mu = 1,\ldots,m$, determine a move $\hat\sigma^i$. P_k chooses move $\hat\sigma^k$ as follows. At every node of a subtree B_μ where it is his turn, he picks move $\hat\sigma_\mu^k$. At A, he picks that edge s_μ for which the "equilibrium pay-off" $A_k^{(\mu)}(\hat\sigma_\mu^1,\ldots,\hat\sigma_\mu^n)$ is maximal. By deviating from this $\hat\sigma^k$ he can only decrease his pay-off if the other players choose $\hat\sigma^i$. And the same holds true for the remaining players.

<u>Example</u>. <u>A variant of Nim</u>.

Three players, P_1, P_2, and P_3 partake in a game. Each in turn $(P_1, P_2, P_3, P_1, P_2, P_3,\ldots)$ removes either one or two beans from a pile of M beans. Whoever takes the last bean loses, and must pay one unit to the third from last player. If he should lose, P_3 then pays to P_1, P_2 pays to P_3, and P_1 to P_2. Figure 19.2 depicts the tree for this game in the case $M = 6$.

At each of the nodes, the number of beans remaining has been written in. At the left are the pay-offs which occur if a game ends at the various levels. At every node with a branching, the optimal edge is marked with an arrow. If

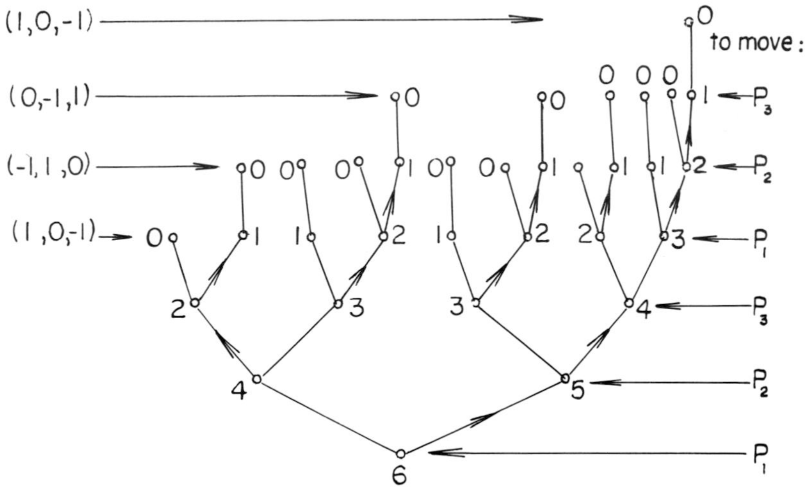

Figure 19.2. Nim variant

all the players follow the marked paths, their moves form an equilibrium point. Whenever a player is faced with five beans, he will lose, regardless of whether he picks one or two beans, as long as the other players choose equilibrium moves.

It is easily seen that an equilibrium point exists for each $M > 6$ if the players' moves are determined by the following rule.

number of beans on hand	number of beans to be removed	equilibrium pay-off
4k	2	1
4k + 1	1 or 2	-1
4k + 2	1	0
4k + 3	1	1

The numbers in the last column represent the pay-off to the player whose turn it is if all players choose equilibrium moves.

Games with non-finite trees have also been investigated. It turns out that for such games, an equilibrium point need not necessarily exist (cf. Burger, 1959, p. 33 ff).

19.3. Cooperative n-Person Zero Sum Games

We begin with a discussion of a simple example of a three person game. Let the three players be P_1, P_2, and P_3. We allow any two of them to form a coalition. If this happens, we have a two person game where one player is the coalition of two and the other is the remaining third original player. The only requirement is that player P_k, if he is the single player, must pay an amount a_k to the coalition of the other two players. All three players know the amounts a_1, a_2, and a_3. Discussions on the means of dividing the winnings are permitted.

We now assume that P_1 demands an amount z_1 in return for joining a coalition. If P_1 joins with P_2, this means that P_2 will be left with the amount $a_3 - z_1$; if P_1 joins with P_3, P_3 receives $a_2 - z_1$. Finally, if P_2 and P_3 form a coalition against P_1, they win a_1 together. If $a_1 > (a_3-z_1) + (a_2-z_1)$, P_2 and P_3 obviously will ally against P_1. Consequently, P_1 must make sure that $a_1 \leq (a_3-z_1) + (a_2-z_1)$; the greatest amount he can demand is

$$z_1 = (1/2)(a_2 + a_3 - a_1) \tag{19.7}$$

since he will not find a coalition partner otherwise. Similarly we find that the largest amount which players P_2 and P_3 can demand in return for joining a coalition, respectively is

$$\left.\begin{aligned} z_2 &= (1/2)(a_3 + a_1 - a_2), \\ z_3 &= (1/2)(a_1 + a_2 - a_3). \end{aligned}\right\} \tag{19.8}$$

If player P_i declines to participate in a coalition and allows the other two players to ally against him, his winnings will be $-a_i$ (where the a_i perhaps are negative). Depending on the sign of the difference

$$s_i = z_i - (-a_i) = (1/2)(a_1 + a_2 + a_3)$$

it is or is not advantageous for P_i to join a coalition. s_i is independent of i; thus

$$s = (1/2)(a_1 + a_2 + a_3) \tag{19.9}$$

alone determines the desirability of entering into a coalition, for all three players.

If $s \leq 0$, no player can demand more in a coalition than he could have obtained as a single player. However, if $s > 0$, it is desirable for every player to join a coalition. The most that P_i then can demand is $z_i = s - a_i$, $i = 1, 2, 3$. For example, if $a_i = i$, $i = 1, 2, 3$, then $s = 3$ and player P_i can demand at most $z_i = 3 - i$ in a coalition. The player who would lose the least can demand the most.

Example. Three cavemen, P_1, P_2, and P_3, get into an argument over the amber in their possession. P_1 has 100 pieces, P_2 has 200 pieces, and P_3 has 300 pieces of amber.

P_1 to P_3: Let's do a number on P_2 and rip off his 200 pieces.

P_3: Cool. We'll each have another 100 pieces.

P_1: No way. I get all 200.

P_3: ... and I get zero? Hey man, you gotta be kiddin.

P_1: Uh, Uh, 'cause otherwise, P_2 and I will do you, and you'll be out 300.

P_3: Mmmm. I dig. Let's do it.

19.4. The Characteristic Function of a Game

We now discuss the general case of an n-person game. Let the participating players be denoted by $1, 2, 3, \ldots, n$. Every player has a finite set of possible moves, and the payoff functions are defined as in §19.1, so that we have a zero sum game. We allow coalitions, and a coalition can consist of any subset of the players, i.e., any subset S of $\{1, 2,\ldots,n\}$. If some of the players have joined into coalition S, the remaining players form a coalition S^-. Thus S and S^- are sets with the properties $S \cap S^- = \emptyset$ (the empty set) and $S \cup S^- = \{1, 2,\ldots,n\}$. This turns an n-person game into a two person game with players S and S^-. The sets of moves are finite, and we still have a zero sum game, and hence a matrix game as in §18. By the minimax theorem, §18.2, the game has a value of v (the expected value of the pay-off to S when both sides pursue optimal strategies). In this way, we obtain a value $v = v(S)$ for every coalition S, and therefore, a function $v(S)$ defined for all the subsets of $\{1, 2,\ldots,n\}$. The function $v(S)$ is called the <u>characteristic function</u> of the game and has the following properties:

$$\left.\begin{array}{ll} \text{(a)} & v(\emptyset) = 0, \\ \text{(b)} & v(S^-) = -v(S), \\ \text{(c)} & v(S \cup T) \geq v(S)+v(T) \quad \text{for} \quad S \cap T = \emptyset. \end{array}\right\} \qquad (19.10)$$

Since there are no members in coalition $\emptyset$, there can be no pay-off; this implies (a). (b) follows because we have a zero sum game. To prove (c), we denote the counter-coalition to $S \cup T$ by R; thus $(S \cup T)^- = R$, $S^- = T \cup R$, and $T^- = S \cup R$. Coalition S wins an amount of $v(S)$ if both the coalitions S and $T \cup R$ pursue an optimal strategy. If the members of T deviate from a strategy which is optimal against S and if S keeps to an optimal strategy, then S will win an amount which is $\geq v(S)$. The analogous remark holds for T. Thus if S pursues a strategy which is optimal against $T \cup R$ and if T pursues a strategy which is optimal against $S \cup R$, then S and T together will win an amount which is $\geq v(S) + v(T)$. If S and T now change to a strategy which is optimal for $S \cup T$, the win for $S \cup T$ will be enlarged, or at least the same as it was.

Several consequences can be derived from properties (a) through (c) of the characteristic function:

$$v(\{1,\dots,n\}) = v(\emptyset^-) = -v(\emptyset) = 0; \tag{19.11}$$

$$v(S_1 \cup S_2 \cup \dots \cup S_r) \geq \sum_{\rho=1}^{r} v(S_\rho), \tag{19.12}$$

if the sets S are pairwise disjoint; in particular, then,

$$v(\{1\}) + v(\{2\}) + \dots + v(\{n\}) \leq v(\{1,2,\dots,n\}) = 0. \tag{19.13}$$

Properties (a) through (c) characterize the characteristic functions of n-person zero sum games, as the following theorem shows.

Theorem: If $w(S)$ is a function defined for all subsets S of $\{1, 2,\dots,n\}$ which satisfies properties (a) through (c) of (19.10), then there exists a game with a characteristic function $v(S) = w(S)$.

Proof: Let $w(S)$ be a set function with properties (a) through (c). Then consequences (19.11) through (19.13) hold for $w(S)$. Let every player k, $k = 1,\dots,n$, choose a subset S_k of $\{1, 2,\dots,n\}$ which contains him. This defines the finite sets of moves. Pay-offs are made according to the following rule. Every set of players, S, such that $S_k = S$ for all $k \in S$ is called a ring. Two rings are either disjoint or identical. The set $\{1, 2,\ldots,n\}$ thus consists of a number of rings together with the remaining players who belong to no ring. Let the rings and the remaining players (considered as one element sets) be denoted by $T_1, T_2,\ldots,T_t$. Let the number of elements of T_q be denoted by n_q, $q = 1,\ldots,t$. Since the sets T_q are pairwise disjoint and since their union is $\{1, 2,\ldots,n\}$, we have

$$\sum_{q=1}^{t} n_q = n. \tag{19.14}$$

Let the pay-off to a player $k \in T_q$ be

$$z_q = \frac{1}{n_q} \cdot w(T_q) - \frac{1}{n} w, \tag{19.15}$$

where $w = \sum_{q=1}^{t} w(T_q)$. It follows from (19.12) and (19.11) that

$$w \le w(T_1 \cup \ldots \cup T_t) = w(\{1,\ldots,n\}) = 0,$$

i.e., that $w \le 0$. We have a zero sum game, for the sum of the pay-offs to all the players is

$$\sum_{q=1}^{t} n_q z_q = \sum_{q=1}^{t} w(T_q) - w = 0.$$

Let $v(S)$ be the characteristic function of this game. It remains to show that $v(S) = w(S)$.

I. Player $k \in T_q$ receives the amount z_q, so all of the players in T_q together receive $n_q \cdot z_q$. Since $w \le 0$, (19.15) implies that $n_q \cdot z_q \ge w(T_q)$.

II. (a), (b), and (c) hold for $v(S)$ and $w(S)$. That implies that $v(S) \ge w(S)$. For $S = \emptyset$, this follows from (a). So suppose $S \neq \emptyset$. If the players in S join to form a ring, then by I the sum of the pay-offs to them is $\geq w(S)$. If they follow an optimal strategy as a coalition, the pay-off will be larger or at least the same. Therefore, $v(S) \geq w(S)$.

III. Also therefore, $v(S^-) \geq w(S^-)$ and hence $v(S) = -v(S^-) \leq -w(S^-) = w(S)$, so that $v(S) = w(S)$.

19.5. Strategically Equivalent Games. Essential Games

It is possible to alter the pay-offs in an n-person game in a certain manner, thereby obtaining an equivalent game with a different characteristic function. Let every player k be paid an additional amount α_k which is independent of his choice of move. In order to keep the game a zero sum game, we require

$$\sum_{k=1}^{n} \alpha_k = 0. \tag{19.16}$$

The fixed supplementary payments have no affect on the strategies and hence do not influence the formation of coalitions. The new game therefore is to be regarded as _strategically equivalent_ to the original. The characteristic function $\hat{v}(S)$ of the new game is

$$\hat{v}(S) = v(S) + \sum_{k\varepsilon S} \alpha_k.$$

It is now possible to find α_k satisfying (19.16) such that

$$\hat{v}(\{1\}) = \hat{v}(\{2\}) = \cdots = \hat{v}(\{n\}); \tag{19.17}$$

indeed,

$$\alpha_k = -v(\{k\}) + \frac{1}{n} \sum_{j=1}^{n} v(\{j\}) \quad (k = 1,\ldots,n).$$

Whenever (19.17) holds, the characteristic function is called reduced. If we then set $\hat{v}(\{k\}) = -\gamma$, $k = 1,\ldots,n$, then $\gamma \geq 0$ by (19.13). If S is a set with $n - 1$ elements, $\hat{v}(S) = \gamma$. If S is a set with r elements, $2 \leq r \leq n - 2$, we can find bounds for $\hat{v}(S)$. For $\hat{v}(S) \geq r(-\gamma)$ and $\hat{v}(S^-) = -\hat{v}(S) \geq -(n-r)\gamma$, so

$$-r\gamma \leq \hat{v}(S) \leq (n-r)\gamma. \tag{19.18}$$

There are now two cases to be distinguished.

I. $\gamma = 0$. By (19.18), $\hat{v}(S) = 0$ for all S; the game is inessential because each player may just as well play for himself alone, since there is no benefit in joining a coalition.

II. $\gamma > 0$. The game is essential. Every player who plays for himself alone receives a pay-off of $-\gamma$, thus losing a positive amount. Every coalition of $n - 1$ players wins the positive amount γ. Hence there is an incentive to form coalitions.

Even if the characteristic function of a game is not in reduced form, it is not difficult to determine whether the game is essential or inessential. Let $V = \sum_{k=1}^{n} v(\{k\})$, recall (19.16), and observe that

$$\begin{aligned} \gamma &= -\hat{v}(\{k\}) = -\frac{1}{n} \sum_{k=1}^{n} \hat{v}(\{k\}) = \\ &= -\frac{1}{n} \sum_{k=1}^{n} v(\{k\}) - \sum_{k=1}^{n} \alpha_k = -\frac{1}{n} V. \end{aligned} \tag{19.19}$$

Thus a game is essential iff $V < 0$.

Since multiplication of all pay-offs by a fixed factor clearly transforms a game into a strategically equivalent one, there is no loss of generality in assuming that $\gamma = 1$ for essential games with reduced characteristic functions. If S is a coalition of r players in such a game, and if $2 \leq r \leq n - 2$, then by (19.18) we have $-r \leq v(S) \leq n - r$, where we again write $v(S)$ for the characteristic function instead of $\hat{v}(S)$. We can deduce from Figure 19.3 how the values of $v(S)$ are restricted by the number of elements, r, in S.

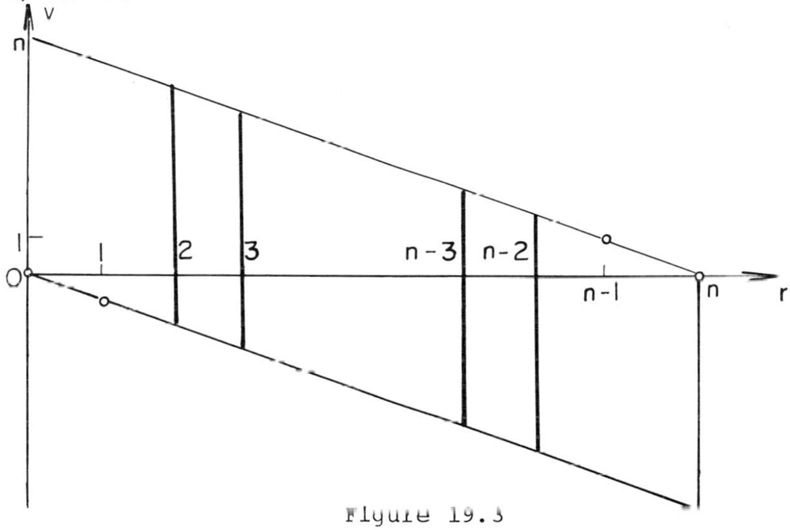

Figure 19.3

For $r = 0, 1, n-1$, or n, $v(S)$ is uniquely determined. For $2 \leq r \leq n - 2$, any of the points on the bold vertical segments in Figure 19.3 may be a value of $v(S)$. By constructing examples, as in von Neumann-Morgenstern, 1953, one can show that each of these points can be a value of $v(S)$ (so that (19.18) cannot be sharpened).

For essential three person games with reduced characteristic function and with $\gamma = 1$, all $v(S)$ are uniquely determined. Thus all essential three person games are strategically equivalent, and hence equivalent to the three person game of the introductory example.

Now phenomena occur when $n \geq 4$. Here we will discuss the case $n = 4$. For a four person game with reduced characteristic function and $\gamma = 1$, $v(S)$ is not uniquely determined iff S contains two elements. For the three coalitions $S_1 = \{1,4\}$, $S_2 = \{2,4\}$, and $S_3 = \{3,4\}$, let

$$v(\{1,4\}) = 2a_1, \quad v(\{2,4\}) = 2a_2,$$
$$v(\{3,4\}) = 2a_3;$$

then by condition (b) of (19.10), $v(S)$ is determined for the countercoalitions $\bar{S_i}$, $i = 1, 2, 3$, and

$$v(\{2,3\}) = -2a_1, \quad v(\{1,3\}) = -2a_2, \quad v(\{1,2\}) = -2a_3.$$

This determines $v(S)$ for all two player coalitions S. (19.18) with $\gamma = 1$ and $r = 2$ yields $-2 \leq v(S_j) \leq 2$, $j = 1, 2, 3$, and hence

$$|a_j| \leq 1 \qquad (j = 1,2,3). \tag{19.20}$$

Therefore we can assign to any such game a point of the cube in Figure 19.4.

Conversely, there is a four person game for every point of this cube; the proof and an extensive discussion may be found in von Neumann-Morgenstern, 1953. As an example, we briefly discuss the game corresponding to the vertex

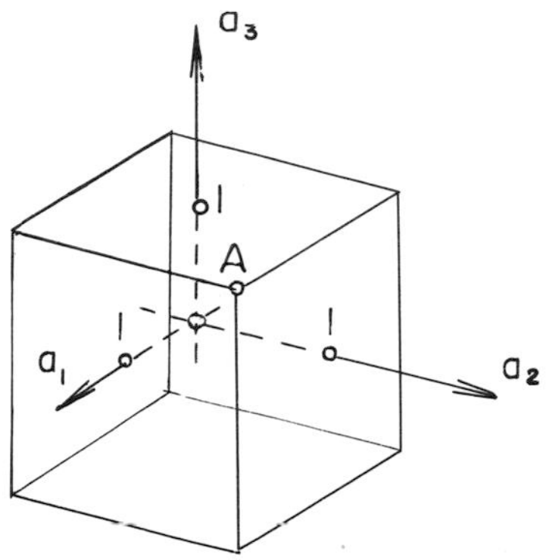

Figure 19.4

$A = (a_1, a_2, a_3) = (1, 1, 1)$. For this game,

$$v(\{1,4\}) = v(\{2,4\}) = v(\{3,4\}) = 2, \quad v(\{1,2,3\}) = 1. \qquad (19.21)$$

If any of the coalitions $\{i,4\}$ where $i = 1, 2,$ or 3, were augmented by an additional player j, the winnings for the coalition would decrease from 2 to 1. The desirable coalitions, S, therefore, are precisely the ones for which $v(S)$ has been given in (19.21). Player 4 finds himself in a favored position. He needs only one coalition partner in order to win an amount 2 with his partner. Should players 1, 2, and 3 ally against 4, they would win an amount 1. It is also easy to determine the amount z_j which player j can demand in return for joining a winning coalition. For

$$z_1 + z_4 = z_2 + z_4 = z_3 + z_4 = 2$$
$$z_1 + z_2 + z_3 = 1$$

implies

$$z_1 = z_2 = z_3 = \frac{1}{3}, \quad z_4 = \frac{5}{3} .$$

A player belonging to a losing coalition must pay off an amount 1 in every case.

19.6. Symmetric n-Person Games

We conclude with some remarks on symmetric games. An n-person zero sum game with characteristic function $v(S)$ is called a symmetric game if $v(S)$ depends only on the number, r, of elements in S, i.e., $v(S) = v_r$. In this case, conditions (19.10) become

$$\left.\begin{array}{lll} \text{(a)} & v_0 = 0, & \\ \text{(b)} & v_{n-r} = -v_r & (r = 0,\dots,n), \\ \text{(c)} & v_{p+q} \geq v_p + v_q & \text{for } p + q \leq n. \end{array}\right\} \qquad (19.22)$$

(c) may be rewritten as $v_p + v_q + v_r \leq 0$ for $p + q + r = n$. The characteristic function for a symmetric game is always in reduced form. If we set the normalization factor $\gamma = 1$, we obtain

$$\left.\begin{array}{l} v_0 = v_n = 0, \quad v_1 = -1, \quad v_{n=1} = 1 \\ -r \leq v_{n-r} \leq n - r \quad (r = 2,\dots,n - 2). \end{array}\right\} \qquad (19.23)$$

for essential symmetric games.

For a symmetric four person game with characteristic function normalized as above, condition (b) of (19.22) implies that

$$v_2 = -v_2,$$

i.e., that $v_2 = 0$. All essential symmetric four person games therefore are strategically equivalent (the center of the cube in Figure 19.4). For $n \geq 5$ and an essential symmetric n-person game with $-v_1 = v_{n-1} = 1$, the numbers

$$v_2, v_3, \ldots, v_{\left[\frac{n-1}{2}\right]}$$

(where $[x]$ means, as usual, the greatest integer not exceeding x, so e.g., $[2.5] = [2] = 2$), may be chosen arbitrarily within the bounds prescribed by (19.22). Thus the number of free parameters determining such a game is $[\frac{n-3}{2}]$.

It is also easy to find the number of free parameters in the general case of a not necessarily symmetric game (with reduced characteristic function and $\gamma = 1$). There are 2^n subsets, S, of $\{1, 2, \ldots, n\}$. Since $v(S) = -v(S^-)$, $v(S)$ is fixed for all of these sets once it is fixed for $(1/2)2^n = 2^{n-1}$ of these. Since $v(\emptyset) = 0$ and $v(\{k\}) = -1$, $k = 1, \ldots, n$, $n + 1$ of the parameters are fixed. The remaining $2^{n-1} - n - 1$ parameters may be chosen freely within the bounds determined by (19.10). The following table gives the number of parameters which may be chosen freely for various n.

n	general games	symmetric games
3	0	0
4	3	0
5	10	1
6	25	1
7	56	2
...	....	...
n	$2^{n-1}-n-1$	$\left[\frac{n-3}{2}\right]$

APPENDIX

1. The Separation Theorem

The proof of the Kuhn-Tucker Theorem in §7 uses the following intuitively obvious theorem.

Separation Theorem. Let B_1 and B_2 be proper convex subsets of R^n which have no points in common. Let B_1 be open. Then there exists a hyperplane, $\underset{\sim}{a}'\underset{\sim}{x} = \beta$, separating B_1 and B_2, i.e., there exists a vector $\underset{\sim}{a} \neq 0$ and a real number β such that

$$\underset{\sim}{a}'\underset{\sim}{x} \leq \beta < \underset{\sim}{a}'\underset{\sim}{y} \quad \text{for} \quad \underset{\sim}{x} \varepsilon B_1 \text{ and } \underset{\sim}{y} \varepsilon B_2.$$

Remark 0. The separation theorem is true without the condition that B_2 be open. In that case, $\underset{\sim}{a}'\underset{\sim}{x} \leq \beta \leq \underset{\sim}{a}'\underset{\sim}{y}$ for $\underset{\sim}{x} \varepsilon B_1$ and $\underset{\sim}{y} \varepsilon B_2$. The proof, however, is more difficult. We use the theorem only in the formulation above.

Remark 1. In speaking of "separation" we allow the case where the set B_1 lies entirely in the hyperplane $\underset{\sim}{a}'\underset{\sim}{x} = \beta$. Example: B_2 consists of the interior of a disk in R^2 and B_1, of a point in the bounding circle. The separating hyperplane is the tangent to the circle at this point.

Remark 2. The separation theorem is a theorem from affine geometry. We make use of this in always choosing the most suitable coordinate system in the course of the proof.

We begin by proving the separation theorem for the case where B_1 is the (one point) set containing only the origin.

<u>Lemma</u>. Let B be an open convex set in R^n which does not contain the origin. Then there exists a vector $\underset{\sim}{a} \neq \underset{\sim}{0}$ with the property that $\underset{\sim}{x} \in B$ implies $\underset{\sim}{a}'\underset{\sim}{x} > 0$.

<u>Proof</u> (by induction): The case $n = 1$ is trivial, since B is then an open interval which does not contain the origin.

$n = 2$. Choose the coordinate system so that no points of B lie on the negative part of the x_1-axis. For $-\pi \leq \phi < \pi$, let h_ϕ be the ray from the origin which forms an angle of ϕ with the positive x_1-axis.

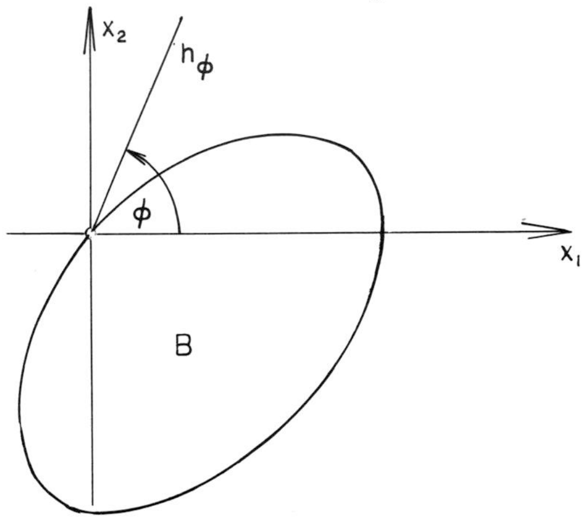

Figure A/1

Let Φ be the set of ϕ for which h_ϕ contains points of B. Φ is a (one dimensional) open convex set and does not contain either $\phi = -\pi$ or $\phi = \pi$; hence an open subinterval (ϕ_0, ϕ_1) of $[-\pi, \pi]$. It follows that

$\phi_1 - \phi_0 < \pi$. Otherwise, there would be a line through the origin consisting of two rays, each of which contains points of B; since B is convex, the origin would have to belong to B. Set

$$\underset{\sim}{a} = \begin{pmatrix} \sin \phi_1 \\ -\cos \phi_1 \end{pmatrix}.$$

Then $\underset{\sim}{x} \varepsilon B$ implies $\underset{\sim}{a}'\underset{\sim}{x} > 0$. For if $\underset{\sim}{x} \varepsilon B$, then $\underset{\sim}{x} = c\begin{pmatrix} \cos \phi \\ \sin \phi \end{pmatrix}$ where $\phi_0 < \phi < \phi_1$ and $c > 0$; therefore, $\underset{\sim}{a}'\underset{\sim}{x} = c \cdot \sin(\phi_1 - \phi) > 0$.

$n \geq 3$. We assume that the Lemma holds for R^{n-1}. Regard the hyperplane $x_1 = 0$ as R^{n-1}. Its intersection with B, denoted B', is open (in this R^{n-1}) and convex (and is perhaps empty). By the induction hypothesis, there exists a vector $\underset{\sim}{a}^1$ (εR^n) whose first component is zero such that $\underset{\sim}{x} \varepsilon B'$ implies $\underset{\sim}{a}^{1'}\underset{\sim}{x} > 0$. Initially, the coordinate system can be chosen so that $\underset{\sim}{a}^1 = (0, 1, 0, \ldots, 0)'$.

Now let B" be the projection of B into the (x_1,x_2)-plane, i.e., the set of all points $\underset{\sim}{x}^* = (x_1^*,x_2^*)'$ such that there is an

$$\underset{\sim}{x} = (x_1^*,x_2^*,x_3,\ldots,x_n)' \ \varepsilon \ B.$$

B" is an open convex set in R^2 which does not contain the origin; for $\underset{\sim}{x} \varepsilon B$ and $x_1 = 0$ implies $\underset{\sim}{a}^{1'}\underset{\sim}{x} = x_2 > 0$.

Since the lemma is true for $n = 2$, there is a vector $\underset{\sim}{a}^{*'} = (a_1^*,a_2^*)$ such that $\underset{\sim}{a}^{*'}\underset{\sim}{x}^* > 0$ for $\underset{\sim}{x}^* \varepsilon B"$. The vector $\underset{\sim}{a}' = (a_1^*, a_2^*, 0,\ldots, 0)$ then satisfies the statement of the lemma for the case of R^n.

Proof of the separation theorem:

$B = \{\underset{\sim}{y} - \underset{\sim}{x} | \underset{\sim}{x} \in B_1, \underset{\sim}{y} \in B_2\}$ is a convex set. It is open because it is the union of open sets: $B = \bigcup_{\underset{\sim}{x} \in B_1} \{\underset{\sim}{y} - \underset{\sim}{x} | \underset{\sim}{y} \in B_2\}$. It does not contain the origin because B_1 and B_2 do not intersect. The conditions of the lemma are fulfilled. Thus there exists a vector $\underset{\sim}{a}$ such that $\underset{\sim}{a}'(\underset{\sim}{y}-\underset{\sim}{x}) > 0$ for $\underset{\sim}{x} \in B_1$ and $\underset{\sim}{y} \in B_2$. If neither B_1 nor B_2 are empty, we can let $\beta = \inf_{\underset{\sim}{y} \in B_2} \underset{\sim}{a}'\underset{\sim}{y}$, and then $-\infty < \beta < \infty$ and $\underset{\sim}{a}'\underset{\sim}{x} \leq \beta < \underset{\sim}{a}'\underset{\sim}{y}$ for $\underset{\sim}{x} \in B_1$ and $\underset{\sim}{y} \in B_2$.

The theorem is of course also true if B_1 or B_2 or both are the empty set. Thus if B_1 is empty and B_2 is not, replace B_1 by a non-empty set $\overline{B}_1$ which does not intersect the (proper) subset B_2 of R^n. The separation theorem holds for $\overline{B}_1$ and B_2 and therefore for B_1 and B_2.

2. An Existence Theorem for Quadratic Optimization Problems

The proof of theorem 2, §12.2, uses a theorem of Barankin and Dorfman on the existence of a solution for quadratic optimization problems. For the sake of completeness, we include here a proof of this theorem which uses only the material presented in this book, in particular, the specialization of the Kuhn-Tucker theorem to quadratic optimization problems developed in §12.1. As in §11 and §12, we consider a quadratic optimization problem of type (11.1).

Theorem: If the set

$$M = \{x \mid x \in R^n,\ Ax \leq b,\ \ x \geq 0\}$$

of feasible points is not empty and if the objective function

$$Q(x) = p'x + x'Cx$$

(where C is a positive semi-definite matrix) is bounded below on M, then $Q(x)$ attains its minimum on M.

Proof: Let $e \in R^n$ be the vector whose components are all 1: $e = (1, 1, \ldots, 1)'$. For $x \in R^n$, we then have

$$e'x = \sum_{i=1}^{n} x_i .$$

The sets $M_s = \{x \mid Ax \leq b,\ e'x \leq s,\ x \geq 0\}$ are bounded and are not empty for sufficiently large, real s, say $0 < s_0 \leq s < \infty$. $s < s'$ implies $M_s \subset M_{s'} \subset M$. The continuous function $Q(x)$ does attain its minimum on each such (bounded, closed, and non-empty) set M_s. Let x^s be the minimal point of $Q(x)$ with respect to M_s for $s_0 \leq s < \infty$. By theorem 1, §12.1, there exist vectors v^s, u^s, and y^s and real numbers η_s and ζ_s, all corresponding to x^s, such that the Kuhn-Tucker conditions are satisfied:

$$\left.\begin{aligned} Ax^s + y^s &= b, \\ e'x^s + \eta_s &= s, \\ v^s - 2Cx^s - A'u^s - e\zeta_s &= p, \\ x^{s'}v^s + u^{s'}y^s + \eta_s\zeta_s &= 0, \\ x^s, v^s \geq 0,\quad u^s, y^s \geq 0,\quad & \eta_s, \zeta_s \geq 0. \end{aligned}\right\} \qquad (K)$$

We must now distinguish between two cases.

(a) There is an s such that $\zeta_s = 0$. Then x^s, v^s, u^s, and y^s also satisfy conditions (12.2) and (12.3); by theorem 1, §12.1, x^s is a minimal point with respect to M.

(b) $\zeta_s > 0$ for $s_0 \leq s < \infty$. Since $\eta_s \zeta_s = 0$, $\eta_s = 0$ for all these s and hence $e'x^s = s$. Let

$$t^s = \frac{1}{e'x^s} x^s = \frac{1}{s} x^s,$$

and then $t^s \geq 0$ and $e't^s = 1$. Now the set of $t \in R^n$ with $t \geq 0$ and $e't = 1$ is closed and bounded, so that the t^s contain a convergent sequence with $s \to \infty$; let this sequence of t^s be the one with $s = s_1, s_2, s_3, \ldots$, or briefly, with $s \in S$. Let the vector to which this sequence converges be

$$t = \lim_{s \in S} t^s.$$

This vector t has the following properties.

1) $$e't = 1, \quad t \geq 0.$$

2) $$At \leq 0;$$

because $At^s = \frac{1}{s} Ax^s \leq \frac{1}{s} b$ for all $s \in S$. This implies $At \leq 0$.

3) $$Ct = 0, \quad p't = 0;$$

because $s < s'$ implies that the set M_s is contained in $M_{s'}$, so that $Q(x^s)$ does not increase with increasing s. Thus $Q(x^s)$ is bounded above for $s_0 \leq s < \infty$ and below by assumption. But

$$Q(x^s) = sp't^s + s^2 t^{s'} Ct^s,$$

and the boundedness of the right-hand side for $s \to \infty$ and $s \, \varepsilon \, S$ implies that $\underset{\sim}{p}'\underset{\sim}{t} = \underset{\sim}{t}'\underset{\sim}{Ct} = 0$, and this in turn implies $\underset{\sim}{Ct} = \underset{\sim}{0}$ (cf. §6.2).

We now define two index sets, $I \subset \{1, 2, \ldots, n\}$ and $J \subset \{1, 2, \ldots, m\}$. Let I be the set of indices i for which the components t_i of vector $\underset{\sim}{t}$ are positive, and let J be the set of indices j for which the components $(\underset{\sim}{At})_j$ of vector $\underset{\sim}{At}$ are negative. Thus,

$$t_i > 0 \quad \text{for} \quad i \, \varepsilon \, I, \quad t_i = 0 \quad \text{for} \quad i \notin I;$$
$$(\underset{\sim}{At})_j < 0 \quad \text{for} \quad j \, \varepsilon \, J, \quad (\underset{\sim}{At})_j = 0 \quad \text{for} \quad j \notin J.$$

Now choose $\bar{s}$ so large that for $s \, \varepsilon \, S$ and $s \geq \bar{s}$,

$$t_i^s > 0 \quad \text{for} \quad i \, \varepsilon \, I,$$

$$\left. \begin{array}{l} (\underset{\sim}{At}^s)_j < \frac{1}{2}(\underset{\sim}{At})_j < 0 \\[2ex] \frac{s}{2}(\underset{\sim}{At})_j \leq b_j \end{array} \right\} \quad \text{for} \quad j \, \varepsilon \, J.$$

Then also

$$x_i^s = s t_i^s > 0 \quad \text{for} \quad i \, \varepsilon \, I,$$

$$(\underset{\sim}{Ax}^s)_j = s(\underset{\sim}{At}^s)_j < \frac{s}{2}(\underset{\sim}{At})_j \leq b_j \quad \text{for} \quad j \, \varepsilon \, J.$$

Choose now a fixed $s \, \varepsilon \, S$ with $s \geq \bar{s}$. For every real $\lambda \geq 0$, the vector $\underset{\sim}{x}^s + \lambda\underset{\sim}{t}$ will satisfy the conditions

(I) $\underset{\sim}{x}^s + \lambda\underset{\sim}{t} \geq \underset{\sim}{0}$,

(II) $\underset{\sim}{A}(\underset{\sim}{x}^s + \lambda\underset{\sim}{t}) + (\underset{\sim}{y}^s - \lambda\underset{\sim}{At}) = \underset{\sim}{b}$,

(III) $\underset{\sim}{e}'(\underset{\sim}{x}^s + \lambda\underset{\sim}{t}) = \underset{\sim}{e}'\underset{\sim}{x}^s + \lambda\underset{\sim}{e}'\underset{\sim}{t} = s + \lambda$

        (since $\eta_s = 0$),

(IV) $\underset{\sim}{v}^s - 2\underset{\sim}{C}(\underset{\sim}{x}^s + \lambda\underset{\sim}{t}) - \underset{\sim}{A}'\underset{\sim}{u}^s - \underset{\sim}{e}\zeta_s = \underset{\sim}{p}$

(since $\underset{\sim}{C}\underset{\sim}{t} = \underset{\sim}{0}$),

(V) $(\underset{\sim}{x}^s + \lambda\underset{\sim}{t})'\underset{\sim}{v}^s + \underset{\sim}{u}^{s'}(\underset{\sim}{y}^s - \lambda\underset{\sim}{A}\underset{\sim}{t}) + \eta_s\zeta_s = 0$

($(\underset{\sim}{x}^s + \lambda\underset{\sim}{t})'\underset{\sim}{v}^s = 0$ since for components $t_i > 0$, i.e., for $i \in I$, $x_i^s > 0$ and hence $v_i^s = 0$; similarly, $\underset{\sim}{u}^{s'}(\underset{\sim}{y}^s - \lambda\underset{\sim}{A}\underset{\sim}{t}) = 0$),

(VI) $Q(\underset{\sim}{x}^s + \lambda\underset{\sim}{t}) = Q(\underset{\sim}{x}^s)$ (since $\underset{\sim}{p}'\underset{\sim}{t} = 0$ and $\underset{\sim}{C}\underset{\sim}{t} = \underset{\sim}{0}$).

By theorem 1, §12.1, and (I) through (V), $\underset{\sim}{x}^s + \lambda t$ is a minimal point of $Q(\underset{\sim}{x})$ with respect to $M_{s+\lambda}$; then by (VI) $\underset{\sim}{x}^s$ is also a minimal point with respect to $M_{s+\lambda}$ and for all $\lambda \geq 0$. If we now choose an arbitrary $\lambda > 0$, then $\underset{\sim}{x}^s$ satisfies the Kuhn-Tucker conditions (K) for $s+\lambda$ instead of s with the appropriate $\underset{\sim}{v}^{s+\lambda}$, $\underset{\sim}{u}^{s+\lambda}$, $\underset{\sim}{y}^{s+\lambda}$, $\eta_{s+\lambda}$, and $\zeta_{s+\lambda}$; in particular,

$$\underset{\sim}{e}'\underset{\sim}{x}^s + \eta_{s+\lambda} = s + \lambda.$$

Since $\underset{\sim}{e}'\underset{\sim}{x}^s = s$, $\eta_{s+\lambda} = \lambda > 0$, and hence $\zeta_{s+\lambda} = 0$. Thus in case (b) it is still possible to find a minimal point for which case (a) applies and which therefore is a minimal point with respect to M.

PROBLEMS

1. x five story and y two story buildings are to be constructed on a swampy plot, where the need for adequate foundations greatly increases the cost of a taller building. The work produced by one person in one month will be denoted a "person-month". The remaining information is contained in the following table.

number of stories	costs in $	person-months	area in yd^2	number of occupants per bldg.	number of bldgs.
5	600,000	120	800	30	x
2	200,000	60	600	12	y
available:	18,000,000	4,500	42,000		

How should x and y be chosen if we want to maximize the number of people who can live on this plot of land?

Solution: $x = 15$, $y = 45$; for this solution 3,000 yd^2 are left open.

2. A cabinet maker wants to maximize his profit on the production of x_1 tables and x_2 chairs. He can sell at most 20 tables, so that $x_1 \leq 20$. Details are contained in the table.

	per table	per chair	total available
hours of labor	6	1.5	240
cost of material, labor, etctera, in $	180	30	5,400
profit in $	80	15	

Solution: $x_1 = 10$, $x_2 = 120$, total profit $Q = \$2,600$.

3. A farm is to be planted in rye and potatoes. For each acre of arable land, the data is as follows.

	costs of cultivation in $	labor required in hrs.	profit in $
for potatoes	5	2	20
for rye	10	10	60

We want to divide the acreage, with x_1 of potatoes and x_2 of rye, so as to maximize the profit. We have 1200 acres available, as well as $7,000 and 5,200 hours of labor. A problem of this type is treated in detail in Stiefel, 1961, p. 28.

Solution: Plant 600 acreas of potatoes, 400 of rye, and leave 200 fallow; the maximal profit will be $36,000.

4. Which of the point-sets described by the following conditions represent polyhedra? (We use x, y, and z for

the coordinates instead of x_1, x_2, and x_3.)

a) $|x| \leq 1, \quad |y| \leq 1, \quad |z| \leq 1,$

b) $|x| + |y| + |z| \leq 1,$

c) $-1 \leq x \leq y \leq z \leq 1,$

d) $|x| \leq |y| \leq |z|,$

e) $|x| \leq |y| \leq |z| \leq 1,$

f) $|x + y + z| \leq 1$

g) $|x + y + z| \leq 1, \quad |x| \leq 2, \quad |y| \leq 2, \quad |z| \leq 2.$

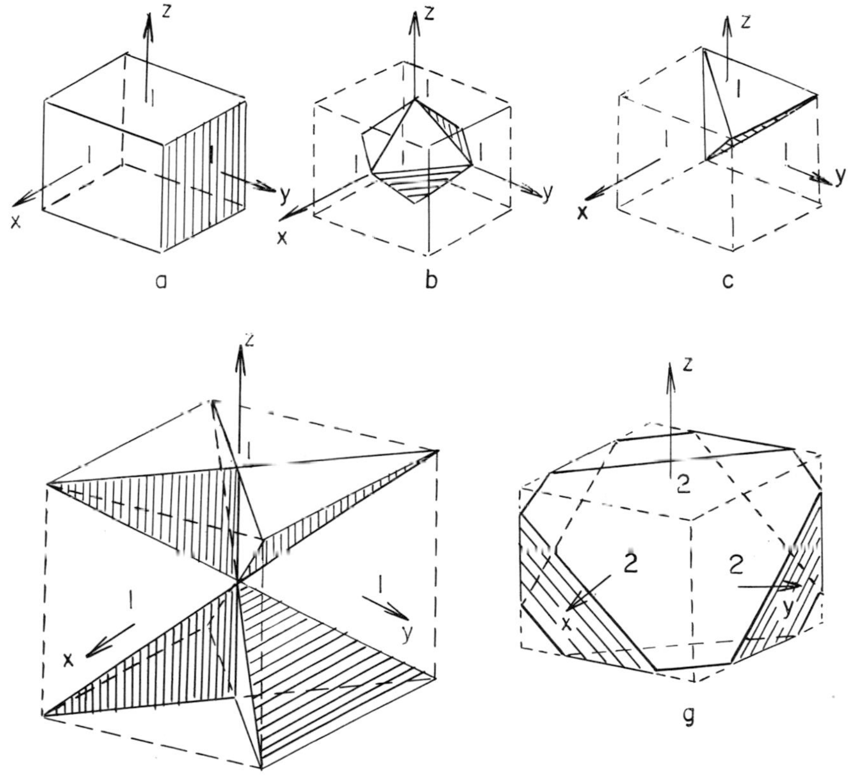

Figure A/2

Solution: Only the point-sets defined by a), b), c),

and g) represent polyhedra; see Figure A/2. For e) we get four congruent tetrahedra. For d), four infinite "cones", of which the part with $|z| \leq 1$ is shown in figure e). For f) we get an unbounded point-set which lies between the two parallel planes $x + y + z = \pm 1$. If we slice a finite piece out of this plate-like point set, by $|x| \leq 2$, $|y| \leq 2$, and $|z| \leq 2$, we obtain the object in figure g), namely a polyhedron with eight sides (two congruent 6-edged and six congruent 4-edged figures) and 12 edges.

5. The regular polyhedra in R^3 are the tetrahedron, the cube, the octahedron, the dodecahedron, and the icosahedron. Which of these have degenerate vertices?

6. Find the dual problem for the transportation problem (4.15). One solution of the dual problem can be found from the T-tableau of the solution of the original problem. What is the solution of the problem dual to the example considered in §4.8?

7. a) Find the limit load $P^*(x,y)$ (see §5.4) for a square plate ($|x| \leq 1$, $|y| \leq 1$), supported at (1, 1), (-1, 1), (1, -1), and (-1, -1). The permissible loads are $P_j \leq 1$, $j = 1,\ldots,4$.

Solution: $$P^*(x,y) = \mathrm{Min}\left(\frac{4}{|x|+1}, \frac{4}{|y|+1}\right).$$

b) The same as a) except that $0 \leq P_j \leq 1$, $j = 1,\ldots,4$.

Solution: $$P^*(x,y) = \min\left(\frac{4}{|x|+1}, \frac{4}{|y|+1}, \frac{2}{|x|+|y|}\right).$$

(See Figure 5.2)

8. (The cellar problem) A very large coal cellar with given outline B and area F is to be subdivided for n users into n individual cellars of area F/n in such a way that the additional walls which have to be drawn have the least possible total length L; see Figure A/3. The problem is idealized by regarding the walls as very thin (curves) and by disregarding the need for corridors leading to the individual cellars. For example, let B be the square with sides of length 1, or the rectangle with sides of length 1 and 2. Since we do not know in advance which arrangements of the walls should be considered, the problem is barely tractable for large values of n, either mathematically or with a computer. The examples should be tried with n = 2,3,...,7.

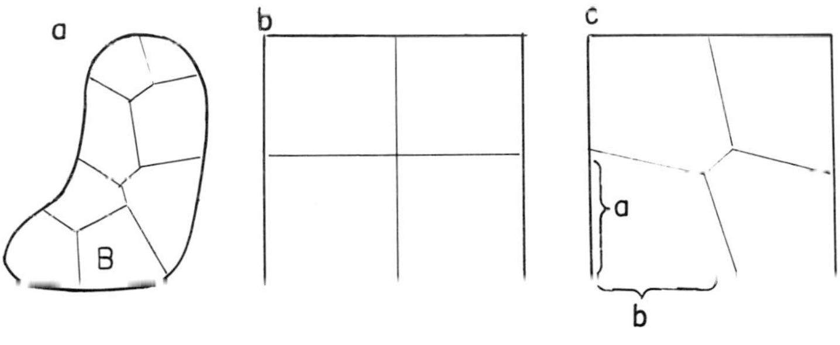

Figure A/3. Subdivision of a coal cellar.

Hints for a solution: We know from the isoperimetric problem that the only curves we need to consider are either straight line segments or arcs of circles. If we restrict the problem to polygons, then we must allow for a subdivision which, in approximating a circular arc, has vertices with only two edges leading out, and hence, is not necessarily composed of convex polygons only. At first glance, the example of the square and $n = 4$ is already surprising, in that the subdivision into four congruent subsquares (Figure A/3b) with $L = 2$ is not the optimal one. An arrangement such as in Figure A/3c, and with length a and b as indicated there, yields an optimal value of $L \approx 1.981$ when $a \approx 0.53$ and $b \approx 0.47$.

9. For Example 3, §6.9, find other sets $\{n_i\}$ which are solutions for $s(20, 3)$ and $s(20, 4)$.

Solution: $s(20, 3) = 4$ if $\{n_i\} = \{1, 4, 6, 7\}$ or $\{1, 3, 7, 12\}$ or $\{1, 3, 8, 12\}$. Sample solutions for $s(20, 4)$ are $\{n_i\} = \{1, 4, c\}$ where $c = 5, 6,$ or 9, and $\{n_i\} = \{1, 5, d\}$ where $d = 6$ or 8.

10. Four localities in the (x,y)-plane have center coordinates $P_1 = (0, 0)$, $P_2 = (1, 0)$, $P_3 = (1, 2)$, and $P_4 = (0, 1)$; see Figure A/4. Find that location, S, for a factory which minimizes the sum of the distances, $\sum_{j=1}^{4} \overline{P_jS}$, from the four localities.

Solution: $S = (\frac{1}{3}, \frac{2}{3})$, $\sum_{j=1}^{4} \overline{P_jS} = \sqrt{2} + \sqrt{5}$.

Show generally that whenever the four points P_1, P_2, P_3, and P_4 define a convex four sided figure, the optimal point S is the intersection of the diagonals.

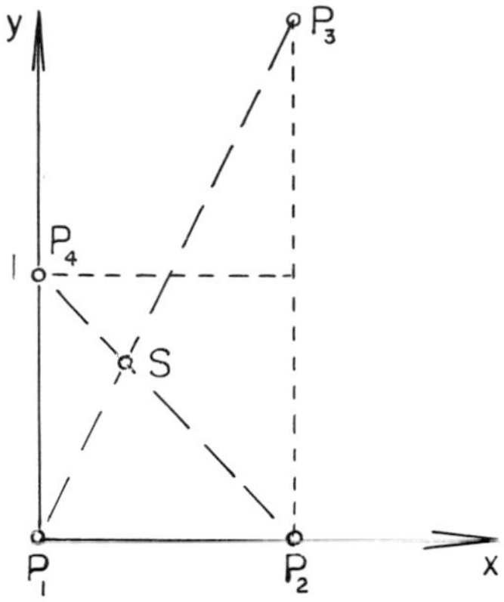

Figure A/4. Optimal location of a factory.

11. (Gustav-Victor game) Gustav and Victor each place a die with one side up so that the other cannot see it. Each then makes his choice known to the other. Thus, each chooses for himself one of the numbers 1 through 6. Victor bets that the two numbers they have chosen are different. If this is the case, Gustav must pay Victor the amount \$ v, where v is the difference between the two numbers, in absolute value. Gustav bets that the two numbers are the same, and if this is the case, Victor must pay Gustav the amount \$ a. Is the game fair if they agree that a = 12?

Solution: The pay-off matrix which shows the amounts which Victor must pay to Gustav looks like

G \ V	1	2	3	4	5	6
1	a	-1	-2	-3	-4	-5
2	-1	a	-1	-2	-3	-4
3	-2	-1	a	-1	-2	-3
4	-3	-2	-1	a	-1	-2
5	-4	-3	-2	-1	a	-1
6	-5	-4	-3	-2	-1	a

The game is almost fair, but not completely fair. The value of the game is $v = -5/266 \approx -0.018797$. We may not conclude that the value, v, is positive just because the sum of all the matrix elements, namely 2, is positive.

In order to provide some concept of the influence of a on the value, we note that for $a = 10$, the value of the game is $v = -35/96 \approx -0.3646$.

12. (Three person Nim as a cooperative game) Each of the players P_1, P_2, P_3 in turn removes some beans from a pile of M beans. The number, z, of beans removed at any turn must belong to a set, K, of possible numbers which the players have fixed by agreement before the beginning of the game. The player who removes the last bean or who cannot make a move allowed by the rules, loses the game.

The simplest case is the one with $K = \{1, 2\}$, i.e., each player removes either one or two beans. P_1 begins the game by removing z beans, where $z \in K$.

Suppose now that the game is a "covert cooperative

game", i.e., P_2 and P_3 will attempt to play in such a way that P_1 is forced to lose. For which numbers M can P_1 avoid a loss, and for which numbers can't he? Consider the special cases where: a) $K = \{1, 2\}$; b) $K = \{2, 3\}$; c) $K = \{3, 4\}$.

Solution: Call the number M favorable if P_1 can prevent a loss, and unfavorable otherwise. In case a), the numbers $M = 2, 3, 4, 7$, and 8 are favorable, and all others are unfavorable. In case b), e.g., the numbers 27 and 28 are favorable, and the number 26 as well as all numbers ≥ 29 are unfavorable. In case c), e.g., the numbers 59 and 60 are favorable, while 58 and all numbers ≥ 61 are unfavorable.

13. (Approximation problems) Categorize the following approximations according to the types of optimization problems listed in Figure 6.7. The function $f(x) = 2 - x^{1/2}$ is to be approximated on the interval $[0,1]$ by a function of the form $w(x) = (a + x)^{-1}$ where $a > 0$ is to be chosen to give the best approximation

1) in the Tchebychev sense, i.e., to minimize

$$\Phi(a) = \operatorname*{Max}_{x \varepsilon I} |w(x) - f(x)|,$$

2) in the mean, i.e., to minimize

$$\Psi(a) = [\int_0^1 (w - f)^2 dx]^{1/2}.$$

Solution: 1. Strict quasiconvex optimization. The function $\Phi(a)$ has two vertices in the region $a > 0$ and hence is not differentiable; see Figure A/5.

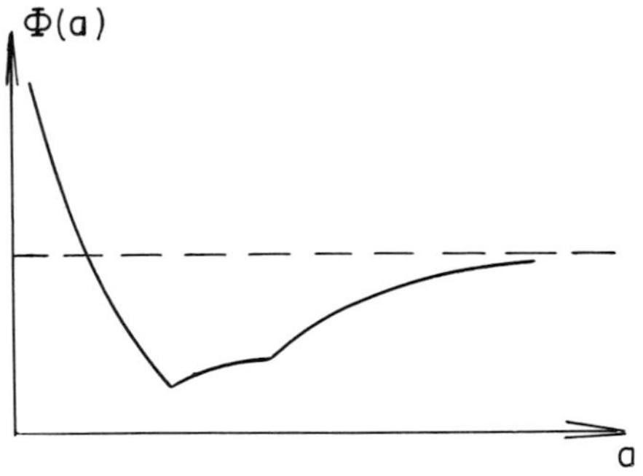

Figure A/5

2. Pseudoconvex optimization. The function $\Psi(a)$ is differentiable for $a > 0$; see Figure A/6.

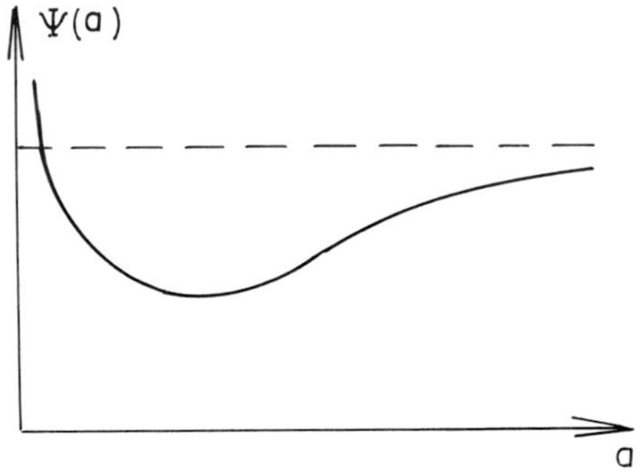

Figure A/6

14. (Shortest path) In the (x,y)-plane we have "streets" along the lines $y = 1$ and $y = 2$ along which a person P can move with velocity v. P would like to get from the point $P_0 = (0, 0)$ to the point $P_1 = (a,b)$ in the least possible time, where, e.g., $(a,b) = (4, 3)$; see Figure A/7. The speed in each of the three regions, $y < 1$, $1 < y < 2$, and $2 < y$, respectively is v/α_1, v/α_2, and v/α_3,

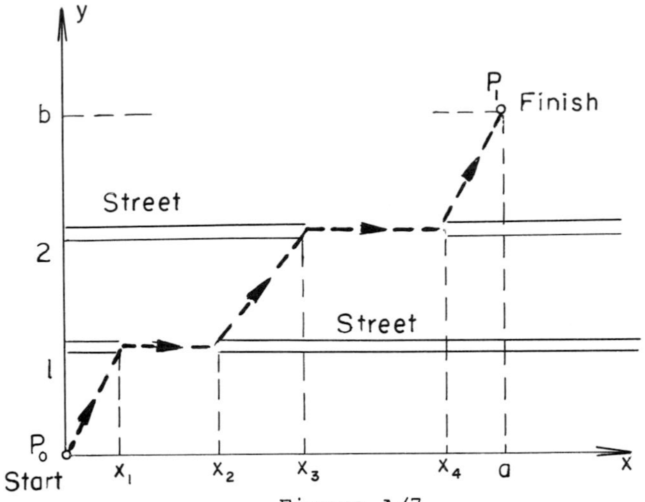

Figure A/7

where $\alpha_j > 1$, $j = 1, 2, 3$. In view of Figure A/7, we have an optimization problem of what sort?

Solution: A non-linear separable optimization.

Let the path we want have the intervals $[x_1, x_2]$ and $[x_3, x_4]$, respectively in common with the lines $y = 1$ and $y = 2$. Set $x_1 = c_1$, $x_3 - x_2 = c_2$, and $a - x_4 = c_3$. Then for a sufficiently large we want to minimize the function

$$z = z(c_1, c_2, c_3)$$

$$= [\alpha_1 \sqrt{1+c_1^2} + \alpha_2 \sqrt{1+c_2^2} + \alpha_3 \sqrt{(b-2)^2+c_3^2} +$$

$$+ a - c_1 - c_2 - c_3]/v$$

Analogously, one can easily consider more complicated optimization problems. For example, in Figure A/8 a person is to get to a circular lake along intervening, partially curved streets in the least possible time.

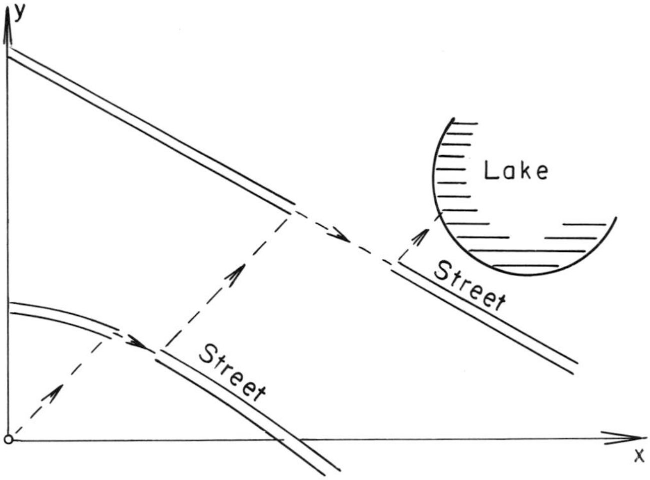

Figure A/8

BIBLIOGRAPHY

Abadie, J. (ed.): Nonlinear programming. Amsterdam: North Holland Publ. Company 1967.

________ (ed.): Integer and Nonlinear Programming, Amsterdam, North-Holland Publ. Company 1970.

Altman, M.: Bilinear programming. Serie des sciences math. astr. et phys. Vol. 16, Nr. 9, 741-746 (1968).

Arrow, K. J., L. Hurwicz, and H. Uzawa: Studies in linear and non-linear programming. Stanford: University Press 1964, 2nd printing.

Barankin, E., and R. Dorfman: On quadratic programming. University of California Publications in Statistics 2, 258-318 (1958).

Bonnesen, T., and W. Fenchel: Theorie der konvexen Körper. Berlin: Springer 1934.

Boot, J. C. G.: Quadratic programming. Amsterdam: North Holland Publishing Company 1964.

Bracken, J., and G. P. McCormick: Selected applications of nonlinear programming. New York: John Wiley & Sons 1968.

Burger, E.: Einführung in die Theorie der Spiele. Berlin: de Gruyter 1966, 2nd edition.

Burkard, R. E.: Methoden der ganzzahligen Optimierung, Berlin: Springer 1972.

Cheney, E. W., and A. A. Goldstein: Newtons method for convex programming and Tchebycheff approximation. Numerische Math. 1, 253-268 (1959).

Collatz, L.: Aufgaben monotoner Art. Arch. Math. 3, 366-376 (1952).

________. Approximation in partial differential equations. Proc. Symposium on Numerical Approximation. Edited by R. E. Langer. Madison 1959. 413-422.

________: Functional Analysis and Numerical Mathematics. Berlin: Springer 1966.

________: Tschebyscheffsche Approximation, Randwertaufgaben und Optimierungsaufgaben. Wissenschaftliche Zeitschrift der Hochschule für Architektur und Bauwesen Weimar 12, 504-509 (1965).

________: Applications of nonlinear optimization to approximation problems. In: Integer and nonlinear programming. Amsterdam: North Holland Publ. Comp. 1970, p. 285-308.

Converse, A. O.: Optimization. New York-Chicago: Holt, Rinehart, Winston 1970.

Danø, S.: Linear programming in industry, theory and applications. Wien: Springer 1960.

Dantzig, G. B.: Linear Programming and Extensions. Princeton, New Jersey: University Press 1963.

Dieter, U.: Optimierungsaufgaben in topologischen Vektorraümen I: Dualitätstheorie. Z. Wahrscheinlichkeitstheorie verw. Geb. 5, 89-117 (1966).

Duffin, R. J., E. L. Peterson, and C. M. Zener: Geometric programming. New York-London-Sidney: John Wiley & Sons 1967.

Eggleston, H. G.: Convexity. Cambridge: University Press 1966.

Elsner, L.: Konvexe Optimierung beim Einschließungssatz für Eigenwerte von Matrizen, private communication (1971).

Ferguson, Th. S.: Mathematical Statistics, A Decision Theoretic Approach. New York: Academic Press 1967.

Fiacco, A. V., McCormick, G. P.: Nonlinear Programming: Sequential Unconstrained Minimization Techniques. New York: Wiley 1968.

Fletcher, R. (ed.): Optimization. Symposium of the Institute of Mathematics and its Applications, University of Keele, 1968. London: Academic Press 1969.

Gale, D.: The theory of linear economic models. New York: McGraw-Hill 1960.

Gass, S. I.: Linear Programming. New York: McGraw-Hill 2nd ed., 1964.

Goldman, A. J., and A. W. Tucker: Theory of linear programming. Ann. Math. Studies 38, 53-97 (1956).

Gomory, R. E.: An algorithm for integer solutions to linear programs. 269 – 302 in Graves-Wolfe 1963.

Graves, R. L., and Ph. Wolfe: Recent advances in mathematical programming. New York-San Francisco-Toronto-London: McGraw-Hill 1963.

Hinderer, K.: Foundations of Non-stationary Dynamic Programming with Discrete Time Parameter. Berlin: Springer 1970.

Holmes, R. B.: A Course on Optimization and Best Approximation. Berlin: Springer 1972.

Hordijk, A.: Dynamic Programming and Markov Potential Theory. Amsterdam: Math. Center Tracts Nr. 51, 1974.

Junginger, W.: Über die Lösung des dreidimensionalen Transportproblems. Diss. Univ. Stuttgart 1970.

Karlin, S.: Mathematical methods and theory in games, programming and economics, Vol. I, II. London-Paris: Pergamon 1959.

Kelley, J. E. Jr.: The cutting plane method for solving convex programs. J. Soc. Indust. Appl. Math. 8, 703-712 (1960).

Kirchgässner: Graphentheoretische Lösung eines nichtlinearen Zuteilungsproblems. Unternehmensforschung 9, 217-229 (1965).

Knödel, W.: Lineare Programme und Transportaufgaben. Zeitschrift für moderne Rechentechnik und Automation 7, 63-68 (1960).

König, D.: Theorie der endlichen und unendlichen Graphen. Leipzig: Akad. Verlagsgesellschaft 1936.

Krabs, W.: Fehlerquadrat-Approximation als Mittel zur Lösung des diskreten Linearen Tschebyscheff-Problems. Z. Angew. Math. Mech. 44, T 42-45 (1964).

________: Lineare Optimierung in halbgeordneten Vektorräumen. Num. Math. 11, 220-231 (1968).

Kuhn, W.: Das Skin-Spiel ist zitiert bei GASS, 1964, Kap. 12.1.

Künzi, H. P., und W. Krelle: Nichtlineare Programmierung. Berlin-Göttingen-Heidelberg: Springer 1962.

Künzi, H. P., W. Krelle, H. Tzschach and C. A. Zehnder: Numerical Methods of Mathematical Optimization with Algol and Fortran Programs. New York: Academic Press 1968.

Kushner, H.: Introduction to Stochastic Control. New York: Holt, Rinehart and Winston 1971.

Laurent, P. J.: Approximation et Optimisation, Paris: Hermann 1972.

Lempio, F.: Separation und Optimierung in Linearen Räumen. Hamburg: Dissertation 1971.

________: Lineare Optimierung in unendlichdimensionalen Vektorräumen, Computing, 8, 284-290 (1971).

Luenberger, D. G.: Optimization by Vector Space Methods. New York: Wiley 1969.

________: Introduction to Linear and Nonlinear Programming. Reading (Mass.): Addison-Wesley 1973.

Mangasarian, O. L.: Nonlinear Programming, New York: McGraw-Hill 1969.

McCormick, G. P.: Second order conditions for constrained minima. SIAM J. Appl. Math. 15, 641-652 (1967).

Meinardus, G.: Approximation of Functions: Theory and Numerical Treatment. Berlin: Springer 1967.

Melsa, J. L., and D. G. Schultz: Linear control systems. New York: McGraw-Hill 1970.

Neumann, J. von and O. Morgenstern: Theory of Games and Economic Behaviour. Princeton: University Press 1953.

Owen, G.: Game Theory. Philadelphia: Saunders 1968.

Parthasarathy, T., and Raghavan, T. E. S.: Some Topics in Two-Person Games. New York: Elsevier 1971.

Pfanzagl, J.: Allgemeine Methodenlehre der Statistik, Bd. II, 2. Auflage. Berlin: de Gruyter 1966.

Prager, W.: Lineare Ungleichungen in der Baustatik. Schweiz. Bauzeitung 80, 19 (1962).

Ross, S. M.: Applied Probability Models with Optimization Applications. San Francisco: Holden-Day 1970.

Schröder, J.: Das Iterationsverfahren bei allgemeinerem Abstandsbegriff. Math. Z. 66, 111-116 (1956).

Sengupta, J. K., and G. Tintner: A review of stochastic linear programming. Review of the Internat. Statistic Institut 39, 197-223 (1971).

Stiefel, E.: Über diskrete und lineare Tschebyscheff-Approximationen. Numerische Math. 1, 1-28 (1959).

________: Note on Jordan elimination, linear programming and Tschebyscheff approximation. Numerische Math. 2, 1-17 (1960).

________: An Introduction to Numerical Mathematics. New York: Academic Press 1971.

Stoer, J.: Duality in nonlinear programming and the minmax theorem. Numerische Math. 5, 371-379 (1963).

________: Über einen Dualitätssatz der nichtlinearen Programmierung. Numerische Math. 6, 55-58 (1964).

________ and C. Witzgall: Convexity and optimization in finite dimensions. I. Berlin-Heidelberg-New York: Springer 1970.

Tolle, H: Optimization Methods (with Ordinary Differential Equations as Constraints). New York/Heidelberg/Berlin, Springer 1975.

Uzawa, H.: The Kuhn-Tucker theorem in concave programming. In: Arrow, Hurwicz, Uzawa 1958.

Vajda, S.: Mathematical programming. Reading, Mass.: Addison-Wesley 1961.

Valentine, F. A.: Convex Sets. New York: McGraw-Hill 1964.

Vogel, W.: Lineares Optimieren. Leipzig: Akademische Verlagsgesellschaft Geest + Portig 1967.

Wetterling, W.: Lösungsschranken beim Differenzenverfahren zur Potentialgleichung. International Series of Numerical Mathematics. 9, 209-222 (1968).

________: Lokal optimale Schranken bei Randwertaufgaben. Computing 3, 125-130 (1968).

________: Definitheitsbedingungen für relative Extrema bei Optimierungs- und Approximationsaufgaben. Numerische Math. 15, 122-136 (1970).

________: Über Minimalbedingungen und Newton-Iteration bei nichtlinearen Optimierungsaufgaben. International Series of Numerical Mathematics 15, 93-99 (1970a).

Wolfe, Ph.: The simplex method for quadratic programming. Econometrica 27, 382-398 (1959).

________. Accelerating the cutting plane method for nonlinear Programming. J. Soc. Indust. Appl. Math. 9, 481-488 (1961).

________: Methods of nonlinear programming. In: Graves and Wolfe 1963, p. 67-86.

Zangwill, W. I.: Nonlinear Programming, a Unified Approach. Englewood Cliffs, New Jersey: Prentice Hall 1969.

Zoutendijk, G.: Methods of feasible directions. Amsterdam: Elsevier Publishing Company 1960.

INDEX